Chemical Composition and Biological Activities of Essential Oils

Chemical Composition and Biological Activities of Essential Oils

Editors

Edoardo Marco Napoli
Maura Di Vito

MDPI • Basel • Beijing • Wuhan • Barcelona • Belgrade • Manchester • Tokyo • Cluj • Tianjin

Editors
Edoardo Marco Napoli
Institute of Biomolecular Chemistry
National Research Council
ICB-CNR
Italy

Maura Di Vito
University of Bologna
Italy
Università Cattolica del Sacro Cuore
Italy

Editorial Office
MDPI
St. Alban-Anlage 66
4052 Basel, Switzerland

This is a reprint of articles from the Special Issue published online in the open access journal *Antibiotics* (ISSN 2079-6382) (available at: https://www.mdpi.com/journal/antibiotics/special_issues/Chemical_Composition).

For citation purposes, cite each article independently as indicated on the article page online and as indicated below:

LastName, A.A.; LastName, B.B.; LastName, C.C. Article Title. *Journal Name* **Year**, *Volume Number*, Page Range.

ISBN 978-3-0365-1250-1 (Hbk)
ISBN 978-3-0365-1251-8 (PDF)

Cover image courtesy of Edoardo Napoli and Maria Grazia Bellardi.

Contents

About the Editors

Edoardo Marco Napoli, Ph.D. is a Staff Research Scientist of Italian National Research Council (C.N.R.—Italy). Prior to joining the "Natural substances and food chemistry" team at the Institute of Biomolecular Chemistry (ICB-CNR), Dr. Napoli graduated summa cum laude in chemistry from the University of Catania (Italy). He continued his studies in the chemistry of natural compounds, earning his Ph.D. at the same university in 2003. After several years in a multinational pharmaceutical company as Regulatory Affairs Manager, he became a permanent research staff member at ICB-CNR in 2008. He is also a biotechnologist, with a bachelor's degree achieved in 2014. His research activities are focused on natural products' chemistry, phytochemistry, cosmetics, natural ingredients, flavors and fragrances and the exploitation of agroindustrial wastes. At the moment, the pharmacological and biological activities of natural compounds and phytocomplexes from different origins have become the main topic of his research. He has authored more than 60 publications in peer-reviewed journals, several scientific dissemination articles, and book chapters.

Maura Di Vito achieved a Ph.D. with honors, and both a degree in Biology, Specialization in Clinic Pathology and Ph.D. in Human Pathology at the University of Rome "Sapienza", and higher education in Herbal Medicine at the University of Viterbo "La Tuscia".

She has expertise in molecular biology, microbiology, protein analysis, microscopy, histology, and cell cultures in rapport to research in microbiology, oncology, and pathophysiology.

After her Ph.D., she started her studies in the field of Complementary and Alternative Medicines (CAM) research at different national institutions. In 2013, she worked at the Network of Integrated Medicine of Tuscan Region for the joint action of European Partnership for Action Against Cancer (EPAAC), and she collaborated with research groups of the Dept. of Infectious, Parasitic and Immune-mediated Diseases of National Health Institute of Rome. Actually, she is active in the study of the antimicrobial activity of natural products at the Dept. of Agricultural and Food Sciences (DISTAL) of Università of Bologna, and at the Dept. of "Scienze biotecnologiche di base, cliniche intensivologiche e perioperatorie", Catholic University of Sacred Heart of Rome, Italy. She has authored several publications, scientific dissemination articles, and book chapters.

antibiotics

Editorial

Toward a New Future for Essential Oils

Edoardo Napoli [1,*,†] and Maura Di Vito [2,3,*,†]

1 Istituto Chimica Biomolecolare—C.N.R., Via Paolo Gaifami 18, 95126 Catania, Italy
2 Dipartimento di Scienze e Tecnologie Agro-Alimentari, Università of Bologna, Viale G. Fanin 42, 40127 Bologna, Italy
3 Dipartimento di Scienze Biotecnologiche di Base, Cliniche Intensivologiche e Perioperatorie, Università Cattolica del Sacro Cuore, Largo A. Gemelli 8, 00168 Rome, Italy
* Correspondence: edoardo.napoli@icb.cnr.it (E.N.); wdivit@gmail.com (M.D.V.)
† The two authors contributed equally to this work.

Citation: Napoli, E.; Di Vito, M. Toward a New Future for Essential Oils. *Antibiotics* **2021**, *10*, 207. https://doi.org/10.3390/antibiotics10020207

Received: 5 February 2021
Accepted: 16 February 2021
Published: 19 February 2021

Publisher's Note: MDPI stays neutral with regard to jurisdictional claims in published maps and institutional affiliations.

Essential oils (EOs) are peculiar phytocomplexes in the already widely varied world of natural bioactive substances. They represent the volatile and aromatic components of some officinal plants, called aromatic plants [1], obtained through well-regulated extraction techniques [2]. They have been known, used, and studied for therapeutic purposes since ancient times, when, due to their evocative power, they were also considered a means of communication between humans and spiritual entities. However, it is their uses as therapeutic agents in the form of active ingredients belonging to the spices and to preserve food that have driven increasing interest in their use over the years. Their fame as antiseptic substances derive from their use in traditional medicines around the world [3].

Antibiotic multi-resistance has become a global emergency in recent decades. The indiscriminate use of antibiotic substances in both human and veterinary fields has led us from the dream of magic pills to the modern nightmare of the ineffectiveness of many previously active molecules. Despite all efforts, international research has struggled to find new and effective chemical drugs, so is looking to the natural world as a source of new scaffolds with new antibiotic activity. The broad-spectrum antimicrobial activity manifested by EOs, in a similar context, must be noticed. Their nature as a complex chemical mixture, which varies in terms of the quantity of their individual bioactive components, makes them resistant to any mechanism of microbial resistance. However, much remains to be understood about their mechanism of action, although often it is attributed to their ability to interact with the bacterial microbial wall [4]. The control of microbial growth is also an important problem for the food industry (food deterioration and shelf-life extension) and for cultural heritage artworks (indoor and outdoor biodeterioration) [5–8]. In this scenario, EOs can play an important role.

EOs have also shown antiviral activity, a peculiarity that is arousing some interest, especially in recent years [9,10]. The use of these natural compounds for inhalation and their interactions with the central nervous system are still topics of fascination that, however, require more robust scientific confirmation, which we expect in the near future due to the numerous research groups that are lending their attention in this field [11–13].

However, is there another side to this coin? Despite their considerable application potential, EOs present a series of problems that have limited their use on a large scale to date.

The first obstacle to the widespread diffusion of EOs is their potential for acute and chronic toxicity and their role in the development of some allergic reactions. Although most of the components considered individually and many EOs in their entirety are generally recognized as safe (GRAS), their use is often conditioned by restrictions that are not supported by adequate scientific documentation or are based on literature data that are not yet sufficient to determine the correct exposure doses. Therefore, their use is avoided according to the prudence principle. Additional effort by the scientific community in this regard is highly desirable.

Another problem that is important is their chemical variability. Being composed of secondary metabolites, they are considerably affected by external factors (climatic, agronomic, and anthropic), which make them extremely qualitatively and quantitatively variable over time. This problem has been, in part, limited due to the introduction of the chemotyping of EOs. An effort to fix the chemical compositions of EOs through analysis of large plant populations as well as the search for new endemics and new chemotypes must not fail to consider the variability derived from the biodiversity of the various aromatic species.

The extremely lipophilic nature, volatility, and susceptibility to oxidation shown by EOs also represent a limitation, especially technological, to their application. The lipophilicity makes them hardly usable in polar solvents, whereas their volatility and instability in the air has made their application difficult, especially in the agronomic field. The many studies on the applicability of nanotechnologies to overcome these limits has opened new perspectives that, until now, we could not have imagined.

Hydrolates (Hys), hydrophilic co-products of EOs, are increasingly attracting the interest of the scientific community due to their versatile characteristics. Although more perishable than EOs, these natural compounds have fair antimicrobial activity, are generally safe, and do not need to be diluted in a vehicle before use. These characteristics make them suitable not only for human and veterinary medicine (for both topical applications and oral administration), but also for other applications such as environmental, entomological, and agronomic applications [14–16].

In this Special Issue, we selected 13 manuscripts from eminent research groups, each of which provides a significant advance in knowledge about the biological activities and potential applications of EOs. Seven of them deal with antibacterial activity with reference to human, veterinary, and food applications. The potential antibacterial and antibiofilm activities of *Pimenta dioica* and *Pimenta racemosa* EOs in relation to their chemical composition, in addition to their ability to treat *Acinetobacter baumannii* wound infection in a mouse model, were investigated by Ismail et al. [17]. *P. dioica* leaf EO efficiently inhibited and eradicated biofilm formed by *A. baumannii*. Both *P. dioica* and *P. racemosa* leaf EOs also showed bactericidal activity. In addition, a significant reduction in *A. baumannii* microbial load in a mouse wound infection model was found.

Recent evidence suggests that *Staphylococcus* spp. has the ability to colonize the reproductive system and to affect its structure and functions. Kačániová et al. [18] focused on the antibacterial effects of 14 selected EOs against 50 *Staphylococcus* spp. cultures isolated from human semen. The best anti-*Staphylococcus* activities were found for the EOs of *Canarium luzonicum* (Blume) A. Gray, *Amyris balsamifera*, *Cinnamomum camphora*, and *Pogostemon cablin*.

Di Vito et al. [19] studied the in vitro antimicrobial and anti-biofilm effectiveness of both *Origanum vulgare* EO and a commercial product based on EOs on 29 *Salmonella* spp. strains isolated from chicken and pig farms. The authors concluded on the possible use of a commercial formulation both to reduce the meat contamination of *Salmonella* spp. before slaughter and in synergy with low doses of ciprofloxacin against resistant strains diffusion.

Di Stefano et al. [20] focused on the comparative study of the antimicrobial and antifungal activities of different grades of EOs extracted from the resins of three different *Boswellia sacra* cultivars toward relevant pathogens. One EO showed a minimum inhibitory concentration (MIC) of 52 mg/mL toward both *Staphylococcus aureus* and *Pseudomonas aeruginosa* pathogens, while other samples were particularly active against a dermatological strain of *Propionibacterium acnes*. Data obtained from in vitro studies showed that all EOs had a significant antifungal effect against *Candida albicans* and *Malassezia furfur*.

The biological activities of an EO distilled by Brazilian medicinal plant *Aniba rosaeodora* were evaluated by Teles et al. [21]. This EO, with linalool as its major compound, showed activity against all the bacteria strains tested, standard strains, and marine environment bacteria, with the lower minimum inhibitory concentration observed for *S. aureus*. The antitrypanosomal activity of *A. rosaeodora* EO and linalool was observed at high concentrations against epimastigote forms and even higher concentrations against intracellular amastigotes of *Trypanosoma cruzi*. Both *A. rosaeodora* EO and linalool did not exhibit a cytotoxic effect in BALB/c peritoneal macrophages, and both reduced nitrite levels in unstimulated cells, revealing a potential effect on nitric oxide production.

The inhibition of glucosyltransferase activity and glucan production as an antibiofilm mechanism of lemongrass EO against *Escherichia coli* O157:H7 was studied by Ortega-Ramirez et al. [22]. The planktonic growth of *E. coli* was inhibited by EO, citral, and geraniol as per as the bacterial adhesion on stainless steel. All compounds decreased the glucans production. The evidence collected by docking analysis indicated that both terpenes could interact with the helix finger of the glucosyltransferase responsible for polymer production.

Finally, the antimicrobial activity of some hydrolates was evaluated in comparison with the corresponding EOs by Di Vito et al. [23]. The authors highlight that although the minimum inhibitory concentration values of the EOs are lower than the corresponding Hys, the volatiles contained in the Hys are more effective in inhibiting microbial growth because they are active at lower concentrations. These data support the growing interest in Hys, especially when it is necessary to act in hydrophilic environments with products safer than essential oils.

The antifungal potential of EOs was investigated in three studies. An example of the use of EOs as novel alternatives to the application of synthetic fungicides to control seed-borne pathogens is provided by Moumni et al. [24], with their in vitro study conducted on the growth inhibition of seven EOs against *Stagonosporopsis cucurbitacearum* and *Alternaria alternata*. EO with citral, β-myrcene, and geraniol as major components controlled these fungi most effectively, followed by EOs containing terpinen-4-ol or linalool. Kisova et al., using a microarray approach, provide an important contribution to understanding the mechanism of action of EOs against fungi [25]. They evaluated the gene expression of *Penicillium chrysogenum* exposed to the indirect contact (vapors) of eight EOs. A microarray investigation confirmed their main impact on the metabolic processes in *P. rubens* involved vital functions.

Nanoencapsulation of EOs is a promising topic in nanotechnology. Ecofriendly EOs nanosuspensions with their broad-spectrum antimicrobial activity could be a valid alternative to synthetic products. The antifungal activity of a nanoencapsulated EO of oregano and thyme was evaluated by Kapustova et al. [26]. Their results showed that the nanosystems containing both thyme and oregano EOs are active against various fungal strains belonging to natural environments and materials. In particular, the minimum inhibitory concentration and minimum fungicidal concentration values were two to four times lower than EOs alone. This result suggests interesting applications in the agri-food and environmental fields.

Two studies were devoted to the use of EOs as chemotaxonomic markers. This is an approach that finds application both in botany, providing a tool for the enhancement of biodiversity and endemics, and in the quality control of EOs. The differences in the composition of EOs obtained from the aerial parts of six *Ferula* species, determined by also using a chemometric approach, were addressed by Youssef et al. [27]. In this manuscript, the authors evaluate the in vitro antioxidant potential of the EOs together with tyrosinase inhibitory potential using different assays, concluding that *Ferula* species could serve as a promising natural antioxidant drug that could be included in different cosmetics or pharmaceutical products to counteract hyperpigmentation.

The results of gas chromatographic analytical techniques coupled with random amplified polymorphic DNA (RAPD) techniques allowed Kim et al. [28] to obtain a perfect distinction between two Korean thyme cultivars (Wolchul and Odae) and commercial thyme cultivars.

The role of EOs as feeding agents for farm animals and their effect on meat quality is a topic of interest. The beneficial effect of oregano EO administration on animal diet was demonstrated by Garcia-Galicia et al. [29] in a study on lambs. This work demonstrates that oregano EO was beneficial for lambs feeding and could be a natural alternative to replace monensin in lamb diets, with improvements demonstrated in the quality of the meat.

In conclusion, the works published in this Special Issue provide a broad and clearly non-exhaustive vision of the application potential of EOs and show the path that is being followed for their application in the food, human health, veterinary fields, with particular reference to the control of microbial and fungal infections and/or contaminations. We expect this Special Issue to be of great help to those interested in the valorization and exploitation of EOs.

Conflicts of Interest: The authors declare no conflict of interest.

References

1. Napoli, E.M.; Ruberto, G. Sicilian Aromatic Plants: From Traditional Heritage to A New Agro-Industrial Exploitation. In *Spices: Types, Uses and Health Benefits*; Kralis, J.F., Ed.; Nova Publishers: New York, NY, USA, 2012; pp. 1–56.
2. *European Pharmacopeia*, 6th ed.; Supplement 5.8; European Directorate for the Quality of Medicines and Healthcare: Strasbourg, France, 2007.
3. Napoli, E.; Ruberto, G.; Siracusa, L. New tricks for old guys: Recent developments in the chemistry, biochemistry, applications and exploitations of selected species from Lamiaceae family. *Chem. Biodivers.* **2020**, *17*, e1900677. [CrossRef]
4. Sharmeen, J.B.; Shanoo, S.; Kannan, R.R.R.; Fawzi, M.M. Chemistry, bioactivities, mode of action and industrial application of essential oils. *Trend Food Sci. Technol.* **2020**, *101*, 89–105.
5. Di Vito, M.; Bellardi, M.G.; Colaizzi, P.; Ruggiero, D.; Mazzucca, C.; Micheli, L.; Sotgiu, S.; Iannuccelli, S.; Michelozzi, M.; Mondello, F.; et al. Hydrolates and gellan: An eco-innovative synergy for safe cleaning of paper artworks. *Stud. Conserv.* **2017**, *63*, 1–11. [CrossRef]
6. Stupar, M.; Grbić, M.L.; Džamić, A.; Unković, N.; Ristić, M.; Jelikić, A.; Vukojević, J. Antifungal activity of selected EOs and biocide benzalkonium chloride against the fungi isolated from cultural heritage objects. *S. Afr. J. Bot.* **2014**, *93*, 118–124. [CrossRef]
7. Romano, I.; Granata, G.; Poli, A.; Finore, I.; Napoli, E.; Geraci, C. Inhibition of bacterial growth on marble stone of 18th century by treatment of nanoencapsulated essential oils. *Int. Biodeterior. Biodegrad.* **2020**, *148*, 104909. [CrossRef]
8. Palla, F.; Bruno, M.; Mercurio, F.; Tantillo, A.; Rotolo, V. EOs as natural biocides in conservation of cultural heritage. *Molecules* **2020**, *25*, 730. [CrossRef] [PubMed]
9. Schnitzler, P. EOs for the treatment of *Herpes simplex* virus infections. *Chemotherapy* **2019**, *64*, 1–7. [CrossRef]
10. Leigh-de Rapper, S.; van Vuuren, S.F. Odoriferous therapy: A review identifying EOs against pathogens of the respiratory tract. *Chem. Biodivers.* **2020**, *17*, e2000062.
11. Yuan, R.; Zhang, D.; Yang, J.; Wu, Z.; Luo, C.; Han, L.; Yang, F.; Lin, J.; Yang, M. Review of aromatherapy EOs and their mechanism of action against migraines. *J. Ethnopharmacol.* **2021**, *30*, 113326. [CrossRef] [PubMed]
12. Setzer, W.N. EOs and anxiolytic aromatherapy. *Nat. Prod. Commun.* **2009**, *4*, 1305–1316. [PubMed]
13. Zhang, N.; Yao, L. Anxiolytic effect of EOs and their constituents: A review. *J. Agric. Food Chem.* **2019**, *67*, 13790–13808. [CrossRef] [PubMed]
14. Politi, M.; Menghini, L.; Conti, B.; Bedini, S.; Farina, P.; Cioni, P.L.; Braca, A.; Leo, M. Reconsidering Hydrosols as main products of aromatic plants manufactory: The Lavandin (*Lavandula intermedia*) case study in Tuscany. *Molecules* **2020**, *25*, 2225. [CrossRef] [PubMed]
15. Di Vito, M.; Bellardi, M.G.; Sanguinetti, M.; Mondello, F.; Girolamo, A.; Barbanti, L.; Garzoli, S.; Sabatino, M.; Ragno, R.; Vitali, A.; et al. Potent in vitro activity of *Citrus aurantium* EO and *Vitis vinifera* Hydrolate against gut yeast isolates from irritable bowel syndrome patients-The right mix for potential therapeutic use. *Nutrients* **2020**, *12*, 1329. [CrossRef] [PubMed]
16. Marino, A.; Nostro, A.; Mandras, N.; Roana, J.; Ginestra, G.; Miceli, N.; Taviano, M.F.; Gelmini, F.; Beretta, G.; Tullio, V. Evaluation of antimicrobial activity of the hydrolate of *Coridothymus capitatus* (L.) Reichenb. fil. (Lamiaceae) alone and in combination with antimicrobial agents. *BMC Complementary Med. Ther.* **2020**, *20*, 89. [CrossRef]
17. Ismail, M.M.; Samir, R.; Saber, F.R.; Ahmed, S.R.; Farag, M.A. Pimenta oil as a potential treatment for *Acinetobacter baumannii* wound infection: In vitro and in vivo bioassays in relation to its chemical composition. *Antibiotics* **2020**, *9*, 679. [CrossRef]
18. Kačániová, M.; Terentjeva, M.; Štefániková, J.; Žiarovská, J.; Savitskaya, T.; Grinshpan, D.; Kowalczewski, P.L.; Vukovic, N.; Tvrdá, E. Chemical composition and antimicrobial activity of selected essential oils against *Staphylococcus* spp. isolated from human semen. *Antibiotics* **2020**, *9*, 765. [CrossRef]
19. Di Vito, M.; Cacaci, M.; Barbanti, L.; Martini, C.; Sanguinetti, M.; Benvenuti, S.; Tosi, G.; Fiorentini, L.; Scozzoli, M.; Bugli, F.; et al. *Origanum vulgare* essential oil vs. a commercial mixture of essential oils: In vitro effectiveness on *Salmonella* spp. from poultry and swine intensive livestock. *Antibiotics* **2020**, *9*, 763. [CrossRef]
20. Di Stefano, V.; Schillaci, D.; Cusimano, M.G.; Rishan, M.; Rashan, L. In Vitro antimicrobial activity of Frankincense Oils from *Boswellia sacra* grown in different locations of the Dhofar region (Oman). *Antibiotics* **2020**, *9*, 195. [CrossRef]

21. Teles, A.M.; Silva-Silva, J.V.; Pereira Fernandes, J.M.; da Silva Calabrese, K.; Abreu-Silva, A.L.; Carvalho Marinho, S.; Nascimento Mouchrek, A.; Mouchrek Filho, V.E.; Almeida-Souza, F. *Aniba rosaeodora* (Var. *amazonica* Ducke) essential oil: Chemical composition, antibacterial, antioxidant and antitrypanosomal activity. *Antibiotics* **2021**, *10*, 24. [CrossRef]
22. Ortega-Ramirez, L.A.; Gutiérrez-Pacheco, M.M.; Vargas-Arispuro, I.; González-Aguilar, G.A.; Martínez-Téllez, M.A.; Fernando Ayala-Zavala, J.F. Inhibition of glucosyltransferase activity and glucan production as an antibiofilm mechanism of Lemongrass essential oil against *Escherichia coli* O157:H7. *Antibiotics* **2020**, *9*, 102. [CrossRef]
23. Di Vito, M.; Smolka, A.; Proto, M.R.; Barbanti, L.; Gelmini, F.; Napoli, E.; Bellardi, M.G.; Mattarelli, P.; Beretta, G.; Sanguinetti, M.; et al. Is the antimicrobial activity of hydrolates lower than that ofessential oils? *Antibiotics* **2021**, *10*, 88. [CrossRef]
24. Moumni, M.; Romanazzi, G.; Najar, B.; Pistelli, L.; Ben Amara, H.; Mezriouri, K.; Karous, O.; Chaieb, I.; Allagui, M.B. Antifungal activity and chemical composition of seven essential oils to control the main seedborne fungi of cucurbits. *Antibiotics* **2021**, *10*, 104. [CrossRef]
25. Kisova, Z.; Šoltýsová, A.; Bucková, M.; Beke, G.; Puškárová, A.; Pangallo, D. Studying the gene expression of *Penicillium rubens* under the effect of eight essential oils. *Antibiotics* **2020**, *9*, 343. [CrossRef] [PubMed]
26. Kapustova, M.; Granata, G.; Napoli, E.; Puskarova, A.; Buckova, M.; Pangallo, D.; Geraci, C. Nanoencapsulated essential oils with enhanced antifungal activity for potential application on agri-food, material and environmental fields. *Antibiotics* **2021**, *10*, 31. [CrossRef]
27. Youssef, F.S.; Mamatkhanova, M.A.; Mamadalieva, N.Z.; Zengin, G.; Aripova, S.F.; Alshammari, E.; Ashour, M.L. Chemical profiling and discrimination of essential oils from six Ferula species using GC analyses coupled with chemometrics and evaluation of their antioxidant and enzyme inhibitory potential. *Antibiotics* **2020**, *9*, 518. [CrossRef] [PubMed]
28. Kim, M.; Moon, J.; Kim, S.; Sowndharajan, K. Morphological, Chemical, and genetic characteristics of Korean native Thyme Bak-Ri-Hyang (*Thymus quinquecostatus* Celak.). *Antibiotics* **2020**, *9*, 289. [CrossRef] [PubMed]
29. Garcia-Galicia, I.A.; Arras-Acosta, J.A.; Huerta-Jimenez, M.; Renteria-Monterubbio, A.L.; Loya-Olguin, J.L.; Carillo-Lopez, L.M.; Tirado-Gallegos, J.M.; Alarcon-Rojo, A.D. Natural oregano essential oil may replace antibiotics in lamb diets: Effects on meat quality. *Antibiotics* **2020**, *9*, 248. [CrossRef] [PubMed]

antibiotics

MDPI

Article

Antifungal Activity and Chemical Composition of Seven Essential Oils to Control the Main Seedborne Fungi of Cucurbits

Marwa Moumni [1,2], Gianfranco Romanazzi [2,*], Basma Najar [3], Luisa Pistelli [3], Hajer Ben Amara [1], Kaies Mezrioui [1,2], Olfa Karous [4], Ikbal Chaieb [5] and Mohamed Bechir Allagui [1]

1 Laboratory of Plant Protection, National Institute for Agronomic Research of Tunisia, University of Carthage, Rue Hédi Karray, Ariana 2080, Tunisia; m.moumni@staff.univpm.it (M.M.); mzbenamara@gmail.com (H.B.A.); s1087850@studenti.univpm.it (K.M.); allagui.bechir@gmail.com (M.B.A.)
2 Department of Agricultural, Food and Environmental Sciences, Marche Polytechnic University, Via Brecce Bianche, 60131 Ancona, Italy
3 Dipartimento di Farmacia, Università di Pisa, Via Bonanno 33, 56126 Pisa, Italy; basma.najar@farm.unipi.it (B.N.); luisa.pistelli@unipi.it (L.P.)
4 National Agricultural Institute of Tunisia, 43 Avenue Charles Nicolle, Tunis 1082, Tunisia; karous-olfa@hotmail.fr
5 Regional Centre for Research in Horticulture and Organic Agriculture (CRRHAB), Chott Mariem, Sousse 4042, Tunisia; ikbal_c@yahoo.fr
* Correspondence: g.romanazzi@univpm.it; Tel.: +39-071-220-4336

Citation: Moumni, M.; Romanazzi, G.; Najar, B.; Pistelli, L.; Ben Amara, H.; Mezrioui, K.; Karous, O.; Chaieb, I.; Allagui, M.B. Antifungal Activity and Chemical Composition of Seven Essential Oils to Control the Main Seedborne Fungi of Cucurbits. *Antibiotics* **2021**, *10*, 104. https://doi.org/10.3390/antibiotics10020104

Academic Editors: Edoardo Marco Napoli and Maura Di Vito

Received: 30 December 2020
Accepted: 20 January 2021
Published: 22 January 2021

Publisher's Note: MDPI stays neutral with regard to jurisdictional claims in published maps and institutional affiliations.

Abstract: Essential oils represent novel alternatives to application of synthetic fungicides to control against seedborne pathogens. This study investigated seven essential oils for in vitro growth inhibition of the main seedborne pathogens of cucurbits. *Cymbopogon citratus* essential oil completely inhibited mycelial growth of *Stagonosporopsis cucurbitacearum* and *Alternaria alternata* at 0.6 and 0.9 mg/mL, respectively. At 1 mg/mL, *Lavandula dentata, Lavandula hybrida, Melaleuca alternifolia, Laurus nobilis,* and two *Origanum majorana* essential oils inhibited mycelia growth of *A. alternata* by 54%, 71%, 68%, 36%, 90%, and 74%, respectively. *S. cucurbitacearum* mycelia growth was more sensitive to *Lavandula* essential oils, with inhibition of 71% at 1 mg/mL. To determine the main compounds in these essential oils that might be responsible for this antifungal activity, they were analyzed by gas chromatography–mass spectrometry (GC-MS). *C. citratus* essential oil showed cirtal as its main constituent, while *L. dentata* and *L. nobilis* essential oils showed eucalyptol. The *M. alternifolia* and two *O. majorana* essential oils had terpinen-4-ol as the major constituent, while for *L. hybrida* essential oil, this was linalool. Thus, in vitro, these essential oils can inhibit the main seedborne fungi of cucurbits, with future in vivo studies now needed to confirm these activities.

Keywords: *Alternaria alternata*; cucurbits; *Cymbopogon citratus*; GC-MS; *Stagonosporopsis cucurbitacearum*

1. Introduction

Cucurbits are an important source of income for countries in the Mediterranean basin, with a total production of nearly 3,356,669 tonnes in 2018 [1]. Squash (*Cucurbita maxima* Duchesne; *Cucurbita moschata* Duchesne) is one of the major cucurbits grown in tropical and temperate regions. *Cucurbita* spp. can be affected by a number of fungal pathogens, which can cause major economic losses [2]. The majority of these fungi are seedborne, such as gummy stem blight (with foliar symptoms) and black rot (with fruit symptoms), which are caused by *Stagonosporopsis cucurbitacearum* (Fr.) Aveskamp, Gruyter & Verkley (anamorph: *Phoma cucurbitacearum* (Fr.) Sacc.), synonym *Didymella bryoniae* (Fuckel) Rehm, and which represent serious diseases that are a major constraint to cucurbit production worldwide [2–4]. Nuangmek et al. [5] reported that losses in cantaloupe can also reach 100% under conditions conducive to *S. cucurbitacearum*. *Alternaria alternata* (Fr.) Keissl. is the agent of leaf spot, which is a further major factor responsible for low cucurbit production. The genus *Alternaria* affects the plant seedlings, leaves, stalks, stems, flowers and fruit. Yield losses of 50% and more can occur under weather conditions that are conducive to leaf spot, and in particular temperatures of 25 to 32 °C associated with 40% relative humidity during the day and 95% at night [6]. Many other pathogens have been detected on cucurbits seeds, such as *Fusarium solani* [7,8], *Alternaria cucumerina* [9], *Paramyrothecium roridum*, and *Albifimbria verrucaria* [10,11].

The most important unit of the squash crop is the seed, which should be of high quality and pathogen free. The propagation of such seedborne fungi is generally controlled by chemical treatments [7,12,13]. Indeed, there have been many studies on chemical control against seedborne *S. cucurbitacearum* [14–16]. Sudisha et al. [17] reported that seed treatment with the Dithane M-45 (Mancozeb 75% WP) fungicide can reduce the incidence of gummy stem blight in muskmelon crops, although this active ingredient was recently refused approval for use in the European Union, so its use will be banned in few months. Chemical fungicides are generally adopted for disinfestation, disinfection, and protection of seeds and the emerging plantlets. However, these chemicals can also cause environmental pollution due to their high persistence in the soil and water, because of their slow biodegradability [18,19].

In recent years, alternatives to synthetic fungicides have been investigated due to the extensive use of fungicides for plant and seed treatments, the problems of pathogen resistance to fungicides that this causes, and the increased demand for organic and free-of-residue vegetables [20–23]. Natural organic compound, such as plant extracts and essential oils, are among the environmentally friendly alternatives that are being developed and tested for antifungal activities against seedborne pathogens [24]. Essential oils are a rich source of broad-spectrum antifungal plant-derived metabolites that inhibit both fungal growth and their production of toxic metabolites [25]. Tea tree essential oil contains terpinen-4-ol, 1,8-cineole, and γ-terpinene, and at 2%, it has shown potent inhibition of mycelial growth of *Fusarium graminearum*, *Fusarium culmorum*, and *Pyrenophora graminea* [26]. Riccioni and Orzali [27] reported that tea tree essential oil represents a source of sustainable eco-friendly botanical fungicides, because of its efficacy in the control of seedborne fungi. The genus *Cymbopogon* (Poaceae) is known for its essential oils, especially for extracts of lemongrass (*Cymbopogon citratus* (DC.) Stapf). The in vitro evaluation of the effectiveness of this essential oil on the main seedborne pathogens of cucurbits was reported previously [28–30].

The objectives of the present study were to evaluate the inhibitory effects of seven essential oils that differ in their chemical compositions and to determine what the most important compounds in these seven essential oils might be, using gas chromatography–mass spectrometry (GC-MS) analysis.

2. Results

2.1. In Vitro Inhibition of Fungal Growth by the Seven Essential Oils

The effects of increasing concentrations of seven essential oils on mycelial growth of the fungi *A. alternata* and *S. cucurbitacearum* were investigated. These essential oils were from various sources, and are defined as (see Table 1): *C.cit*, *Cymbopogon citratus* (lemon grass); *L.dent*, *Lavandula dentata* (lavender); *L.hyb*, *Lavandula hybrida* (lavandin); *M.alt*, *Melaleuca alternifoglia* (tea tree); *L.nob*, *Laurus nobilis* (bay laurel); *O.maj1/2*, *Origanum majorana* 1/2 (majoram).

As can be seen in Figures 1 and 2, and as summarized in Tables 2 and 3, all of these essential oils inhibited the growth of these two fungi in a dose-dependent manner. The greatest inhibitory activity was shown by the *C.cit* essential oil, with 100% inhibition of mycelial growth of both *A. alternata* and *S. cucurbitacearum* reached at 0.6 mg/mL and 0.9 mg/mL, respectively (Table 2). *A. alternata* was generally more sensitive to these essential oils than *S. cucurbitacearum*, and at 1 mg/mL essential oils, its mycelia growth was inhibited by 55.0%, 71.5%, 68.2%, 36.1%, 74.2%, and 90.5% by *L.dent*, *L.hyb*, *M.alt*, *L.nob*,

O.maj1, and *O.maj2*, respectively (Table 2). At the same essential oil concentration, *S. cucurbitacearum* radial growth was inhibited by 73.5%, 74.0%, 73.7%, 65.3%, 60.0%, and 67.3%, respectively (Table 3). The positive control of the fungicide combination of 25 g/L difenoconazole plus 25 g/L fludioxonil completely inhibited the mycelial growth of *A. alternata* at all concentrations tested. Against *S. cucurbitacearum*, this fungicide combination at 0.1, 0.5, and 1 mg/mL inhibited the mycelial growth by 75.7%, 84.9%, and 86.7%, respectively.

In addition, the *C.cit* essential oil had a fungicidal effect against *S. cucurbitacearum* from 900 µg/mL. Indeed, for *A. alternata*, *C.cit* had fungistatic effects at 0.6 mg/mL and 0.7 mg/mL, and it was fungicidal from 0.8 mg/mL (Table 4). These data show that the *C.cit* had potent antifungal activity against *A. alternata* and *S. cucurbitacearum* with IC_{50} values of 0.315 mg/mL and 0.102 mg/mL, respectively (Figure 3). The essential oils of *L.dent*, *L.hyb*, *M.alt*, *O.maj1*, and *O.maj2* showed moderate antifungal activities against *A. alternata*, with IC_{50} values from 0.473 mg/mL to 0.893 mg/mL, as similarly against *S. cucurbitacearum*, with IC_{50} values from 0.322 mg/mL to 0.884 mg/mL. However, *L.nob* showed only weak antifungal activities against both *A. alternata* and *S. cucurbitacearum*, as seen by its relatively high IC_{50} values of 1.310 mg/mL and 1.248 mg/mL, respectively (Figure 3).

Table 1. Details of the essential oils included in this study.

Code	Species	Family	Common Name	Source
C.cit	*Cymbopogon citratus* (DC.) Stapf	*Poaceae*	Lemongrass	Biopesticides Laboratory, Regional Centre for Research in Horticulture and Organic Agriculture (CRRHAB), Tunisia
L.dent	*Lavandula dentata* L.	*Lamiaceae*	Lavender	CRRHAB, Tunisia
L.hyb	*Lavandula hybrida* E.Rev. ex Briq	*Lamiaceae*	Lavandin	FLORA s.r.l. (Batch N° 161808)
M.alt	*Melaleuca alternifolia* (Maiden & Betche) Cheel	*Myrtaceae*	Tea tree	FLORA s.r.l. (Batch N° 161960)
L.nob	*Laurus nobilis* L.	*Lauraceae*	Bay laurel	Medicinal Plants Laboratory, National Institute of Agronomy of Tunisia (INAT)
O.maj1	*Origanum majorana* L.	*Lamiaceae*	Marjoram	INAT
O.maj2	*Origanum majorana* L.	*Lamiaceae*	Marjoram	CRRHAB, Tunisia

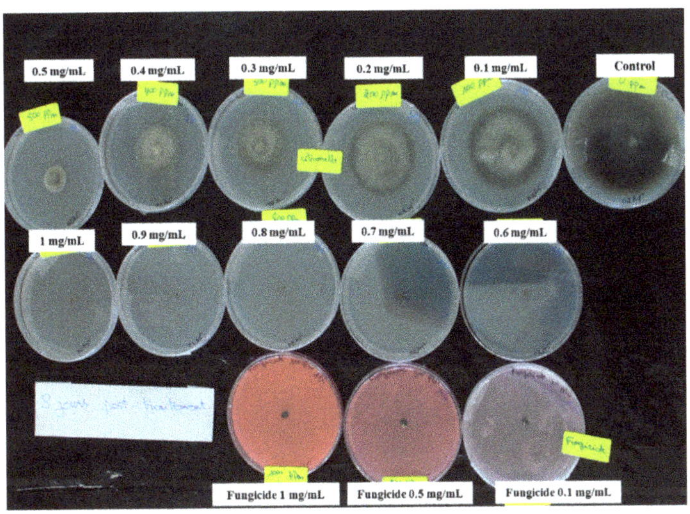

Figure 1. Representative experiment showing inhibition of *Alternaria alternata* mycelial growth by *Cymbopogon citratus* essential oil at 0.1 to 1 mg/mL and by the fungicide combination of 25 g/L difenoconazole plus 25 g/L fludioxonil at 0.1, 0.5 and 1 mg/mL, as seen after 8 days of incubation at 22 ± 2 °C.

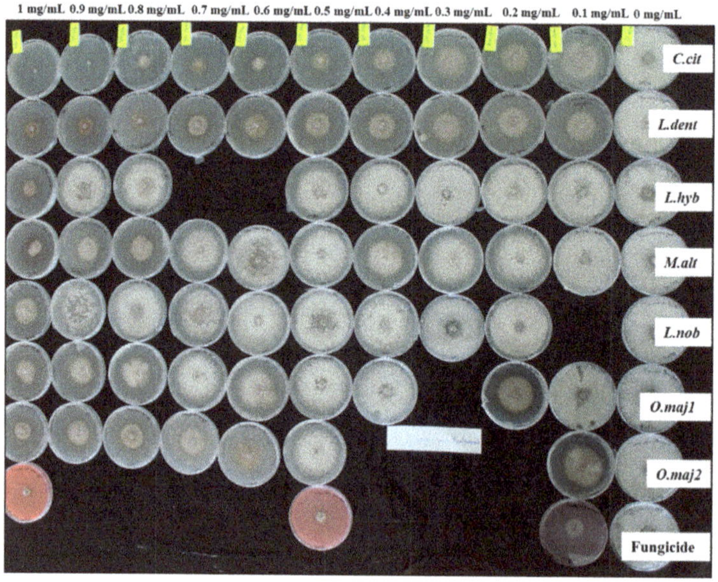

Figure 2. Representative experiment showing inhibition of *Stagonosporopsis cucurbitacearum* mycelial growth by the seven essential oils: *C.cit, Cymbopogon citratus; L.dent, Lavandula dentata; L.hyb, Lavandula hybrida; M.alt, Melaleuca alternifolia; L.nob, Laurus nobilis; O.maj1/2, Origanum majorana* 1/2, at increasing concentrations (right to left; as indicated) from 0 mg/mL (control) to 1 mg /mL, and by the fungicide combination of 25 g/L difenoconazole plus 25 g/L fludioxonil (positive control) at 0.1, 0.5 and 1 mg/mL, after 7 days of incubation at 22 ± 2 °C.

Table 2. Mycelial growth inhibition of *Alternaria alternata* by the seven essential oils. *C.cit*, *Cymbopogon citratus*; *L.dent*, *Lavandula dentata*; *L.hyb*, *Lanvandula hybrid*; *M.alt*, *Melaleuca alternifolia*; *L.nob*, *Laurus nobilis*; *O.maj1/O.maj2*, *Origanum majorana* 1/2, after 7 days of incubation at 22 ± 2 °C.

Essential Oil	Inhibition of Mycelial Growth of *Alternaria alternata* (%) at Increasing Essential Oil Concentrations (mg/mL)										
	0	0.1	0.2	0.3	0.4	0.5	0.6	0.7	0.8	0.9	1
C.cit	0.00	20.28 ± 4.32	25.03 ± 1.56	43.21 ± 2.33	65.92 ± 6.29	79.14 ± 10.97	100	100	100	100	100
L.dent	0.00	9.95 ± 2.30	20.07 ± 3.63	29.99 ± 4.97	44.44 ± 3.13	43.82 ± 1.84	43.00 ± 3.18	37.63 ± 1.65	44.44 ± 0.21	43.00 ± 1.89	54.98 ± 0.83
L.hyb	0.00	31.23 ± 4.40	32.88 ± 4.72	42.38 ± 2.18	50.64 ± 2.89	51.05 ± 2.50	59.31 ± 3.76	63.44 ± 1.64	65.72 ± 1.49	67.37 ± 0.55	71.50 ± 0.95
M.alt	0.00	13.47 ± 5.36	15.74 ± 3.06	21.52 ± 10.17	21.73 ± 5.38	24.00 ± 1.09	25.44 ± 4.42	29.37 ± 3.45	62.00 ± 1.49	67.99 ± 2.33	68.19 ± 2.98
L.nob	0.00	0.58 ± 0.29	0.58 ± 0.29	1.28 ± 0.41	8.72 ± 2.51	12.02 ± 1.56	23.17 ± 2.48	24.00 ± 1.65	25.03 ± 3.41	33.91 ± 4.96	36.08 ± 2.44
O.maj1	0.00	20.07 ± 2.18	37.84 ± 9.68	42.59 ± 2.30	45.48 ± 4.13	47.13 ± 1.80	47.75 ± 1.09	53.12 ± 7.10	60.97 ± 3.41	64.27 ± 3.25	74.18 ± 1.45
O.maj2	0.00	27.10 ± 0.55	31.85 ± 1.99	32.47 ± 1.99	33.91 ± 0.55	37.01 ± 0.41	42.38 ± 1.24	42.59 ± 0.83	46.92 ± 0.55	63.86 ± 1.61	90.50 ± 9.50
Fungicides [a]	0.00	100	-	-	-	100	-	-	-	-	100

Data are means ±SD (n = 3). [a] 25 g/L difenoconazole + 25 g/L fludioxonil. - not tested.

Table 3. Mycelial growth inhibition of *Stagonosporopsis cucurbitacearum* by the seven essential oils. *C.cit*, *Cymbopogon citratus*; *L.dent*, *Lavandula dentata*; *L.hyb*, *Lanvandula hybrid*; *M.alt*, *Melaleuca alternifolia*; *L.nob*, *Laurus nobilis*; *O.maj1/O.maj2*, *Origanum majorana* 1/2, after 7 days of incubation at 22 ± 2 °C.

Essential Oil	Inhibition of Mycelial Growth of *Stagonosporopsis cucurbitacearum* (%) at Increasing Essential Oil Concentrations (mg/mL)										
	0	0.1	0.2	0.3	0.4	0.5	0.6	0.7	0.8	0.9	1
C.cit	0.00	51.76 ± 3.86	53.53 ± 1.22	58.24 ± 0.59	67.25 ± 1.37	74.51 ± 3.08	85.49 ± 7.45	82.16 ± 9.00	87.06 ± 6.51	100	100
L.dent	0.00	29.41 ± 0.00	41.18 ± 6.79	47.06 ± 1.36	51.76 ± 0.68	61.37 ± 0.71	62.75 ± 0.71	66.27 ± 0.39	71.37 ± 0.78	73.33 ± 0.39	73.53 ± 0.59
L.hyb	0.00	5.49 ± 1.68	6.27 ± 2.18	10.39 ± 2.05	12.75 ± 1.04	21.57 ± 5.85	22.75 ± 4.43	27.65 ± 4.75	53.92 ± 0.52	63.73 ± 0.85	73.92 ± 3.63
M.alt	0.00	3.92 ± 1.96	22.16 ± 1.04	20.78 ± 3.03	20.59 ± 3.02	30.78 ± 3.16	31.37 ± 4.11	41.96 ± 3.35	60.20 ± 0.71	63.92 ± 0.20	73.73 ± 0.85
L.nob	0.00	0.98 ± 0.71	3.73 ± 3.73	5.88 ± 2.38	5.88 ± 2.37	5.69 ± 0.52	6.67 ± 0.71	11.37 ± 4.37	15.29 ± 4.57	29.22 ± 8.35	65.29 ± 3.83
O.maj1	0.00	0.00	5.69 ± 1.87	8.24 ± 1.22	9.02 ± 1.19	10.98 ± 1.37	33.73 ± 4.31	33.73 ± 3.74	47.45 ± 6.97	54.12 ± 2.96	60.00 ± 1.56
O.maj2	0.00	34.12 ± 4.42	49.22 ± 1.41	54.31 ± 4.27	56.47 ± 1.96	57.25 ± 1.56	58.82 ± 1.04	59.61 ± 1.80	61.57 ± 1.19	63.53 ± 0.34	67.25 ± 1.74
Fungicides [a]	0.00	75.69 ± 0.00	-	-	-	84.90 ± 0.68	-	-	-	-	86.67 ± 0.39

Data are means ±SD (n = 3). [a] 25 g/L difenoconazole + 25 g/L fludioxonil. - not tested.

Table 4. Fungistatic and fungicidal activities of *Cymbopogon citratus* essential oil on mycelia growth of *Stagonosporopsis cucurbitacearum* and *Alternaria alternata* after 7 days of incubation at 22 ± 2 °C.

Fungus	Fungicidal and Fungistatic Activities of *Cymbopogon citratus* Essential Oil at Increasing Concentrations (mg/mL)									
	0.6		0.7		0.8		0.9		1	
	Fungicidal	Fungistatic	Fungicidal	Fungistatic	Fungicidal	Fungistatic	Fungicidal	Fungistatic	Fungicidal	Fungistatic
Alternaria alternata	NO	Yes	NO	Yes	Yes	NO	Yes	NO	Yes	NO
Stagonosporopsis cucurbitacearum	-	-	-	-	-	-	Yes	NO	Yes	NO

- not tested.

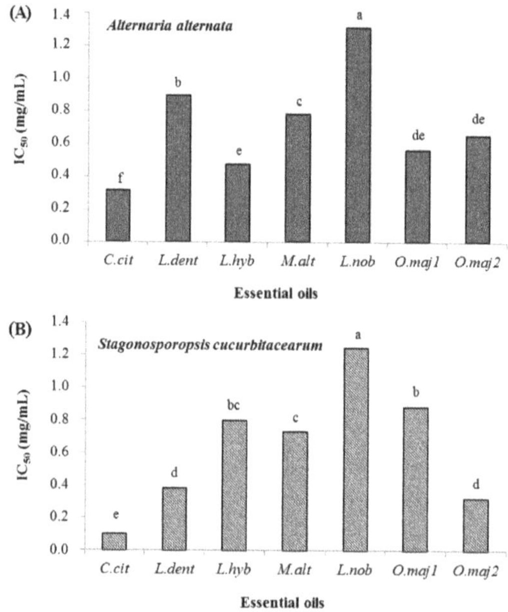

Figure 3. Inhibitory concentration for 50% reduction (IC_{50}) of mycelial growth of *Alternaria alternata* (**A**) and *Stagonosporopsis cucurbitacearum* (**B**) by the seven essential oils: *C.cit, Cymbopogon citratus; L.dent, Lavandula dentata; L.hyb, Lanvandula hybrida; M.alt, Melaleuca alternifolia; L.nob, Laurus nobilis; and O.maj1/2, Origanum majorana* 1/2. Data with different letters (**a–f**) are significantly different between treatments ($p \leq 0.05$; Fisher's LSD).

2.2. Chemical Profiles of the Essential Oils

The 41 components given in Table 5 were identified as comprising from 97.7% (*O.maj1*) to 100% (*L.hyb*) of these essential oils. The oxygenated monoterpenes dominated in all of the essential oils, even though these belonged to different plant families and species. They represented more than two-thirds of the fraction in three of the four *Lamiaceae*: *L.dent* (81.1%), *L.hyb* (90.8%), and *O.maj2* (66.8%). The oxygenated monoterpenes were also the highest proportionally in *C.cit* (88.5%) and *L.nob* (70.3%). On the other hand, the compositions of *M.alt* and *O.maj1* were divided mainly between oxygenated monoterpenes as the main class (48.1%, 49.7%, respectively) and monoterpene hydrocarbons in similar proportions (40.4%, 44.3%, respectively). In more detail, *C.cit* showed α-citral (geranial; 51.6%) and β-citral (neral; 26.0%), whereby these two major oxygenated monoterpenes represented together over three-quarters of the total composition. In the *Lamiaceae*, almost two-thirds of *L.dent* was eucalyptol (63.5%) and β-selinene (4.1%). Instead, the total composition of *O.maj1* and *O.mag2* included around half and over two-thirds as terpenen-4-ol (44.8%) and *p*-cymene (68.2%), respectively, followed by γ-terpinene (12.6%) for *O.maj1* and α-terpineol (5.4%) for *O.maj2*. For the two commercial essential oils, the main compounds of *L.hyb* were linalool (33.7%) and linalyl acetate (27.7%), followed by camphor (9.3%), while *M.alt* showed terpinen-4-ol (41.1%) as 86% of its oxygenated monoterpene, with γ-terpinene (16.0%), p-cymene (9.3%), and α-terpinene (6.1%), together representing 78% of the monoterpene hydrocarbons. Finally, more than half of the identified fractions of the *Lauraceae L.nob* were characterized by the combination of eucalyptol (47.9%) and α-terpinyl acetate (10.2%).

Table 5. Relative levels of the volatile constituents of the different essential oils determined by gas chromatography–mass spectrometry analysis.

	Compound [a]	Class	Linear Retention Index		Relative amount (%)						
			(*n*-alkane) [b]	(Adams, 2007) [c]	*Poaceae*	*Lamiaceae*				*Myrtaceae*	*Lauraceae*
					C.cit	*L.dent*	*L.hyb*	*O.maj1*	*O.maj2*	*M.alt*	*L.nob*
1	α-Pinene	mh [d]	937	932	0.1	0.6	0.5	0.6	0.3	2.7	5.6
2	Sabinene	mh	974	969	0.1	-	0.1	4.5	2.2	-	6.7
3	β-Pinene	mh	979	974	-	3.1	0.5	0.4	0.2	0.7	5.0
4	β-Myrcene	mh	991	988	5.3	-	0.4	1.1	0.6	0.6	1.2
5	α-Terpinene	mh	1017	1014	-	-	-	5.7	0.8	6.1	0.8
6	*p*-Cymene	mh	1025	1020	0.3	0.9	0.2	11.3	17.8	9.3	0.6
7	Limonene	mh	1030	1024	0.4	1.1	0.8	3.5	2.4	1.0	1.7
8	Eucalyptol	om	1032	1026	0.6	63.5	6.5	0.7	0.2	2.8	47.9
9	γ-Terpinene	mh	1060	1054	-	-	-	12.6	3.8	16.0	1.4
10	Terpinolene	mh	1088	1086	-	-	0.2	3.4	1.2	3.0	0.3
11	Linalool	om	1099	1095	0.8	1.8	33.7	1.1	3.6	-	7.4
12	*cis*-*p*-Menth-2-en-1-ol	om	1122	1118	-	-	-	1.3	0.7	0.1	-
13	*trans*-Pinocarveol	om	1139	1135	-	2.9	-	-	-	-	-
14	*trans*-*p*-Menth-2-en-1-ol	om	1141	1136	-	-	-	1.2	0.7	0.1	-
15	Camphor	om	1145	1141	-	0.5	9.3	-	0.1	-	-
16	γ-Terpineol	om	1166	1162	-	1.3	-	-	-	-	0.3
17	*endo*-Borneol	om	1167	1165	-	0.4	4.4	0.1	0.2	-	-
18	*p*-Mentha-1,5-dien-8-ol	om	1170	1166	2.5	-	-	-	-	-	-
19	Terpinen-4-ol	om	1177	1174	0.1	1.3	4.5	32.4	50.1	41.1	1.5
20	*p*-Cymen-8-ol	om	1183	1179	1.1	0.3	-	0.3	0.4	-	-
21	Cryptone	nt	1186	1183	-	1.3	-	-	-	-	-
22	α-Terpineol	om	1189	1186	-	1.8	1.1	6.0	5.4	3.7	1.6
23	Myrtenal	om	1198	1195	-	2.7	-	-	-	-	-
24	*trans*-Piperitol	om	1208	1207	-	-	-	1.0	0.6	-	-
25	β-Citral	om	1240	1235	26.0	-	-	-	-	-	-
26	Carvone	om	1243	1239	0.9	1.6	-	0.2	0.6	-	-
27	Geraniol	om	1253	1249	2.7	-	-	-	-	-	-
28	Linalyl acetate	om	1257	1254	-	-	27.7	2.7	2.2	-	0.2
29	α-Citral	om	1270	1264	51.6	-	-	-	-	-	-
30	2-Undecanone	nt	1294	1293	1.2	-	-	-	-	-	-

Table 5. *Cont.*

	Compound [a]	Class	Linear Retention Index		Relative amount (%)						
			(*n*-alkane) [b]	(Adams, 2007) [c]	*Poaceae*	*Lamiaceae*				*Myrtaceae*	*Lauraceae*
					C.cit	*L.dent*	*L.hyb*	*O.maj1*	*O.maj2*	*M.alt*	*L.nob*
31	4-Terpinyl acetate	om	1300	1300	-	-	-	1.5	1.0	-	-
32	Lavandulyl acetate	om	1304	1288	-	-	3.0	-	-	-	-
33	α-Terpinyl acetate	om	1350	1346	-	-	-	0.1	-	-	10.2
34	Methyleugenol	pp	1404	1403	-	-	-	-	-	-	3.1
35	β-Caryophyllene	sh	1419	1417	-	-	2.0	2.4	1.6	0.5	0.5
36	Aromandendrene	sh	1440	1439	-	-	-	-	-	2.2	-
37	β-Selinene	sh	1486	1489	-	4.1	-	-	-	0.1	-
38	Eremophyllene	sh	1498	1492	-	-	-	-	-	1.5	-
39	δ-Cadinene	sh	1524	1522	-	-	-	-	-	1.7	0.1
40	Caryophyllene oxide	os	1583	1582	0.2	1.9	0.1	0.2	0.3	-	-
41	β-Eudesmol	os	1651	1649	-	2.1	-	-	-	-	-
Monoterpene Hydrocarbons		mh			6.5	5.9	3.6	44.3	29.7	40.4	24.0
Oxygenated Monoterpenes		om			88.5	81.1	90.8	49.7	66.8	48.1	70.3
Sesquiterpene Hydrocarbons		sh			-	5.5	4.0	3.3	2.0	9.3	1.2
Oxygenated Sesquiterpens		os			0.3	4.6	0.2	0.6	1.1	1.8	-
Penylpropanoids		pp			-	-	-	-	-	-	3.6
Non-terpene Derivatives		nt			3.1	2.2	1.4	-	-	-	0.4
Total Identified					98.4	99.3	100.0	97.7	99.6	99.6	99.5

[a] Compounds present at $\geq$1% in at least one of the analyzed essential oils. C.cit: Cympobogon citratus; L.dent: Lavandula dentata; L.hyb: Lavandula hybrida; O.maj1/2: Origanum majorana 1/2; M.alt: Melaleuca alternifolia; L.nob: Laurus nobilis. [b] Linear retention index relative to n-alkane on the DB5 column. [c] Linear retention index reported by Adams (2007). [d] mh: monoterpene hydrocarbons; om: oxygenated monoterpenes; os: oxygenated sesquiterpens; sh: sesquiterpene hydrocarbons; pp: penylpropanoids; nt: nonterpene derivatives.

3. Discussion

In this study, these in vitro assays for the antifungal activities of these seven essential oils on mycelial growth of two fungi showed that the lemongrass (*C.cit*) essential oil was the most effective. The mycelial growth of *A. alternata* was totally inhibited by application of *C.cit* at a moderate concentration, while *S. cucurbitacearum* was completely inhibited at the highest *C.cit* concentration, with fungicidal activity seen in both cases. Only a few studies have investigated lemongrass essential oils and these fungi, with most studies focused on the essential oil activity rather than its composition. Shafique et al. [29] reported total inhibition of *A. alternata* by a *C. citratus* essential oil, with an IC_{50} of 279.13 µL/L, as also reported by Jie et al. [31]. In the same year, Guimarães et al. [32] confirmed in vitro fungitoxic activity on *A. alternata*, with the essential oil rich in citral (69.3%) and myrcene (23.8%). For the second fungus here, *S. cucurbitacearum*, Fiori et al. [30] reported 100% inhibition of mycelial growth and spore germination at a rate of 20 µL *C. citratus* essential oil. Seixas et al. [28] reported the same result at 0.25, 0.5, 0.75, 1, and 1.25 mg/mL *C. citratus* essential oil. A number of studies have reported these high proportions of the two isomers α-citral and β-citral in *C. citratus* essential oils, even when collected from different countries [33–37], as also confirmed by the present study. Brügger et al. [38] reported that in addition to the high proportion of citral, their commercial *C. citratus* essential oil showed relevant amounts of nonan-4-ol (6.5%) and camphene (5.2%). These two compounds were completely absent in the *C.cit* essential oil used in the present study. A Brazilian commercial *C. citratus* essential oil also indicated a different composition, which was rich in nonterpenes, as especially 4,8-dimethyl-3,7-nonadien-2-one (25.0%), 1-heptadec-1-ynyl-cyclopentanol (9.6%), and 7,7-dimethyl-bicycloheptan-2-ol (8.0%); here, the proportion of citral was less than 37% [39]. The antifungal activity of *C. citratus* essential oil has also been reported against other fungi, including *Aspergillus flavus*. This activity can be ascribed to the presence of various components such as citral, geraniol, and β-myrcene [37,40]. According to some studies, citral and geranol can indeed inhibit the mycelial growth of *Fusarium oxysporum*, *Colletotrichum gloeosporioides*, *Bipolaris* sp. and *A. alternata* [41,42]. These major compounds in the *C. citratus* essential oil also have antioxidant and antimicrobial activities [43]. Furthermore, Kurita et al. [44] defined the fungicidal action of citral as due to its ability to receive electrons from the fungus cell, through charge transfer with an electron donor in the cell, which results in death of the fungus. In previous studies, β-myrcene and geraniol were found in *C.cit* essential oil. These compounds with citral can contributed to inhibit the mycelial growth of *A. alternata* and *S. cucurbitacearum*.

The cultivated lavender *L. dentata* essential oil showed eucalyptol (63.5%) as the major compound in the present study, which was higher than that previously reported for both inflorescences (46.3%) and the aerial parts (40.4%) [45]. Iranian lavandin (*L. hybrida*) has also been characterized by high proportions of oxygenated monoterpenes, with eucalyptol (41.1%) as the main component, followed by borneol (20.7%) and camphor (10.8%) [46]. All of these components were present in *L.hyb* in the present study, although in lower amounts (6.5%, 4.4%, 9.3%, respectively), with the main component here being linalool (33.7%) and linalyl acetate (27.7%). The antifungal activity of linalool on *Candida* species was recently studied by Dias et al. [47], who indicated the potential use of this unsaturated monoterpene as a strong candidate with antifungal potency. According to Pitarokili et al. [48], linalyl acetate was inactive against all of the fungi they studied, although it showed weak activity against only *Sclerotinia sclerotiorum*. On the contrary, they confirmed the antifungal effects of linalool.

Good antifungal effects on mycelial growth of *A. alternata* were also seen here using the *Origanum* essential oils. Even though the *O. majorana* essential oils tested here had the same classes of compounds shown in a Brazilian species studied by Chaves et al. [49], they did not show pulegone as the main compound. An Italian species investigated by Della Pepa et al. [50] was in partial agreement with the present study for the high amount of terpinen-4-ol (29.6%), while a Tunisian oregano species were characterized by similar terpinen-4-ol content [51,52]. Moreover, Busatta et al. [53] showed that an Egyptian essential oil that was extracted by hydrodistillation of dried leaves of *O. majorana* showed the same dominance of terpinen-4-ol (31.8%) and γ-terpinene (13.0%). Although most studies on the composition of *Origanum* essential oils have agreed on the main compounds from *O. majorana* [54–57], these have indeed varied. The effectiveness of *Origanum* might be due to its high content of terpinen-4-ol, a monoterpene alcohol that is known to have good antifungal activity, as previously reported against *Fusarium avenaceum*, *Fusarium moniliforme*, *Fusarium semitectum*, *F. solani*, *F. oxysporum*, and *F. graminearum* [58,59]. Its fungicidal activity was also reported by Morcia et al. [60], who analyzed its potency on mycotoxigenic plant pathogens. However, Ebani et al. [61] showed weak activity of a tea tree essential oil on *Aspergillus fumigatus* even though terpinen-4-ol was present at relatively high levels. This might be explained by the synergistic effects among all of the different components in each of the essential oils.

The *M. alternifolia* essential oil in the present study was characterized by high levels of terpinen-4-ol (41.1%) and γ-terpinene (16.0%). These are comparable with the data reported by Elmi et al. [62] and Silva et al. [63]. In their investigations of Italian and Brazilian, commercial essential oils, they confirmed the predominance of terpinen-4-ol (41.5%, 43.1%, respectively) and γ-terpinene (20.6%, 22.8%, respectively). They also reported relatively high levels of both α-terpinene (9.6%, 9.3%, respectively) and α-terpineol (4.4%, 5.2%, respectively), while in the present study these two compounds were present in lower amounts (6.1%, 3.7%, respectively). It is also interesting to note the high proportion of p-cymene (9.3%) in the present study. In a more recent study, α-terpineol (4.4%) and 1,8-cineol (4.0%) were found in relatively high amounts, together with terpinen-4-ol (30.2%) and γ-terpinene (16.9%) [61]. In the present study, 1,8-cineol was also present but at a lower amount (2.8%).

The *L.nob* profile in the present study was in good agreement with that reported by Dhifi et al. [64], where they also showed high proportions of oxygenated monoterpenes (64.3%), with eucalyptol as the main constituent (46.8%) in their Tunisian species.

4. Materials and Methods

4.1. Origin of the Essential Oils

The lemongrass (*Cymbopogon citratus* (DC.) Stapf), lavender (*Lavandula dentata* L.), sweet marjoram (*Origanum majorana* L.), and bay laurel (*Laurus nobilis* L.) essential oils were provided by different laboratories (see Table 1), where the dried aerial parts of the plants were hydrodistilled using a Clevenger apparatus, as recommended by the European Pharmacopeia. The lavandin (*Lavandula hybrida* E.Rev. ex Briq) and tea tree (*Melaleuca alternifolia* (Maiden & Betche) Cheel) essential oils were from Flora Srl (Lorenzana, Pisa, Italy). The selection of these essential oils was based initially on their availability in our laboratory, and then on the studies in the literature that have reported in vitro activities of some of these against pathogen growth [65–67].

4.2. Fungal Strains

The *A. alternata* (GenBank accession: MK497774) and *S. cucurbitacearum* (GenBank accession: MF401569) strains used in the present study were isolated from infected squash seeds [2]. Pure cultures were transferred into Petri dishes (diameter, 90 mm) with potato dextrose agar (PDA; 42 g/L; Liofilchem Srl, Roseto degli Abruzzi, Italy) and incubated at 22 ± 2 °C with a photoperiod of 12/12 h dark/ ultraviolet light (TL-D 36W BLB 1SL; Philips, Dublin, Ireland).

4.3. In Vitro Antifungal Activities on Mycelial Growth

The antifungal activities of these *C.cit*, *L.dent*, *L.hyb*, *O.maj1*, *O.maj2*, *M.alt*, and *L.nob* essential oils were determined according to their contact phase effects on mycelial growth of *A. alternata* and *S. cucurbitacearum*. For these tests, the essential oils were dissolved in sterilized distilled water with 0.1% (*v*/*v*) Tween 20 (Sigma Aldrich, Steinheim, Germany), to obtain homogeneous emulsions. The autoclaved PDA medium (cooled to 40 °C) had the essential oil emulsions added to obtain the final concentrations of 0.1, 0.2, 0.3, 0.4, 0.5, 0.6, 0.7, 0.8, 0.9, and 1 mg/mL. The negative control was PDA containing 0.1% (v/v) Tween 20. The positive control was provided by three concentrations (0.1, 0.5, 1 mg/mL) of fungicides as 25 g/L difenoconazole plus 25 g/L fludioxonil (Celest Extra 50 FS; Cambridge, UK). The PDA was mixed and poured immediately into Petri dishes (diameter, 90 mm; 20 mL/plate), and after medium solidification, each plate was inoculated under aseptic conditions with 6 mm plugs of *A. alternata* or *S. cucurbitacearum*, taken from the edges of actively growing cultures. The experiments were carried out as three replicates per concentration and treatment. The inoculated plates were sealed with Parafilm and incubated for 7 days at 22 ± 2 °C with a photoperiod of 12/12 h dark/ ultraviolet light (TL-D 36W BLB 1SL; Philips, Dublin, Ireland). The orthogonal diameters of the colonies were measured daily until the control plates were completely covered by the mycelia. Mycelial growth inhibition was calculated based on Equation (1):

$$\text{Mycelial growth inhibition (\%)} = [(\text{dc} - \text{dt})/\text{dc}] \times 100 \quad (1)$$

where dc and dt represent the mean diameter of the mycelial growth of the control and treated fungal strains, respectively. Moreover, the IC_{50} for mycelial growth inhibition of the fungi was determined from the linear regression between the essential oil concentrations and the mycelial growth inhibition.

Experiments were performed to differentiate between the fungicidal and fungistatic activities of the elevated essential oil concentrations against fungi. Here, each of the completely inhibited fungal plugs were transferred to fresh PDA plates to note their viability after 7 days of incubation under the same conditions.

4.4. Gas Chromatography-Mass Spectrometry Analysis

The volatile constituents of each essential oil were analyzed by GC-MS as previously reported [68]. They were processed using a gas chromatograph (Agilent 7890B; Agilent Technologies Inc., Santa Clara, CA, USA) equipped with a capillary column (Agilent HP-5MS; 30 m × 0.25 mm; coating thickness, 0.25 μm; Agilent Technologies Inc., Santa Clara, CA, USA) and a single quadrupole mass detector (Agilent 5977B; Agilent Technologies Inc., Santa Clara, CA, USA). The analytical conditions were as follows: injector temperature, 220 °C; transfer line temperature, 240 °C; oven temperature programmed, from 60 °C to 240 °C at 3 °C/min; carrier gas, helium at 1 mL/min; injection volume, 1 μL (in 0.5% HPLC grade n-hexane solution); split ratio, 1:25. The full scan acquisition parameters were as follows: scan range, 30 *m/z* to 300 *m/z*; scan time, 1.0 s (See supplementary materials).

Identification of the constituents was based on comparisons of retention times with those of the authentic standards, with comparisons of their linear retention indices relative to the series of n-hydrocarbons. Computer matching was also used against commercial (NIST 14, Adams) and laboratory developed mass spectra libraries built for pure substances and components of known oils, and against the mass spectrometry literature data [69–74].

4.5. Statistical Analysis

Analysis of variance was calculated using SPSS (version 20, IBM, Armonk, NY, USA). The data were analyzed by analysis of variance (ANOVA). Means were compared using Fisher's test protected least significant difference at $p \leq 0.05$. All of the trials were repeated at least twice, and data are means ± standard error (SE).

5. Conclusions

The management of plant diseases using natural compounds is a great and important need nowadays. The present study has demonstrated the in vitro activities of seven essential oils and their efficacies against the fungi *A. alternata* and *S. cucurbitacearum*. These data show that the chemical compositions of essential oils can affect their antimicrobial activities. These essential oils were characterized by high proportions of oxygenated monoterpenes followed by monoterpene hydrocarbons. Essential oil with citral, β-myrcene, and geraniol as major components (i.e., lemongrass [*C.cit*],) controlled these fungi most effectively, followed by essential oils containing terpinen-4-ol or linalool (i.e., marjoram [*O.maj1/2*], tea tree [*M.alt*], lavandin [*L.hyb*]). Antifungal activity of essential oils can be ascribed to individual effect of major components, and/or due to a synergistic effect of its minor components. Further studies are required to determine the effects of these oils as seed treatments, to evaluate their potential as preventive and curative treatments.

Supplementary Materials: The following are available online https://www.mdpi.com/2079-6382/10/2/104/s1. Figure S1. Chromatogram of *Cympobogon citratus*. Figure S2. Chromatogram of *Lavandula dentata*. Figure S3. Chromatogram of *Lavandula hybrida*. Figure S4. Chromatogram of *Origanum majorana*1. Figure S5. Chromatogram of *Origanum majorana*2. Figure S6. Chromatogram of *Melaleuca alternifolia*. Figure S7. Chromatogram of *Laurus nobilis*.

Author Contributions: Conceptualization, M.M., G.R., K.M., and M.B.A.; methodology, M.M., G.R., B.N., L.P., H.B.A., K.M., O.K., and I.C.; software, M.M.; validation, M.M., G.R., and M.B.A.; formal analysis, M.M., G.R., and M.B.A.; investigation, M.M., G.R., M.B.A., and B.N.; resources, M.M. and B.N.; data curation, M.M. and B.N.; writing—original draft preparation, M.M., K.M., and B.N.; writing—review and editing, M.M., G.R., M.B.A., B.N., and L.P.; visualization, M.M., G.R., and M.B.A.; supervision, G.R. and M.B.A. All authors have read and agreed to the published version of the manuscript.

Funding: This research received no external funding.

Institutional Review Board Statement: Not applicable.

Informed Consent Statement: Not applicable.

Data Availability Statement: Data is contained within the article or supplementary material.

Acknowledgments: The company FLORA s.r.l. is acknowledged for kindly providing essential oils used for experiments.

Conflicts of Interest: The authors declare that they have no conflicts of interest.

References

1. Food and Agriculture Organization of the United Nations. FAOSTAT Data. Available online: http://www.fao.org/faostat/en/#data/QC (accessed on 27 October 2020).
2. Moumni, M.; Allagui, M.B.; Mancini, V.; Murolo, S.; Tarchoun, N.; Romanazzi, G. Morphological and molecular identification of seedborne fungi in squash (*Cucurbita maxima, Cucurbita moschata*). *Plant Dis.* **2020**, *104*, 1335–1350. [CrossRef] [PubMed]
3. Moumni, M.; Mancini, V.; Allagui, M.B.; Murolo, S.; Romanazzi, G. Black rot of squash (*Cucurbita moschata* Duchesne) caused by *Stagonosporopsis cucurbitacearum* reported in Italy. *Phytopathol. Mediterr.* **2019**, *58*, 381–385. [CrossRef]
4. Yao, X.; Li, P.; Xu, J.; Zhang, M.; Ren, R.; Liu, G.; Yang, X. Rapid and sensitive detection of *Didymella bryoniae* by visual loop-mediated isothermal amplification assay. *Front. Microbiol.* **2016**, *7*, 1372. [CrossRef] [PubMed]
5. Nuangmek, W.; Aiduang, W.; Suwannarach, N.; Kumla, J.; Lumyong, S. First report of gummy stem blight caused by *Stagonosporopsis cucurbitacearum* on cantaloupe in Thailand. *Can. J. Plant Pathol.* **2018**, *40*, 306–311. [CrossRef]
6. Töfoli, J.G.; Domingues, R.J. Alternarioses in vegetables: Symptoms, etiology, and integrated management. *Biológico* **2004**, *66*, 23–33. [CrossRef]
7. Farrag, E.S.H.; Moharam, M.H.A. Pathogenic fungi transmitted through cucumber seeds and safely elimination by application of peppermint extract and oil. *Not. Sci. Biol.* **2012**, *4*, 83–91. [CrossRef]
8. Boughalleb, N.; El Mahjoub, M. In vitro determination of *Fusarium* spp. infection on watermelon seeds and their localization. *Plant Pathol. J.* **2006**, *5*, 178–182. [CrossRef]
9. Gannibal, P.B. *Alternaria cucumerina* causing leaf spot of pumpkin newly reported in North Caucasus (Russia). *New Dis. Rep.* **2011**, *23*, 36. [CrossRef]
10. Fish, W.W.; Bruton, B.D.; Popham, T.W. Cucurbit host range of *Myrothecium roridum* isolated from watermelon. *Am. J. Plant Sci.* **2012**, *3*, 353–359. [CrossRef]
11. Sultana, N.; Ghaffar, A. Pathogenesis and control of *Myrothecium* spp., the cause of leaf spot on bitter gourd (*Momordica charantia* Linn.). *Pak. J. Bot.* **2009**, *1*, 429–433.
12. Boughalleb, N.; Tarchoun, N.; Dallagi, W. Effect of fungicides on *in vitro* infestation level of radish, carrot and pepper seeds. *Plant Pathol. J.* **2006**, *3*, 388–392. [CrossRef]

13. Yates, I.E.; Arnold, J.W.; Hinton, D.M.; Basinger, W.; Walcott, R.R. *Fusarium verticillioides* induction of maize seed rot and its control. *Can. J. Bot.* **2003**, *81*, 422–428. [CrossRef]
14. Keinath, A.P. Effect of protectant fungicide application schedules on gummy stem blight epidemics and marketable yield of watermelon. *Plant Dis.* **2000**, *84*, 254–260. [CrossRef] [PubMed]
15. Keinath, A.P.; DuBose, V.B.; May, M.H.; Latin, R.X. Comparison of seven fungicides intervals to control gummy stem blight in a fall watermelon crop. *Fungic. Nematic. Tests* **1998**, *53*, 268.
16. Johnson, C.E.; Payne, J.T.; Buckley, J.B. Evaluation of fungicides for gummy stem blight control on watermelon. *Fungic. Nematic. Tests* **1995**, *50*, 184.
17. Sudisha, J.; Niranjana, S.R.; Umesha, S.; Prakash, H.S.; Shetty, H.S. Transmission of seed-borne infection of muskmelon by *Didymella bryoniae* and effect of seed treatments on disease incidence and fruit yield. *Biol. Control* **2006**, *37*, 196–205. [CrossRef]
18. Barnard, M.; Padgitt, M.; Uri, N.D. Pesticide use and its measurement. *Int. Pest Control* **1997**, *39*, 161–164.
19. Misra, G.; Pavlostathis, S.G. Biodegradation kinetics of monoterpenes in liquid and soil-slurry systems. *Appl. Microbiol. Biotechnol.* **1997**, *47*, 572–577. [CrossRef]
20. Slamet, A.S.; Nakayasu, A.; Bai, H. The determinants of organic vegetable purchasing in Jabodetabek region, Indonesia. *Foods* **2016**, *5*, 85. [CrossRef]
21. Antunes, M.D.C.; Cavaco, A.M. The use of essential oils for postharvest decay control. *Flavour Fragr. J.* **2010**, *25*, 351–366. [CrossRef]
22. Shahi, S.K.; Patra, M.; Shukla, A.C.; Dikshit, A. Use of essential oil as botanical-pesticide against postharvest spoilage in *Malus pumilo* fruits. *Biocontrol* **2003**, *48*, 223–232. [CrossRef]
23. Djioua, T.; Charles, F.; Freire, M., Jr.; Filgueiras, H.; Ducamp-Collin, M.; Sallanon, H. Combined effects of postharvest heat treatment and chitosan coating on quality of fresh-cut mangoes (*Mangifera indica* L.). *Int. J. Food Sci. Technol.* **2010**, *45*, 849–855. [CrossRef]
24. Mancini, V.; Romanazzi, G. Seed treatments to control seedborne fungal pathogens of vegetable crops. *Pest Manage. Sci.* **2014**, *70*, 860–868. [CrossRef] [PubMed]
25. Kishore, G.K.; Pande, S. Natural fungicides for management of phytopathogenic fungi. *Annu. Rev. Plant Pathol.* **2004**, 3, 331–356.
26. Terzi, V.; Morcia, C.; Faccioli, P.; Vale, G.; Tacconi, G.; Malnati, M. *In vitro* antifungal activity of the tea tree (*Melaleuca alternifolia*) essential oil and its major components against plant pathogens. *Lett. Appl. Microbiol.* **2007**, *44*, 613–618. [CrossRef] [PubMed]
27. Riccioni, L.; Orzali, L. Activity of tea tree (*Melaleuca alternifolia*, Cheel) and thyme (*Thymus vulgaris*, Linnaeus.) essential oils against some pathogenic seed borne fungi. *J. Essent. Oil Res.* **2011**, *23*, 43–47. [CrossRef]
28. Seixas, P.T.L.; Castro, H.G.; Cardoso, D.P.; Junior, A.F.C.; do Nascimento, I.R. Bioactivity of essential oils on the fungus *Didymella bryoniae* of the cucumber culture. *Appl. Res. Agrotechnol.* **2012**, *5*, 61–66. [CrossRef]
29. Shafique, S.; Majeed, R.A.; Shafique, S. *Cymbopogon citrates*: A remedy to control selected *Alternaria* species. *J. Med. Plant Res.* **2012**, *6*, 1879–1885. [CrossRef]
30. Fiori, A.C.G.; Schwan-Estrada, K.R.F.; Stangarlin, J.R.; Vida, J.B.; Scapim, C.A.; Cruz, M.E.S.; Pascholati, S.F. Antifungal activity of leaf extracts and essential oils of some medicinal plants against *Didymella bryoniae*. *J. Phytopathol.* **2000**, *148*, 483–487. [CrossRef]
31. Jie, Z.; ChaoYing, Z.; ZhanFang, G.; Lei, Y. Inhibititory activity of *Cymbopogon citratus* essential oil against nine phytopathogens. *J. Shanghai Jiaotong Univ.-Agric. Sci.* **2011**, *29*, 72–74.
32. Guimarães, L.G.L.; dasGraças Cardoso, M.; Souza, P.E.; de Andrade, J.; Vieira, S.S. Antioxidant and fungitoxic activities of the lemongrass essential oil and citral. *Rev. Cienc. Agron.* **2011**, *42*, 464. [CrossRef]
33. Ortega-Ramirez, L.A.; Gutiérrez-Pacheco, M.M.; Vargas-Arispuro, I.; González-Aguilar, G.A.; Martínez-Téllez, M.A.; Ayala-Zavala, J.F. Inhibition of glucosyltransferase activity and glucan production as an antibiofilm mechanism of lemongrass essential oil against *Escherichia coli* O157: H7. *Antibiotics* **2020**, *9*, 102. [CrossRef] [PubMed]
34. Plata-Rueda, A.; Martínez, L.C.; da Silva Rolim, G.; Coelho, R.P.; Santos, M.H.; de Souza Tavares, W.; Zanuncio, J.C.; Serrão, J.E. Insecticidal and repellent activities of *Cymbopogon citratus* (Poaceae) essential oil and its terpenoids (citral and geranyl acetate) against *Ulomoides dermestoides*. *Crop Prot.* **2020**, *137*, 105299. [CrossRef]
35. Bermúdez-Vásquez, M.J.; Granados-Chinchilla, F.; Molina, A. Composición química y actividad antimicrobiana del aceite esencial de *Psidium guajava* y *Cymbopogon citratus*. *Agron. Mesoam.* **2019**, *30*, 147–163. [CrossRef]
36. Dègnon, R.G.; Allagbé, A.C.; Adjou, E.S.; Dahouenon-Ahoussi, E. Antifungal activities of *Cymbopogon citratus* essential oil against *Aspergillus* species isolated from fermented fish products of Southern Benin. *J. Food Qual. Hazards Control* **2019**, *6*, 53–57. [CrossRef]
37. Supardan, M.D.; Misran, E.; Mustapha, W.A.W. Effect of material length on kinetics of essential oil hydrodistillation from lemongrass (*Cymbopogon citratus*). *J. Eng. Sci. Technol.* **2019**, *14*, 810–819.
38. Brügger, B.P.; Martínez, L.C.; Plata-Rueda, A.; de Castro e Castro, B.M.; Soares, M.A.; Wilcken, C.F.; Carvalho, A.G.; Serrão, J.E.; Zanuncio, J.C. Bioactivity of the *Cymbopogon citratus* (Poaceae) essential oil and its terpenoid constituents on the predatory bug, *Podisus nigrispinus* (Heteroptera: *Pentatomidae*). *Sci. Rep.* **2019**, *9*, 1–8. [CrossRef]
39. Macedo, I.T.F.; de Oliveira, L.M.B.; André, W.P.P.; de Araújo Filho, J.V.; dos Santos, J.M.L.; Rondon, F.C.M.; Ribeiro, W.L.C.; Camurça-Vasconcelos, A.L.F.; de Oliveira, E.F.; de Paula, H.C.B.; et al. Anthelmintic effect of *Cymbopogon citratus* essential oil and its nanoemulsion on sheep gastrointestinal nematodes. *Rev. Bras. Parasitol. Vet.* **2019**, *28*, 522–527. [CrossRef]

40. Sonker, N.; Pandey, A.K.; Singh, P.; Tripathi, N.N. Assessment of *Cymbopogon citratus* (DC.) stapf essential oil as herbal preservatives based on antifungal, antiaflatoxin, and antiochratoxin activities and in vivo efficacy during storage. *J. Food Sci.* **2014**, *79*, M628–M634. [CrossRef]
41. Dalcin, M.S.; Cafee-Filho, A.C.; de Almeida Sarmento, R.; do Nascimento, I.R.; de Souza Ferreira, T.P.; de Sousa Aguiar, R.W.; dos Santos, G.R. Evaluation of essential oils for preventive or curative management of melon gummy stem blight and plant toxicity. *J. Med. Plant Res.* **2017**, *11*, 426–432. [CrossRef]
42. Kishore, G.K.; Pande, S.; Harish, S. Evaluation of essential oils and their components for broad-spectrum antifungal activity and control of late leaf spot and crown rot diseases in peanut. *Plant Dis.* **2007**, *91*, 375–379. [CrossRef] [PubMed]
43. Farias, P.K.S.; Silva, J.C.R.L.; Souza, C.N.D.; Fonseca, F.S.A.D.; Brandi, I.V.; Martins, E.R.; Azevedo, A.M.; Almeida, A.C.D. Antioxidant activity of essential oils from condiment plants and their effect on lactic cultures and pathogenic bacteria. *Cienc. Rural* **2019**, *49*, e20180140. [CrossRef]
44. Kurita, N.; Makoto, M.; Kurane, R.; Takahara, Y.; Ichimura, K. Antifungal activity of components of essential oils. *Agric. Biol. Chem.* **1981**, *45*, 945–952. [CrossRef]
45. de Paula Martins, R.; da Silva Gomes, R.A.; Malpass, A.C.G.; Okura, M.H. Chemical characterization of *Lavandula dentata* L. essential oils grown in Uberaba-MG. *Cienc. Rural* **2019**, *49*. [CrossRef]
46. Bajalan, I.; Rouzbahani, R.; Pirbalouti, A.G.; Maggi, F. Chemical composition and antibacterial activity of Iranian lavandula × hybrida. *Chem. Biodivers.* **2017**, *14*, 1–8. [CrossRef] [PubMed]
47. Dias, I.J.; Trajano, E.R.I.S.; Castro, R.D.; Ferreira, G.L.S.; Medeiros, H.C.M.; Gomes, D.Q.C. Antifungal activity of linalool in cases of *Candida* spp. isolated from individuals with oral candidiasis. *Braz. J. Biol.* **2018**, *78*, 368–374. [CrossRef] [PubMed]
48. Pitarokili, D.; Couladis, M.; Petsikos-Panayotarou, N.; Tzakou, O. Composition and antifungal activity on soil-borne pathogens of the essential oil of *Salvia sclarea* from Greece. *J. Agr. Food Chem.* **2002**, *50*, 6688–6691. [CrossRef] [PubMed]
49. Chaves, R.D.S.B.; Martins, R.L.; Rodrigues, A.B.L.; Rabelo, É.D.M.; Farias, A.L.F.; Brandão, L.B.; Santos, L.L.; Galardo, A.K.R.; de Almeida, S.S.M.D.S. Evaluation of larvicidal potential against larvae of *Aedes aegypti* (Linnaeus, 1762) and of the antimicrobial activity of essential oil obtained from the leaves of Origanum majorana L. *PLoS ONE* **2020**, *15*, e0235740. [CrossRef]
50. Della Pepa, T.; Elshafie, H.S.; Capasso, R.; De Feo, V.; Camele, I.; Nazzaro, F.; Scognamiglio, M.R.; Caputo, L. Antimicrobial and phytotoxic activity of *Origanum heracleoticum* and *O. majorana* essential oils growing in Cilento (Southern Italy). *Molecules* **2019**, *24*, 2576. [CrossRef]
51. Khadhri, A.; Bouali, I.; Aouadhi, C.; Lagel, M.C.; Masson, E.; Pizzi, A. Determination of phenolic compounds by MALDI–TOF and essential oil composition by GC–MS during three development stages of *Origanum majorana* L. *Biomed. Chromatogr.* **2019**, *33*, 1–13. [CrossRef]
52. Hajlaoui, H.; Mighri, H.; Aouni, M.; Gharsallah, N.; Kadri, A. Chemical composition and in vitro evaluation of antioxidant, antimicrobial, cytotoxicity and anti-acetylcholinesterase properties of Tunisian *Origanum majorana* L. essential oil. *Microb. Pathog.* **2016**, *95*, 86–94. [CrossRef]
53. Busatta, C.; Barbosa, J.; Cardoso, R.I.; Paroul, N.; Rodrigues, M.; de Oliveira, D.; de Oliveira, J.V.; Cansian, R.L. Chemical profiles of essential oils of marjoram (*Origanum majorana*) and oregano (*Origanum vulgare*) obtained by hydrodistillation and supercritical CO_2. *J. Essent. Oil Res.* **2017**, *29*, 367–374. [CrossRef]
54. Fikry, S.; Khalil, N.; Salama, O. Chemical profiling, biostatic and biocidal dynamics of *Origanum vulgare* L. essential oil. *AMB Express* **2019**, *9*, 41. [CrossRef] [PubMed]
55. Jan, S.; Mir, J.I.; Shafi, W.; Faktoo, S.Z.; Singh, D.B.; Wijaya, L.; Alyemeni, M.N.; Ahmad, P. Divergence in tissue-specific expression patterns of genes associated with the terpeniod biosynthesis in two oregano species *Origanum vulgare* L., and *Origanum majorana*. *Ind. Crops Prod.* **2018**, *123*, 546–555. [CrossRef]
56. Nardoni, S.; Pisseri, F.; Pistelli, L.; Najar, B.; Luini, M.; Mancianti, F. In vitro activity of 30 essential oils against bovine clinical isolates of *Prototheca zopfii* and *Prototheca blaschkeae*. *Vet. Sci.* **2018**, *5*, 45. [CrossRef]
57. Semiz, G.; Semiz, A.; Mercan-Doğan, N. Essential oil composition, total phenolic content, antioxidant and antibiofilm activities of four *Origanum* species from southeastern turkey. *Int. J. Food Prop.* **2018**, *21*, 194–204. [CrossRef]
58. Sahab, A.F.; Aly, S.; Hathout, A.S.; Ziedan, E.S.H.; Sabry, B.A. Application of some plant essential oils to control *Fusarium* isolates associated with freshly harvested maize in Egypt. *J. Essent. Oil Bear. Plants* **2014**, *17*, 1146–1155. [CrossRef]
59. Szczerbanik, M.; Jobling, J.; Morris, S.; Holford, P. Essential oil vapours control some common postharvest fungal pathogens. *Aust. J. Exp. Agric.* **2007**, *47*, 103–109. [CrossRef]
60. Morcia, C.; Malnati, M.; Terzi, V. In vitro antifungal activity of terpinen-4-ol, eugenol, carvone, 1,8-cineole (eucalyptol) and thymol against mycotoxigenic plant pathogens. *Food Addit. Contam.* **2012**, *29 Part A*, 415–422. [CrossRef]
61. Ebani, V.V.; Najar, B.; Bertelloni, F.; Pistelli, L.; Mancianti, F.; Nardoni, S. Chemical composition and *in vitro* antimicrobial efficacy of sixteen essential oils against *Escherichia coli* and *Aspergillus fumigatus* isolated from poultry. *Vet. Sci.* **2018**, *5*, 62. [CrossRef]
62. Elmi, A.; Ventrella, D.; Barone, F.; Carnevali, G.; Filippini, G.; Pisi, A.; Benvenuti, S.; Scozzoli, M.; Bacci, M.L. *In vitro* effects of tea tree oil (*Melaleuca alternifolia* essential oil) and its principal component terpinen-4-ol on swine spermatozoa. *Molecules* **2019**, *24*, 1071. [CrossRef] [PubMed]
63. Silva-Flores, P.G.; Pérez-López, L.A.; Rivas-Galindo, V.M.; Paniagua-Vega, D.; Galindo-Rodríguez, S.A.; Álvarez-Román, R. Simultaneous GC-FID quantification of main components of *Rosmarinus officinalis* L. and *Lavandula dentata* essential oils in polymeric nanocapsules for antioxidant application. *J. Anal. Meth. Chem.* **2019**, *2019*, 1–9. [CrossRef] [PubMed]

64. Dhifi, W.; Bellili, S.; Jazi, S.; Nasr, S.B.; El Beyrouthy, M.; Mnif, W. Phytochemical composition and antioxidant activity of Tunisian *Laurus nobilis*. *Pak. J. Pharm. Sci.* **2018**, *31*, 2397–2402.
65. Alves, F.M.; De, F.; França, K.R.D.S.; Araújo, I.G.D.; Nóbrega, L.P.D.; Xavier, A.L.; Dos, S.; Lima, T.S.; Rodrigues, A.P.M.; Júnior, A.F.; et al. Control of *Alternaria alternata* using melaleuca essential oil (*Melaleuca alternifolia*). *J. Exp. Agric. Int.* **2019**, *40*, 1–10. [CrossRef]
66. Black-Solis, J.; Ventura-Aguilar, R.I.; Correa-Pacheco, Z.; Corona-Rangel, M.L.; Bautista-Baños, S. Preharvest use of biodegradable polyester nets added with cinnamon essential oil and the effect on the storage life of tomatoes and the development of *Alternaria alternata*. *Sci. Hortic.* **2019**, *245*, 65–73. [CrossRef]
67. Soylu, E.M.; Kose, F. Antifungal activities of essential oils against citrus black rot disease agent *Alternaria alternata*. *J. Essent. Oil-Bear. Plants* **2015**, *18*, 894–903. [CrossRef]
68. Najar, B.; Marchioni, I.; Ruffoni, B.; Copetta, A.; Pistelli, L.; Pistelli, L. Volatilomic analysis of four edible flowers from Agastache genus. *Molecules* **2019**, *24*, 4480. [CrossRef]
69. Adams, R.P. *Identification of Essential Oil Components by Gas Chromatography/Quadrupole Mass Spectroscopy*, 4th ed.; Allured Publishing Corporation: Carol Stream, IL, USA, 2007; p. 1902.
70. Davies, N.W. Gas chromatographic retention indices of monoterpenes and sesquiterpenes on methyl silicon and Carbowax 20M phases. *J. Chromatogr. A* **1990**, *503*, 1–24. [CrossRef]
71. Jennings, W.; Shibamoto, T. *Qualitative Analysis of Flavor and Fragrance Volatiles by Glass Capillary Gas Chromatography, Food/Nahrung*; Academic Press: New York, NY, USA; London, UK; Sydney, Australia; Toronto, ON, Canada; San Francisco, CA, USA, 1982.
72. Masada, Y. *Analysis of Essential Oils by Gas Chromatography and Mass Spectrometry*; John Wiley & Sons, Inc.: New York, NY, USA, 1976.
73. Stenhagen, E.; Abrahamsson, S.; McLafferty, F.W. *Registry of Mass Spectral Data*; Wiley & Sons: New York, NY, USA, 1974.
74. Swigar, A.A.; Silverstein, R.M. *Monoterpenes, Aldrich Chemical Company*; Aldrich Chemical Company: Milwaukee, WI, USA, 1981.

antibiotics

Article

Nanoencapsulated Essential Oils with Enhanced Antifungal Activity for Potential Application on Agri-Food, Material and Environmental Fields

Magdaléna Kapustová [1,†], Giuseppe Granata [2,†], Edoardo Napoli [2], Andrea Puškárová [1], Mária Bučková [1], Domenico Pangallo [1,*] and Corrada Geraci [2,*]

[1] Institute of Molecular Biology, Slovak Academy of Sciences, Dúbravská cesta 21, 84551 Bratislava, Slovakia; magdalena.kapustova@savba.sk (M.K.); andrea.puskarova@savba.sk (A.P.); maria.buckova@savba.sk (M.B.)
[2] Istituto Chimica Biomolecolare–Consiglio Nazionale delle Ricerche, Via Paolo Gaifami 18, 95126 Catania, Italy; giuseppe.granata@icb.cnr.it (G.G.); edoardo.napoli@icb.cnr.it (E.N.)
* Correspondence: domenico.pangallo@savba.sk (D.P.); corrada.geraci@icb.cnr.it (C.G.); Tel.: +421-2-5930-7443 (D.P.); +39-095-733-8318 (C.G.)
† These authors contributed equally to this work.

Abstract: Nanotechnology is a new frontier of this century that finds applications in various fields of science with important effects on our life and on the environment. Nanoencapsulation of bioactive compounds is a promising topic of nanotechnology. The excessive use of synthetic compounds with antifungal activity has led to the selection of resistant fungal species. In this context, the use of plant essential oils (EOs) with antifungal activity encapsulated in ecofriendly nanosystems could be a new and winning strategy to overcome the problem. We prepared nanoencapsules containing the essential oils of *Origanum vulgare* (OV) and *Thymus capitatus* (TC) by the nanoprecipitation method. The colloidal suspensions were characterized for size, polydispersity index (PDI), zeta potential, efficiency of encapsulation (EE) and loading capacity (LC). Finally, the essential oil nanosuspensions were assayed against a panel of fourteen fungal strains belonging to the Ascomycota and Basidiomycota phyla. Our results show that the nanosystems containing thyme and oregano essential oils were active against various fungal strains from natural environments and materials. In particular, the minimum inhibitory concentration (MIC) and minimum fungicidal concentration (MFC) values were two to four times lower than the pure essential oils. The aqueous, ecofriendly essential oil nanosuspensions with broad-spectrum antifungal activity could be a valid alternative to synthetic products, finding interesting applications in the agri-food and environmental fields.

Keywords: antifungal activity; essential oils; nanoencapsulation; poly(ε-caprolactone); *Origanum vulgare*; *Thymus capitatus*

Citation: Kapustová, M.; Granata, G.; Napoli, E.; Puškárová, A.; Bučková, M.; Pangallo, D.; Geraci, C. Nanoencapsulated Essential Oils with Enhanced Antifungal Activity for Potential Application on Agri-Food, Material and Environmental Fields. *Antibiotics* **2021**, *10*, 31. https://doi.org/10.3390/antibiotics10010031

Received: 6 November 2020
Accepted: 24 December 2020
Published: 1 January 2021

1. Introduction

Fungi are ubiquitous microorganisms that can colonize several natural environments, equipped with enzymatic machinery enabling the degradation of various types of materials and matrices [1]. Members of the genus *Aspergillus*, such as *A. fumigatus* and *A. flavus*, together with *Penicillium rubens*, can be isolated from indoor environments and can colonize and damage cellulolytic substrates such as archival documents and books [2,3]. The fungi of the genus *Cladosporium* are well known as phytopathogenic microorganisms able to infect different kinds of plants [4]. *Penicillium citrinum* is recognized mainly as a citrus fruit pathogen, but occasionally it also occurs in tropical spices and cereals [5]. Other fungi, such as members of the genera *Geotrichum*, *Mucor* and *Fusarium*, can be isolated from various foods where they can release dangerous mycotoxins [6,7].

There are fungi typically from soil, such as *Purpureocillium* and *Exophiala*, and certain types of them can successfully inhabit stones and be responsible for bioweathering phenomena [1,8]. Different species, mainly belonging to the phylum Basidiomycota, are considered to be wood-decay fungi such as *Pleurotus eryngii*, *Bjerkandera adusta* and *Phanerochaete chrysosporium* [9]. The wood-decay fungi can also lead to the biodeterioration of wood components commonly used in construction and may impose serious problems on building stability [10]. Finally, many of the fungi mentioned above can cause various fastidious health complications and, in some cases, also serious pathogenicity [11]. Therefore, it is necessary to implement a series of precautions for their control and inhibition on various kinds of surfaces (especially in an indoor environment) and in food processing.

Over the decades, the massive use of synthetic antifungal products induced resistance phenomena in a large number of fungal species. In this direction, the use of substances derived from plants, such as essential oils, can be considered a valid alternative that better meets the wishes of consumers increasingly oriented toward natural remedies [12]. Essential oils (EOs) are phytocomplexes obtained from aromatic plants by hydrodistillation or steam distillation. Their chemical composition is very complex due the presence of a large variety of volatile compounds, mainly terpenes. Monoterpenes and sesquiterpenes are present in the form of hydrocarbons, alcohols, aldehydes, ketones, esters, ethers, peroxides and phenols. Phenylpropane derivatives occur less frequently than the aforesaid terpenes [13]. The EO chemical composition is influenced by the geographical position, environment condition, stage of ripening and extraction technique. More than half of the essential oils from Lamiaceae family plants have good antifungal activity (minimum inhibitory concentrations (MICs) < 1000 μg/mL) [14]. As reported by Rao et al. [15], terpenes and terpenoids are known to exhibit intense antifungal activity. Their action mechanism is multitarget and does not favor the appearance of resistance fungal strains [16,17].

Unfortunately, EOs are lipophilic compounds, easily degradable by the effects of oxygen, light, moisture and temperature. Nanoencapsulation is a valid strategy to overcome these obstacles. This technology allows for the protection the essential oils from thermal and photodegradation phenomena, increasing their solubility in aqueous environments, masking their flavor and improving their bioaccessibility and bioavailability [18]. The subcellular size and relative larger surface area per unit volume enhances EO concentration in the zone where the microorganisms are preferentially located, such as water-rich phases or liquid–solid interfaces [19].

The nanocapsules, prepared by interfacial deposition of the preformed polymer (nanoprecipitation), represent an effective method to obtain robust nanosystems suitable (by a feasible scale-up process) for applications in various fields, ranging from medicine, health and agri-food to the environment [20]. In these systems, the EO is located in the inner core, surrounded by a polymeric wall.

In this study, nanocapsules based on the biodegradable and biocompatible poly(ε-caprolactone) (PCL) polymer were prepared. The nanocapsules were loaded with commercial EOs of *Origanum vulgare* (OV) and *Thymus capitatus* (TC), both having known antifungal activity [21–23]. The antifungal ability of the encapsulated and free EOs were assayed against a panel of fourteen different fungi belonging to the Ascomycota and Basidiomycota phyla. These fungi are usually responsible for the biodeterioration and biodegradation of different materials and for the contamination of food, causing damage to human health.

The aim of this work is the preparation of aqueous, ecofriendly nanosuspensions with effective and broad-spectrum antifungal activity for potential applications on different natural matrices and materials.

2. Results

2.1. Chemical Composition of Essential Oils

Thyme and oregano essential oils are, generally, characterized by a large amount of monoterpenes (both hydrocarbons and oxygenated), reaching almost 90% of the whole oil. The two main phenolic monoterpenes, thymol and carvacrol, occur more frequently, often accompanied by their biogenetically correlated compounds *p*-cymene and γ-terpinene [13,24]. Thymus is probably one of the most taxonomically complex genera of its family. Several studies confirmed that the two most common chemotypes are thymol and carvacrol [25], followed by less common non-phenolic chemotypes [26]. As reported in materials and methods section, the two EO samples of this study were commercial. The chemical composition of our oregano sample has already been published in a previous study with different purposes from this one [27], while the chemical composition of our commercial thyme EO is published here for the first time. The two analytical results are listed together in Table 1 for easier comparison by readers. The samples subjected to this study had chemical profiles quite typical of the *Origanum vulgare* thymol and carvacrol chemotype and the *Thymus capitatus* carvacrol chemotype of essential oils. Gas chromatography (GC) techniques, coupled with a flame ionization detector (FID) and a mass spectrometer (MS), allowed the identification of more than 35 components, covering up to 99% and 98% of the total oil compositions, respectively. Table 1 shows the details of the chemical compositions, listing only the components with a % > 0.05. The most represented class for both samples was that of oxygenated monoterpenes (68.7% and 72.9%, respectively), followed by hydrocarbon monoterpenes (30.0% and 22.3%, respectively). The sum of these classes in the oregano sample reached 96.7%, while in the thyme sample it reached 95.1%. For both samples, sesquiterpenes and other components were below 4%. The common feature of these two oils was to have three main components which alone covered more than 80% of the whole composition. The oregano essential oil profile was characterized by the presence of carvacrol (36%) and thymol (25%) as the main compounds, followed by *p*-cymene (22%). All other compounds were below 5%. The thyme essential oil profile was dominated by a high amount of carvacrol (almost 70%), followed by *p*-cymene (9%) and γ-terpinene (8%). The thymol percentage was negligible (0.5%), as were the percentages of all other compounds, with the exception of β-caryophyllene, which reached 2.6%.

Table 1. Chemical composition of commercial *Origanum vulgare* and *Thymus capitatus* essential oils (EOs).

# [a]	RI Lit [b]	RI Exp [c]	Class or Compound [d]	*O. vulgare* [e] % [f]	*T. capitatus* % [f]
			Monoterpene Hydrocarbons	**27.99**	**22.26**
1	930	924	α-Thujene	0.07	0.61
2	939	931	α-Pinene	1.37	0.79
3	954	947	Camphene	0.56	0.16
4	979	974	β-Pinene	0.43	N.D.
7	991	986	β-Myrcene	0.66	1.29
9	1002	999	Δ2-Carene	0.06	N.D.
10	1003	1003	α-Phellandrene	N.D.	0.18
11	1004	1009	*p*-Mentha-1(7),8-diene	N.D.	0.07
12	1017	1013	α-Terpinene	0.36	1.03
13	1025	1027	*p*-Cymene	21.54	9.26
14	1029	1029	Limonene	0.78	0.46
16	1060	1058	γ-Terpinene	2.16	8.22
18	1089	1085	Terpinolene	N.D.	0.19
			Oxygenated Monoterpenes	**68.66**	**72.88**
15	1031	1031	1,8-Cineole	0.84	0.13
17	1070	1067	*cis* Sabinene hydrate	N.D.	0.14
19	1097	1098	Linalool	4.26	0.86
20	1146	1144	Camphor	0.36	N.D.
21	1156	1158	Isoborneol	0.27	N.D.
22	1169	1167	Borneol	0.90	0.41
23	1177	1177	Terpinen-4-ol	0.35	0.59
24	1189	1190	α-Terpineol	0.61	N.D.
25	1208	1238	*trans*-Piperitol	N.D.	0.06
26	1245	1243	Carvacrol methyl ether	0.16	0.06
27	1253	1259	Geraniol	N.D.	0.08
28	1290	1303	Thymol	25.02	0.58
29	1299	1319	Carvacrol	35.95	69.91
31	1373	1364	Carvacrol acetate	N.D.	0.06

Table 1. *Cont.*

# [a]	RI Lit [b]	RI Exp [c]	Class or Compound [d]	*O. vulgare* [e] % [f]	*T. capitatus* % [f]
			Sesquiterpenes	**1.94**	**3.19**
32	1419	1425	β-Caryophyllene	1.70	2.56
33	1455	1459	α-Humulene	0.24	0.10
34	1506	1492	β-Bisabolene	N.D.	0.20
35	1507	1525	(Z)-α-Bisabolene	N.D.	0.16
36	1583	1569	Caryophyllene oxide	N.D.	0.23
			Others	**0.47**	**0.22**
5	979	976	1-Octen-3-ol	0.21	0.16
6	984	982	3-Octanone	0.10	N.D.
8	978	993	3-Octanol	0.05	N.D.
30	1359	1361	Eugenol	0.11	0.06
			TOTAL	**99.06**	**98.55**

[a] The numbering refers to the elution order. [b] Literature retention index (RI). [c] Retention index (RI) relative to the standard mixture of *n*-alkanes on the SPB-5 column. [d] Identified compounds (those < 0.05% have not been reported). [e] Data previously reported (see [27]). [f] Relative peak area percent, representing the averages of three determinations. N.D. = not detected.

2.2. Physicochemical Characterization of Essential Oil-Loaded Nanocapsules (EO-NCs)

In a previous work, we reported the encapsulation of thyme and oregano essential oils in PCL nanocapsules [28]. Unlike the previous work, here we used commercial essential oils having different compositions of volatile components. This offered numerus advantages, including the lower cost of preparation and the standardization in the compositions of the EOs. In this manner, the obtained nanocapsules could potentially be more attractive for industrial sectors. Table 2 shows the reported values of the physicochemical parameters (z-average diameter, polydispersity index (PDI), zeta potential (ζ), encapsulation efficiency (EE) and loading capacity (LC)) which allowed us to characterize the nanocapsules containing thyme and oregano essential oils (TC-NCs and OV-NCs, respectively).

Table 2. Physicochemical Characterization of essential oil (EO)-NCs.

EO-NCs	Z-Average (nm)	PDI	ζ (mV)	EE%	LC%
TC-NCs	198 ± 3	0.09 ± 0.02	−11 ± 1	84 ± 6	52 ± 3
OV-NCs	200 ± 3	0.05 ± 0.03	−10 ± 2	80 ± 9	51 ± 4

The z-average diameters of 198 ± 3 nm and 200 ± 3 nm for the TC-NCs and OV-NCs, respectively, were in agreement with the nanometric structures of the nanocapsules. Low PDI values for both nanocapsules pointed out a narrow size distribution of nanoparticles and the presence in aqueous solution of monodisperse nanosystems (Figure 1). Negative zeta potentials of −11 and −10 mV for the TC-NCs and OV-NCs, respectively, were very similar to those obtained for other stable PCL nanocapsules [28,29]. The percentages of EE and LC were high, with values of 84 ± 6 and 52 ± 3 for the TC-NCs and 80 ± 9 and 51 ± 4 for the OV-NCs. The prepared TC-NC and OV-NC aqueous nanosuspensions showed a total essential oil content of 5.7 ± 0.3 and 5.8 ± 0.4 mg/mL, respectively.

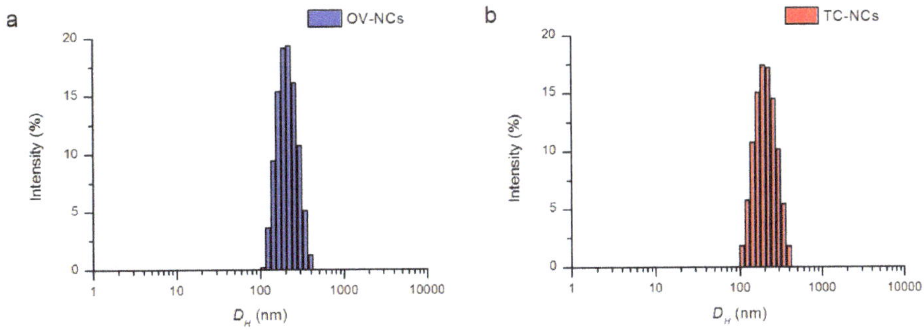

Figure 1. I-weighted distribution of the hydrodynamic diameter (D_H) of (**a**) OV-NCs and (**b**) TC-NCs.

2.3. Antifungal Activities of Pure EOs and EO-NCs

The screening of encapsulated essential oils (EO-NCs) against a panel of fungal strains (Table 3) showed marked antifungal activity (Table 4 and Figure 2). The obtained results evidenced that EO-NCs inhibited the growth of assayed fungi at s MIC values ranging from 0.125 to 0.25 mg/mL. These concentrations were two to four times lower than those observed for pure EOs. For both EO-NCs, an MIC of 0.125 mg/mL was effective for inhibiting the growth of *A. fumigatus*, *C. aggregatocicatricatum*, *C. herbarum* and *P. eryngii*. In addition, this concentration for OV-NCs was also enough to inhibit the fungus *Bjerkandera adusta*.

Table 3. Characteristics of the assayed fungal strains.

Fungal Strain	Environment of Isolation	Source
Aspergillus fumigatus	Indoor air	IMB-SAS
Aspergillus flavus	Indoor air	IMB-SAS
Penicillium rubens	Indoor air	IMB-SAS
Penicillium citrinum	Fruit	IMB-SAS
Cladosporium aggregatocicatricatum	Wax seal	IMB-SAS
Cladosporium herbarum	Wax seal	IMB-SAS
Fusarium oxysporum	Cheese	IMB-SAS
Geotrichum candidum	Cheese	IMB-SAS
Mucor circinelloides	Cheese	IMB-SAS
Exophiala xenobiotica	Soil	IMB-SAS
Purpureocillium lilacinum	Stone	IMB-SAS
Pleurotus eryngii	Tree	CCBAS
Bjerkandera adusta	Tree	CCBAS
Phanerochaete chrysosporium	Tree	CCBAS

IMB-SAS = fungal collection of the Institute of Molecular Biology (Slovak Academy of Sciences). CCBAS = Culture collection of Basidiomycetes, Institute of Microbiology, Academy of Sciences of the Czech Republic.

We also evaluated the fungicidal effect of EO-NCs versus the panel of fourteen fungal strains. In particular, both the TC-NCs and OV-NCs had minimum fungicidal concentrations (MFCs) of 0.25 mg/mL against *A. fumigatus*, *C. aggregatocicatricatum*, *C. herbarum* and *P. eryngii*, while having an MFC of 0.5 mg/mL against *A. flavus*, *P. rubens*, *P. citrinum*, *F. oxysporum*, *G. candidum*, *M. circinelloides*, *E. xenobiotica*, *P. lilacinum* and *P. chrysosporium*. The MFC values of the EOs were two to four times higher than the MFCs of the EO-NCs on the tested strains. The empty NCs revealed no activity on the fungal growth.

Table 4. Minimum inhibitory concentration (MIC) and minimum fungicidal concentration (MFC) values of encapsulated essential oils and pure essential oils.

Fungal Strain	TC-NCs		OV-NCs		TC-EO		OV-EO	
	MIC	MFC	MIC	MFC	MIC	MFC	MIC	MFC
Aspergillus fumigatus	0.125	0.25	0.125	0.25	0.25	0.5	0.25	0.5
Aspergillus flavus	0.25	0.5	0.25	0.5	0.5	1	0.5	1
Penicillium rubens	0.25	0.5	0.25	0.5	0.5	1	0.5	1
Penicillium citrinum	0.25	0.5	0.25	0.5	0.5	1	0.5	1
Cladosporium aggregatocicatricatum	0.125	0.25	0.125	0.25	0.25	0.5	0.25	0.5
Cladosporium herbarum	0.125	0.25	0.125	0.25	0.5	1	0.25	0.5
Fusarium oxysporum	0.25	0.5	0.25	0.5	0.5	1	0.5	1
Geotrichum candidum	0.25	0.5	0.25	0.5	0.5	1	0.5	1
Mucor circinelloides	0.25	0.5	0.25	0.5	0.5	1	0.5	1
Exophiala xenobiotica	0.25	0.5	0.25	0.5	0.5	1	0.5	1
Purpureocillium lilacinum	0.25	0.5	0.25	0.5	0.5	1	0.5	1
Pleurotus eryngii	0.125	0.25	0.125	0.25	0.25	0.5	0.25	0.5
Bjerkandera adusta	0.25	0.5	0.125	0.5	0.5	1	0.5	1
Phanerochaete chrysosporium	0.25	0.5	0.25	0.5	0.5	1	0.5	1

MIC and MFC are expressed in mg/mL. TC-NCs = nanoencapsulated thyme essential oil; OV-NCs = nanoencapsulated oregano essential oil; TC-EO = thyme essential oil; and OV-EO = oregano essential oil.

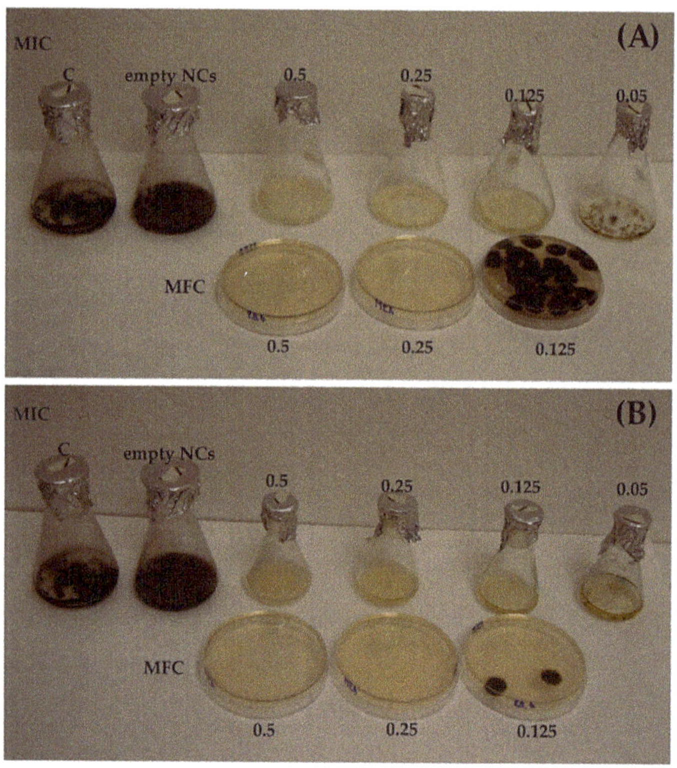

Figure 2. Example of MIC and MFC assays with the isolate *Cladosporium herbarum*. (**A**) Results using a nanoencapsulated oregano EO. (**B**) Results using a nanoencapsulated thyme EO. C = control. The range of nanoencapsulated EO concentrations used was 0.05–0.5 mg/mL.

3. Discussion

The current study evaluated the antifungal activity of TC-NCs and OV-NCs against selected fungal strains that contaminate different natural matrices, indoor air and materials (Table 3). The results evidenced that both TC-NCs and OV-NCs showed in vitro antifungal activity against all tested isolates with an MIC range of 0.125–0.25 mg/mL (Table 4). This activity is two to four times greater than that of pure oils. These results highlight the effectiveness of EO encapsulation in a nanometric structure. In fact, the nanocapsules were able to protect essential oils from degradation phenomena and increase the transport and diffusion mechanisms of essential oils through the cell membrane [30].

Moreover, the same MIC and MFC values were observed for both nanocapsule suspensions in respect to the different strains, with the one exception of the *Bjerkandera adusta* strain, against which the OV-NCs showed more activity than the TC-NCs. The results were imputable to the chemical compositions of commercial essential oils. Although the main bioactive oxygenated monoterpenes (thymol and carvacrol) were present in different quantities in the two essential oils, the sum of their relative percentages was of the same order of magnitude (Table 1). Generally, the growth-inhibiting effects were attributable to the major components, thymol and carvacrol, but the bioactive monoterpene hydrocarbons, *p*-cymene and γ-terpinene present in both essential oils in non-negligible quantities could have given a valuable contribution to the antifungal properties [31].

Regarding the physicochemical characterization of the nanocapsules, the TC-NCs and OV-NCs showed nanodimensions and good PDIs indicating the presence of monodisperse species. This characteristic is crucial for achieving optimal biological activity results [32]. Acidic residues on the PCL polymer and the presence of tween 80 (a neutral surfactant covering the nanoparticle wall) were in agreement with the observed low and negative values of the zeta potential. Moreover, these values were very similar to those previously reported by us [28,29] and by other authors [33,34] for stable poly(ε-caprolactone) nanocapsules.

The good encapsulation in the range of 80–84% and the loading capacity of 51–52% highlighted the ability of the PCL to effectively encapsulate essential oils. Furthermore, the ease of preparation of these nanosystems and their stability in aqueous environments could make them useful in different application fields.

PCL is a biocompatible and biodegradable polymer approved by the Food and Drug Administration for drug delivery application. It is a polymer with a low cost and high hydrophobicity and stability. Moreover, PCL is characterized by slow degradation, which makes it potentially more suitable than other polymers, such as polyglycolide (PGA) and poly D, L-lactide (PDLA), when a slow release of the bioactive is required [35].

We have reported here the first example of PCL-based nanocapsules containing essential oils with more enhanced antifungal activity than pure essential oils against a panel of fourteen fungal strains. In the literature, the antifungal activity of essential oils encapsulated in different nanocarriers was also reported. There are many examples of encapsulated EOs in chitosan nanostructures with antifungal activity [36], the majority of which were obtained by the emulsion ionic gelation technique. In a recent work, Hasheminejad et al. [37] demonstrated that clove essential oil encapsulated by chitosan nanoparticles inhibited the growth of *Aspergillus niger*, isolated from spoiled pomegranate, at a concentration of 1.5 mg/mL. Instead, against *Aspergillus flavus*, Khalili et al. [38] reported an MIC value of 300 mg/L by using a chitosan benzoic acid nanogel containing thyme essential oil under sealed conditions. *Fusarium graminearum* was inhibited by a chitosan-encapsulating *Cymbopogon martini* essential oil [39], with an MIC value of around 0.42 mg/mL.

Nanoemulsions of oregano EO, prepared by Bedoia-Serna et al. [40], showed antifungal effects against *Cladosporium* sp., *Fusarium* sp. and *Penicillium* sp. The authors found that for *Fusarium* sp., unlike with *Penicillium* sp., the encapsulation of the oregano essential oil enhanced its antifungal effect.

β-Cyclodextrin was also used to entrap essential oils with antifungal activity through the formation of an inclusion complex. Capsules of β-cyclodextrin containing clove and Mexican oregano EO were active against *Fusarium oxysporum* [41]. An inclusion complex based on β-cyclodextrin and *Litsea cubeba* EO presented effective antifungal activity against three main fungi in citrus, including *Penicillium digitatum*, *Penicillium italicum* and *Geotrichum citri-aurantii* [42].

The antifungal activity of inorganic nanoparticles loaded with essential oils was reported by Weisany et al. In particular, the authors showed that encapsulation of *Thymus daenensis* and *Anethum graveolens* in silver nanoparticles enhanced their fungicidal activity against the plant pathogen *Colletotrichum nymphaeae* [43]. The same authors proved that copper nanoparticles containing thyme and dill EOs were able to reduce the mycelium growth and strongly inhibit the germination of conidia of *Colletotrichum nymphaeae* [44].

Unexpectedly, in the literature, no example of nanostructured systems containing essential oils effective against the fungi *Mucor circinelloides*, *Exophiala xenobiotica*, *Purpureocillium lilacinum*, *Pleurotus eryngii*, *Bjerkandera adusta* or *Phanerochaete chrysosporium* has been reported.

Although the antibacterial activity of EO nanocapsules based on PCL have been evaluated [28,45,46], less interest has been paid to their antifungal activity. The co-encapsulation of chloramphenicol with lemongrass essential oil in PCL-Pluronic composite nanocapsules was reported by Wang et al. [47]. This nanosystem showed considerably enhanced activity against methicillin-resistant *Staphylococcus aureus* and *Candida* sp. Tea tree essential oil loaded in PCL nanocapsules exhibited antifungal activity against *Trichophyton rubrum*, an etiological agent of superficial human mycosis [48].

In light of this, our results could provide a valuable contribution to the use of encapsulated essential oils as effective natural fungicides, as an alternative to the synthetic ones responsible for increasing resistant fungal strains.

4. Materials and Methods

All solvents were of analytic grade. In order to prepare the essential oil-loaded nanocapsules, sorbitan monostearate (SM) and poly(ε-caprolactone) (PCL) (Mn 45000) were obtained from Sigma-Aldrich (Milan, Italy) and polysorbate 80 (Tween 80) was obtained from Fisher Chemical (Fisher Scientific, Geel, Belgium). Water Chromasolv Plus for HPLC (Honeywell Riedel-de-Haën, Seelze, Germany) was used. Commercial essential oils of oregano and thyme were provided by Esperis S.p.A., (Milan, Italy) and by Flora s.r.l. (Lorenzana, Pisa, Italy), respectively. A standard mix of *n*-alkanes C_9–C_{22} was purchased by Alltech (Italy).

4.1. Characterization of Essential Oils by Gas Chromatography–Flame Ionization Detection (GC–FID) and Gas Chromatography–Mass Spectrometry (GC–MS)

GC–FID and GC–MS analysis were performed on a Shimadzu GC-17A and a Shimadzu GCMS-QP5050A, respectively. For both analyses, the same fused silica capillary column (Supelco SPBTM-5 15 m, 0.1 mm, 0.1 mm) was used.

The operating conditions for both runs were as follows: injector temperature 250 °C; detector temperature 280 °C; carrier gas helium (1 mL/min); split mode (1:200); and volume of injection 1 mL (4% essential oil/CH_2Cl_2 v/v). The heating ramp was as follows: 60 °C for 1 min, 60–280 °C at 10 °C/min, then 280 °C for 1 min. The relative percentages of compounds in each essential oil were determined from the peak areas in the GC–FID profiles. The mass spectrometer operating conditions were as follows: ionization at 70 eV and an ion source temperature of 180 °C. Mass spectral data were acquired in the scan mode in the m/z range of 40–400. Oil solutions were injected with the split mode (1:96) [49].

The identity of components was determined, comparing their retention indices relative to the C_9–C_{22} *n*-alkanes on the SPB-5 column, computer matching of spectral MS data with those from National Institute of Standard and Technology (NIST) Mass Spectral Library (MS) 107 and 21 [50] and the comparison of the fragmentation patterns with those reported in the literature [51].

4.2. Preparation of Essential Oil-Loaded Nanocapsules (EO-NCs)

The EO-loaded nanocapsules were prepared according to the preparation described in Granata et al. [28], but with slight changes, above all concerning scaling up by about three times. A solution of sorbitan monostearate (112 mg), PCL (320 mg), and EO (1.0 g) in acetone (80 mL) at 30 °C was poured into an aqueous solution of polysorbate 80 (275 mg in 160 mL of pure water) while stirring. The suspension was stirred for another 10 min at 25 °C. Then, the organic solvent was removed carefully under vacuum conditions (bath at 30 °C, pressure gradually reduced from 550 to 100 mbar in 1 h, and then reduced at 90 mBar and kept constant for 10 min). Finally, in order to complete the solvent evaporation, the mixture was fluxed by N_2 (1 h, atmospheric pressure), obtaining the EO-NC suspension (160 mL). The empty NC suspension was achieved under the same conditions but without essential oil.

4.3. Physicochemical Characterization of EO-NCs

4.3.1. Encapsulation Efficiency (EE) and Loading Capacity (LC) of EO-NCs

The total amount of essential oil in the EO-NC suspensions was determined by UV−vis spectroscopy (8453 UV-Visible Spectrophotometer, Agilent Technologies, Milan, Italy) over wavelengths ranging from 250 to 450 nm (λ_{max} 274). A 20 µL EO-NC suspension was diluted with 2 mL of acetonitrile, and the absorbance at 274 nm was registered and corrected from the small absorbance due to the other components. The OV or TC concentration was determined from the value of the correct absorbance compared to the linear calibration curve of respective EO, obtained by plotting the absorbance at 274 nm of twelve solutions containing different concentrations of OV (from 11 to 118 µg/mL, R^2 = 0.9999) or eleven solutions containing TC (from 10 to 102 µg/mL, R^2 = 0.9999). The ultrafiltration centrifugation technique (Nanosep 30K Omega, Pall Life Science, Milan, Italy; 90 min at 3500× *g*) was used to estimate the free essential oil. The suspension (500 µL) of the EO-NC was ultrafiltrated, and then an aliquot (60 µL) of the filtrate was diluted with 2 mL of acetonitrile. As described above, the free essential oil amount was determined by the absorbance at 274 nm of the resulting solution. From the total and free amounts of essential oil, the encapsulation efficiency was calculated:

$$EE\ (\%) = ([EO]_{tot} - [EO]_{free})/[EO]_{tot} \times 100 \quad (1)$$

where $[EO]_{tot}$ and $[EO]_{free}$ are the total and free amounts of essential oil in the EO-NC suspensions, respectively.

The loading capacity was calculated by the following equation:

$$LC\ (\%) = (\text{mass of loaded EO})/(\text{mass of loaded nanocapsules}) \times 100 \quad (2)$$

4.3.2. Particle Size, Polydispersity and Zeta Potential Measurements

The mean diameter (z-average), the polydispersity index (PDI) and the I-weighted distribution of the EO-NCs were obtained at 25 °C by dynamic light scattering (DLS). The zeta potential (ζ) values were determined by electrophoretic light scattering (ELS). DLS and ELS experiments were performed on a Zetasizer Nano ZS-90 (Malvern Instruments, Cambridge, UK), and data were analyzed using Zetasizer Version 7.02 software. For these purposes, the EO-NC suspensions were previously diluted (1:200, *v*/*v*) with pure water or with a pre-filtered (0.45 µm) 10 mM NaCl aqueous solution to carry out DLS or ELS, respectively.

Each experiment was replicated at least twice, and measurements performed at least three times. All data are expressed as mean ± standard deviation (SD).

4.4. Microorganisms and Growth Conditions

The fungal strains used in this study, their environment of isolation and sources are detailed described in Table 3 (*Aspergillus fumigatus, Aspergillus flavus, Penicillium rubens, Penicillium citrinum, Cladosporium aggregatocicatricatum, Cladosporium herbarum, Fusarium oxysporum*, *Geotrichum candidum*, *Mucor circinelloides*, *Exophiala xenobiotica*, *Purpureocillium lilacinum*, *Pleurotus eryngii* CCBAS471, *Bjerkandera adusta* CCBAS232 and *Phanerochaete chrysosporium* CCBAS570). The fungal strains were grown at 26 °C on malt extract agar (MEA) for 5 days.

4.5. Minimum Inhibitory Concentration (MIC) and Minimum Fungicidal Concentration (MFC)

The MIC and MFC values of the nanoencapsulated essential oils (EO-NCs) and pure EOs were evaluated with a panel of fourteen fungal strains (Table 3). Fungal suspensions were prepared according to de Lira Mota et al. [52] by washing the surface of the MEA slant culture with 5 mL of sterile saline and shaking the suspensions for 5 min. The resulting mixture of sporangiospores and hyphal fragments was withdrawn and transferred to a sterile tube. After heavy particles were allowed to settle for 3–5 min, the upper suspension was collected and vortexed for 15 s. Fungal suspensions were adjusted to a final concentration of 10^6 conidia mL^{-1} in malt extract broth (MEB). Nine milliliters (9 mL) of MEB containing one hundred microliters of the fungal suspension was distributed into incubation flasks. Encapsulated essential oils of oregano (OV-NC) and thyme (TC-NC) at different concentrations (0.05 to 0.5 mg/mL) and pure oregano (OV-EO) and thyme (TC-EO) at concentrations ranging from 0.05 to 5 mg/mL were added individually to the flasks. The flasks were then incubated at 26 °C for 7 days. After incubation, those fungal strains that did not show any growth were transferred to fresh MEA plates without EO-NCs for an additional 7 days at 26 °C to confirm which concentration had a fungicidal effect. The concentration unfavorable for growth revival during the transfer experiment was taken as the MFC, and this effect was identified as fungicidal. The lowest concentration of EOs or EO-NCs that prevented visible fungal growth and allowed a revival of fungal growth during the transfer experiment was considered the MIC.

5. Conclusions

We prepared essential oil-loaded nanocapsules with broad-spectrum antifungal activity. The obtained results suggest the efficacy of nanoencapsulation for enhancing pure essential oil activity. These ecofriendly nanosytems could be a valid alternative to synthetic antifungals and could find effective applications in various industrial sectors concerning health, agri-food and the environment.

Author Contributions: Conceptualization, C.G. and D.P.; methodology, G.G., A.P. and M.B.; formal analysis, G.G., E.N. and M.K.; investigation, G.G., E.N., M.K., A.P. and M.B.; data curation, E.N., M.K., A.P. and M.B.; writing—original draft preparation, C.G., E.N., G.G., M.K. and D.P.; writing—review and editing, C.G. and D.P.; visualization, G.G. and M.K.; supervision, C.G. and D.P.; project administration, C.G. and D.P.; funding acquisition, D.P. and M.B. All authors have read and agreed to the published version of the manuscript.

Funding: This research was funded by VEGA projects 2/0061/17 (Innovative disinfection strategies: the essential oils effect on microflora and materials of cultural heritage objects) and 2/0059/19 (Combination of nanoparticles and essential oils for mitigating the biodeterioration on various types of building materials).

Institutional Review Board Statement: Not applicable.

Informed Consent Statement: Not applicable.

Data Availability Statement: The data presented in this study are available in this article.

Acknowledgments: We thank Tonia Strano for the GC and GC–MS technical support.

Conflicts of Interest: The authors declare no conflict of interest.

References

1. Sterflinger, K. Fungi: Their role in deterioration of cultural heritage. *Fungal Biol. Rev.* **2010**, *24*, 47–55. [CrossRef]
2. Skóra, J.; Sulyok, M.; Nowak, A.; Otlewska, A.; Gutarowska, B. Toxinogenicity and cytotoxicity of Alternaria, Aspergillus and Penicillium moulds isolated from working environments. *Int. J. Environ. Sci. Technol.* **2016**, *14*, 595–608. [CrossRef]

3. Pinheiro, A.C.; Sequeira, S.O.; Macedo, M.F. Fungi in archives, libraries, and museums: A review on paper conservation and human health. *Crit. Rev. Microbiol.* **2019**, *45*, 686–700. [CrossRef] [PubMed]
4. Bensch, K.; Braun, U.; Groenewald, J.; Crous, P. The genus Cladosporium. *Stud. Mycol.* **2012**, *72*, 1–401. [CrossRef]
5. Reddy, K.R.; Abbas, H.K.; Abel, C.A.; Shier, W.T.; Oliveira, C.; Raghavender, C.R. Mycotoxin contamination of commer-cially important agricultural commodities. *Toxin Rev.* **2009**, *28*, 154–168. [CrossRef]
6. Richard, E.; Heutte, N.; Sage, L.; Pottier, D.; Bouchart, V.; LeBailly, P.; Garon, D. Toxigenic fungi and mycotoxins in mature corn silage. *Food Chem. Toxicol.* **2007**, *45*, 2420–2425. [CrossRef]
7. Hymery, N.; Vasseur, V.; Coton, M.; Mounier, J.; Jany, J.-L.; Barbier, G.; Coton, E. Filamentous Fungi and Mycotoxins in Cheese: A Review. *Compr. Rev. Food Sci. Food Saf.* **2014**, *13*, 437–456. [CrossRef]
8. Topalović, O.; Hussain, M.; Heuer, H. Plants and associated soil microbiota cooperatively suppress plant-parasitic nema-todes. *Front. Microbiol.* **2020**, *11*, 313. [CrossRef]
9. Lundell, T.K.; Mäkelä, M.R.; Hildén, K. Lignin-modifying enzymes in filamentous basidiomycetes-ecological, functional and phylogenetic review. *J. Basic Microbiol.* **2010**, *50*, 5–20. [CrossRef]
10. Schmidt, O. Indoor wood-decay basidiomycetes: Damage, causal fungi, physiology, identification and characterization, preven-tion and control. *Mycol. Prog.* **2007**, *6*, 261–279. [CrossRef]
11. Gostinčar, C.; Grube, M.; Gunde-Cimerman, N. Evolution of Fungal Pathogens in Domestic Environments? *Fungal Biol.* **2011**, *115*, 1008–1018. [CrossRef] [PubMed]
12. Stević, T.; Berić, T.; Šavikin, K.; Soković, M.; Gođevac, D.; Dimkić, I.; Stanković, S. Antifungal activity of selected essential oils against fungi isolated from medicinal plant. *Ind. Crop. Prod.* **2014**, *55*, 116–122. [CrossRef]
13. Napoli, E.M.; Ruberto, G. Sicilian aromatic plants: From traditional heritage to a new agro-industrial exploitation. In *Spices: Types, Uses and Health Benefits*; Kralis, J.F., Ed.; Nova Science Publisher: Hauppauge, NY, USA, 2012; pp. 1–56.
14. Karpiński, T.M. Essential Oils of Lamiaceae Family Plants as Antifungals. *Biomolecules* **2020**, *10*, 103. [CrossRef] [PubMed]
15. Rao, A.; Zhang, Y.; Muend, S.; Rao, R. Mechanism of Antifungal Activity of Terpenoid Phenols Resembles Calcium Stress and Inhibition of the TOR Pathway. *Antimicrob. Agents Chemother.* **2010**, *54*, 5062–5069. [CrossRef] [PubMed]
16. D'Agostino, M.; Tesse, N.; Frippiat, J.; Machouart, M.; Debourgogne, A.; Agostino, D. Essential Oils and Their Natural Active Compounds Presenting Antifungal Properties. *Molecules* **2019**, *24*, 3713. [CrossRef]
17. Gucwa, K.; Milewski, S.; Dymerski, T.; Szweda, P. Investigation of the Antifungal Activity and Mode of Action of Thymus vulgaris, Citrus limonum, *Pelargonium graveolens, Cinnamomum cassia, Ocimum basilicum*, and *Eugenia caryophyllus* Essential Oils. *Molecules* **2018**, *23*, 1116. [CrossRef]
18. Gupta, S.; Variyar, P.S. Nanoencapsulation of essential oils for sustained release: Application as therapeutics and antimicrobials. In *Encapsulations*; Elsevier: Amsterdam, The Netherlands, 2016; pp. 641–672.
19. Weiss, J.; Gaysinsky, S.; Davidson, M.; McClements, J. Nanostructured encapsulation systems: Food antimicrobials. In *Global Issues in Food Science and Technology*; Academic Press: Cambridge, MA, USA, 2009; pp. 425–479.
20. Ephrem, E.; Greige-Gerges, H.; Fessi, H.; Charcosset, C. Optimisation of rosemary oil encapsulation in polycaprolactone and scale-up of the process. *J. Microencapsul.* **2014**, *31*, 746–753. [CrossRef]
21. Kohiyama, C.Y.; Ribeiro, M.M.; Mossini, S.A.; Bando, E.; da Silva Bomfim, N.; Nerilo, S.B.; Rocha, G.H.; Grespan, R.; Mikcha, J.M.; Machinski, M., Jr. Antifungal properties and inhibitory effects upon aflatoxin production of *Thymus vulgaris* L. by Aspergillus flavus Link. *Food Chem.* **2015**, *173*, 1006–1010. [CrossRef]
22. Puškárová, A.; Bučková, M.; Kraková, L.; Pangallo, D.; Kozics, K. The antibacterial and antifungal activity of six essential oils and their cyto/genotoxicity to human HEL 12469 cells. *Sci. Rep.* **2017**, *7*, 8211. [CrossRef]
23. Božik, M.; Cisarova, M.; Tančinová, D.; Kouřimská, L.; Hleba, L.; Klouček, P. Selected essential oil vapours inhibit growth of *Aspergillus* spp. in oats with improved consumer acceptability. *Ind. Crop. Prod.* **2017**, *98*, 146–152. [CrossRef]
24. Saija, A.; Speciale, A.; Trombetta, D.; Leto, C.; Tuttolomondo, T.; La Bella, S.; Licata, M.; Virga, G.; Bonsangue, G.; Gennaro, M.C.; et al. Phytochemical, Ecological and Antioxidant Evaluation of Wild Sicilian Thyme: *Thymbra capitata* (L.) Cav. *Chem. Biodivers.* **2016**, *13*, 1641–1655. [CrossRef] [PubMed]
25. Napoli, E.; Siracusa, L.; Ruberto, G. New tricks for old guys: Recent developments in the chemistry, biochemistry, appli-cations and exploitation of selected species from the Lamiaceae family. *Chem. Biodiver.* **2020**, *17*, e1900677. [CrossRef] [PubMed]
26. György, Z.; Incze, N.; Pluhár, Z. Differentiating *Thymus vulgaris* chemotypes with ISSR molecular markers. *Biochem. Syst. Ecol.* **2020**, *92*, 104118. [CrossRef]
27. Avola, R.; Granata, G.; Geraci, C.; Napoli, E.M.; Graziano, A.C.E.; Cardile, V. Oregano (*Origanum vulgare* L.) essential oil provides anti-inflammatory activity and facilitates wound healing in a human keratinocytes cell model. *Food Chem. Toxicol.* **2020**, *144*, 111586. [CrossRef] [PubMed]
28. Granata, G.; Stracquadanio, S.; Leonardi, M.; Napoli, E.; Consoli, G.M.L.; Cafiso, V.; Stefani, S.; Geraci, C. Essential oils encapsulated in polymer-based nanocapsules as potential candidates for application in food preservation. *Food Chem.* **2018**, *269*, 286–292. [CrossRef] [PubMed]
29. Granata, G.; Consoli, G.M.; Nigro, R.L.; Geraci, C. Hydroxycinnamic acids loaded in lipid-core nanocapsules. *Food Chem.* **2018**, *245*, 551–556. [CrossRef]
30. Bilia, A.R.; Guccione, C.; Isacchi, B.; Righeschi, C.; Firenzuoli, F.; Bergonzi, M.C. Essential Oils Loaded in Nanosystems: A Developing Strategy for a Successful Therapeutic Approach. *Evid. Based Complement. Altern. Med.* **2014**, *2014*, 1–14. [CrossRef]

31. Nazzaro, F.; Fratianni, F.; Coppola, R.; De Feo, V. Essential Oils and Antifungal Activity. *Pharmaceuticals* **2017**, *10*, 86. [CrossRef]
32. Danaei, M.; Dehghankhold, M.; Ataei, S.; Davarani, F.H.; Javanmard, R.; Dokhani, A.; Khorasani, S.; Mozafari, M.R. Impact of Particle Size and Polydispersity Index on the Clinical Applications of Lipidic Nanocarrier Systems. *Pharmaceuticals* **2018**, *10*, 57. [CrossRef]
33. Dos Santos, P.P.; Paese, K.; Guterres, S.S.; Pohlmann, A.R.; Costa, T.M.H.; Jablonski, A.; Flôres, S.H.; Rios, A.D.O. Development of lycopene-loaded lipid-core nanocapsules: Physicochemical characterization and stability study. *J. Nanoparticle Res.* **2015**, *17*, 1–11. [CrossRef]
34. Lobato, K.B.D.S.; Paese, K.; Forgearini, J.C.; Guterres, S.S.; Jablonski, A.; Rios, A.D.O. Characterisation and stability evaluation of bixin nanocapsules. *Food Chem.* **2013**, *141*, 3906–3912. [CrossRef] [PubMed]
35. Woodruff, M.A.; Hutmacher, D.W. The return of a forgotten polymer—Polycaprolactone in the 21st century. *Prog. Polym. Sci.* **2010**, *35*, 1217–1256. [CrossRef]
36. Detsi, A.; Kavetsou, E.; Kostopoulou, I.; Pitterou, I.; Pontillo, A.R.N.; Tzani, A.; Christodoulou, P.; Siliachli, A.; Zoumpoulakis, P. Nanosystems for the Encapsulation of Natural Products: The Case of Chitosan Biopolymer as a Matrix. *Pharamceutics* **2020**, *12*, 669. [CrossRef] [PubMed]
37. Hasheminejad, N.; Khodaiyan, F.; Safari, M. Improving the antifungal activity of clove essential oil encapsulated by chitosan nanoparticles. *Food Chem.* **2019**, *275*, 113–122. [CrossRef]
38. Khalili, S.T.; Mohsenifar, A.; Beyki, M.; Zhaveh, S.; Rahmani-Cherati, T.; Abdollahi, A.; Bayat, M.; Tabatabaei, M. En-capsulation of Thyme essential oils in chitosan-benzoic acid nanogel with enhanced antimicrobial activity against Aspergillus flavus. *Lebensm Wiss Technol.* **2015**, *60*, 502–508. [CrossRef]
39. Kalagatur, N.K.; Nirmal Ghosh, O.S.; Sundararaj, N.; Mudili, V. Antifungal activity of chitosan nanoparticles encapsu-lated with *Cymbopogon martinii* essential oil on plant pathogenic fungi Fusarium graminearum. *Front. Pharmacol.* **2018**, *9*, 610. [CrossRef]
40. Bedoya-Serna, C.M.; Dacanal, G.C.; Fernandes, A.M.; Pinho, S.C. Antifungal activity of nanoemulsions encapsulating oregano (*Origanum vulgare*) essential oil: In vitro study and application in Minas Padrão cheese. *Braz. J. Microbiol.* **2018**, *49*, 929–935. [CrossRef]
41. Estrada-Cano, C.; Castro, M.A.; Muñoz-Castellanos, L.N.; García-Triana, N.A.; Hernández-Ochoa, L. Antifungal activity of microcapsulated clove (*Eugenia caryophyllata*) and Mexican oregano (*Lippia berlandieri*) essential oils against Fusarium ox-ysporum. *J. Microb. Biochem. Technol.* **2017**, *9*, 567–571. [CrossRef]
42. Wang, Y.; Yin, C.; Cheng, X.; Li, G.; Yang, S.; Zhu, X. β-Cyclodextrin Inclusion Complex Containing *Litsea cubeba* Essential Oil: Preparation, Optimization, Physicochemical, and Antifungal Characterization. *Coatings* **2020**, *10*, 850. [CrossRef]
43. Weisany, W.; Amini, J.; Samadi, S.; Hossaini, S.; Yousefi, S.; Struik, P.C. Nano silver-encapsulation of *Thymus daenensis* and *Anethum graveolens* essential oils enhances antifungal potential against strawberry anthracnose. *Ind. Crop. Prod.* **2019**, *141*, 111808. [CrossRef]
44. Weisany, W.; Samadi, S.; Amini, J.; Hossaini, S.; Yousefi, S.; Maggi, F. Enhancement of the antifungal activity of thyme and dill essential oils against *Colletotrichum nymphaeae* by nano-encapsulation with copper NPs. *Ind. Crop. Prod.* **2019**, *132*, 213–225. [CrossRef]
45. Jummes, B.; Sganzerla, W.G.; Da Rosa, C.G.; Noronha, C.M.; Nunes, M.R.; Bertoldi, F.C.; Barreto, P.L.M. Antioxidant and antimicrobial poly-ε-caprolactone nanoparticles loaded with *Cymbopogon martinii* essential oil. *Biocatal. Agric. Biotechnol.* **2020**, *23*, 101499. [CrossRef]
46. da Silva, N.P.; Pereira, E.D.; Duarte, L.M.; de Oliveira Freitas, J.C.; de Almeida, C.G.; da Silva, T.P.; Melo, R.C.; Apolônio, A.C.; de Oliveira, M.A.; de Mello Brandão, H.; et al. Improved an-ti-Cutibacterium acnes activity of tea tree oil-loaded chitosan-poly (ε-caprolactone) core-shell nanocapsules. *Colloids Surf. B Biointerfaces* **2020**, *196*, 111371. [CrossRef] [PubMed]
47. Kalita, S.; Kandimalla, R.; Devi, B.; Kalita, B.; Kalita, K.; Deka, M.; Kataki, A.C.; Sharma, A.; Kotoky, J. Dual delivery of chloramphenicol and essential oil by poly-ε-caprolactone–Pluronic nanocapsules to treat MRSA-Candida co-infected chronic burn wounds. *RSC Adv.* **2017**, *7*, 1749–1758. [CrossRef]
48. Flores, F.C.; De Lima, J.A.; Ribeiro, R.F.; Alves, S.H.; Rolim, C.M.B.; Beck, R.C.R.; Da Silva, C.B. Antifungal Activity of Nanocapsule Suspensions Containing Tea Tree Oil on the Growth of Trichophyton rubrum. *Mycropatology* **2013**, *175*, 281–286. [CrossRef]
49. Napoli, E.M.; Curcuruto, G.; Ruberto, G. Screening of the essential oil composition of wild Sicilian thyme. *Biochem. Syst. Ecol.* **2010**, *38*, 816–822. [CrossRef]
50. NIST, National Institute of Standard and Technology, Mass Spectral Library. 1998. Available online: http://chemdata.nist.gov (accessed on 6 December 2019).
51. Adams, R.P. *Identification of Essential Oil Components by Gas Chromatographic/Quadrupole Mass Spectrometry*; Allured Publishing Corporation: Carol Stream, IL, USA, 2007.
52. Mota, K.S.D.L.; Pereira, F.D.O.; De Oliveira, W.A.; Lima, I.O.; Lima, E.D.O. Antifungal Activity of *Thymus vulgaris* L. Essential Oil and Its Constituent Phytochemicals against *Rhizopus oryzae*: Interaction with Ergosterol. *Molecules* **2012**, *17*, 14418–14433. [CrossRef]

antibiotics

MDPI

Article

Aniba rosaeodora (Var. amazonica Ducke) Essential Oil: Chemical Composition, Antibacterial, Antioxidant and Antitrypanosomal Activity

Amanda Mara Teles [1], João Victor Silva-Silva [2], Juan Matheus Pereira Fernandes [2], Kátia da Silva Calabrese [2,*], Ana Lucia Abreu-Silva [3], Silvio Carvalho Marinho [4], Adenilde Nascimento Mouchrek [5], Victor Elias Mouchrek Filho [5] and Fernando Almeida-Souza [2,3,*]

1 Pós-Graduação em Saúde do Adulto, Universidade Federal do Maranhão, 65080-805 São Luís, Brazil; damarateles@hotmail.com
2 Laboratório de Imunomodulação e Protozoologia, Instituto Oswaldo Cruz, Fiocruz, 21040-900 Rio de Janeiro, Brazil; jvss2906@gmail.com (J.V.S.-S.); juan.fernandes222@gmail.com (J.M.P.F.)
3 Pós-graduação em Ciência Animal, Universidade Estadual do Maranhão, 65055-310 São Luís, Brazil; abreusilva.ana@gmail.com
4 Laboratório de Óleos Essenciais, Universidade Federal do Maranhão, 65065-545 São Luís, Brazil; silviomarinho@yahoo.com.br
5 Laboratório de Controle de Qualidade de Alimentos e Água, Universidade Federal do Maranhão, 65065-545 São Luís, Brazil; adenild@bol.com.br (A.N.M.); victor.mouchrek@ufma.br (V.E.M.F.)
* Correspondence: calabrese@ioc.fiocruz.br (K.d.S.C.); fernandoalsouza@gmail.com (F.A.-S.)

Abstract: *Aniba rosaeodora* is one of the most widely used plants in the perfumery industry, being used as medicinal plant in the Brazilian Amazon. This work aimed to evaluate the chemical composition of *A. rosaeodora* essential oil and its biological activities. *A. rosaeodora* essential oil presented linalool (93.60%) as its major compound. The *A. rosaeodora* essential oil and linalool showed activity against all the bacteria strains tested, standard strains and marine environment bacteria, with the lower minimum inhibitory concentration being observed for *S. aureus*. An efficient antioxidant activity of *A. rosaeodora* essential oil and linalool (EC_{50}: 15.46 and 6.78 μg/mL, respectively) was evidenced by the inhibition of the 2,2-azinobis-(3-ethylbenzothiazoline-6-sulfonic acid) (ABTS) radical. The antitrypanosomal activity of *A. rosaeodora* essential oil and linalool was observed at high concentrations against epimatigote forms (inhibitory concentration for 50% of parasites (IC_{50}): 150.5 ± 1.08 and 198.6 ± 1.12 μg/mL, respectively), and even higher against intracellular amastigotes of *T. cruzi* (IC_{50}: 911.6 ± 1.15 and 249.6 ± 1.18 μg/mL, respectively). Both *A. rosaeodora* essential oil and linalool did not exhibit a cytotoxic effect in BALB/c peritoneal macrophages, and both reduced nitrite levels in unstimulated cells revealing a potential effect in NO production. These data revealed the pharmacological potential of *A. rosaeodora* essential oil and linalool, encouraging further studies.

Keywords: rosewood; linalool; marine bacteria; ABTS; *Trypanosoma cruzi*; cytotoxicity; nitrite; nitric oxide

Citation: Teles, A.M.; Silva-Silva, J.V.; Fernandes, J.M.P.; Calabrese, K.d.S.; Abreu-Silva, A.L.; Marinho, S.C.; Mouchrek, A.N.; Filho, V.E.M.; Almeida-Souza, F. *Aniba rosaeodora* (Var. amazonica Ducke) Essential Oil: Chemical Composition, Antibacterial, Antioxidant and Antitrypanosomal Activity. *Antibiotics* **2021**, *10*, 24. https://doi.org/10.3390/antibiotics10010024

Received: 27 November 2020
Accepted: 28 December 2020
Published: 30 December 2020

1. Introduction

In recent decades, many studies have concentrated on the search for potential antimicrobials, with an increase in the worldwide spending on finding new antimicrobial agents. Faced with bacterial resistance to antibiotic treatment as well as the discovery of new pathogens, the need for new antimicrobials arises [1].

There are promising reports of different plant-derived natural phytochemicals, and a growing interest in exploring their potential [2]. Medicinal plants have been used around the world for various purposes, just as their active chemical compounds have been used to combat various diseases. Essential oils are well known for their pharmacological activities, including antibacterial [3] and trypanocidal activity [4], and may represent a promising source of new natural drugs.

Aniba rosaeodora Ducke, Lauraceae is a large tree that reaches 30 m in height, with yellow-brown bark (hence the name rosewood) and grows in the Amazon region (Figure 1A) [5]. Rosewood essential oil (Figure 1B) is widely used in the perfumery industry and its extraction for industrial purposes began in the interior of the state of Pará, Brazil, around 1930 [6]. All parts of the tree are fragrant, although only the trunk wood is harvested and hydrodistilled to obtain rosewood oil, a valuable product. The species has been used in the Amazon as a medicinal plant to control epileptic seizures, to compose regional fragrances and as an ornamental plant [7].

Figure 1. *Aniba rosaeodora* tree (**A**) and essential oil (**B**).

Rosewood oil is a colorless to pale yellow liquid with a woody floral fragrance, containing monoterpenic alcohol linalool as the main constituent. It is of interest to the flavor and fragrance industries because it is transformed into several valuable derivatives [8]. There are few reports of pharmacological properties of Rosewood oil associated with its chemical profile. These properties are the result of the synergism of all the molecules present in the oil or reflect the activity of its major compound linalool, presenting antibacterial and antifungal [9,10], antioxidant [11] and antiprotozoan properties [12].

Available data on the biological activity of rosewood essential oil are quite limited, because most studies have focused on linalool, its main constituent [13]. Thus, this work aims to evaluate the chemical composition of *Aniba rosaeodora* essential oil extracted in the Amazon biome, as well as its antibacterial, antioxidant and trypanocidal activity.

2. Results

2.1. Physical and Chemical Characterization of A. rosaeodora Essential Oil

The yield of *A. rosaeodora* essential oil obtained from dried leaves thin branches was 2.8%. The oil showed a yellow color and a clean appearance (Figure 1B), and presented a density of 0.89 g/mL at 25 °C, and a refractive index (ND 25) of 1.459. It was soluble in ethanol 70% in a ratio of 1:2. Chemical compounds identified and quantified in *A. rosaeodora* essential oil are presented in the chromatogram (Figure 2) and Table 1. Three compounds were identified and enumerated accordingly with elution order and retention time. The major constituent of *A. rosaeodora* essential oil was linalool with 93.60%. In addition, α-terpinolene and linalool cis-oxide were identified and quantified at 3.37% and 3.03%, respectively.

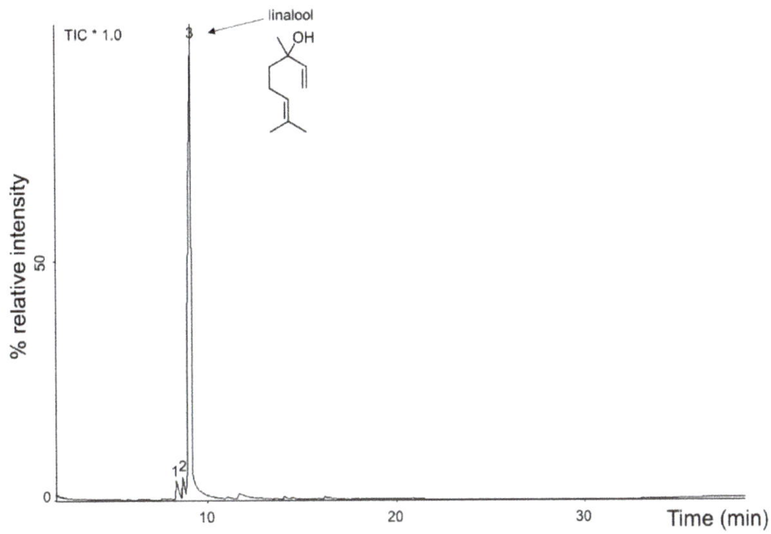

Figure 2. Chromatogram of *Aniba rosaeodora* essential oil. * TIC: total ion chromatogram.

Table 1. Chemical composition of *Aniba rosaeodora* essential oil.

Peak	Compounds	Retention Time (min)	Peak Area (%) [1]
1	α-terpinolene	8.361	3.37
2	linalool cis-oxide	8.702	3.03
3	linalool	9.177	93.60

[1] Peak area percentage in relation to peak total area.

2.2. Antimicrobial Activity of *A. rosaeodora* Essential Oil and Linalool

Bacteria from the marine environment were evaluated by the disc-diffusion method against *A. rosaeodora*, linalool and several reference antibiotics (Table 2). The bacteria culture displayed inhibition halos ranging from 7 to 25 mm, besides the non-inhibition presented in some cultures. Comparing linalool with *A. rosaeodora* essential oil, we found that *A. rosaeodora* was more efficient against *Aeromonas caviae* and *Enterococcus faecalis* than the standard linalool. Linalool exhibited greater activity against *Klebsiella pneumonia* and *Providencia stuartii* than *A. rosaeodora* essential oil, while both compounds presented the same activity against *Aeromonas hydrophila*. The susceptibility test performed with antibiotics showed that *E. faecalis* was the strain that presented sensibility for all the antibiotics analyzed, while the other four strains displayed mixed sensibility to the antibiotics.

The preliminary antibacterial activity against standard strain bacteria evaluated by the disc-diffusion method showed a growth inhibitory halo on *A. rosaeodora* essential oil and linalool disks against Gram-positive (*Staphylococcus aureus*) and Gram-negative strains (*Escherichia coli*, *Pseudomonas aeruginosa* and *Salmonella* sp.) (Table 3). Gram-positive bacteria exhibited the highest inhibition halo from both essential oil and linalool. The minimum inhibitory concentration (MIC) of *A. rosaeodora* essential oil ranged from 250 to 450 µg/mL against the tested strains. As observed in the disc-diffusion method, *S. aureus* was the strain more sensible to *A. rosaeodora* essential oil and linalool activity by MIC analysis. Likewise, both methodologies showed that *A. rosaeodora* essential oil presented better antimicrobial activity than linalool to all strains analyzed.

Table 2. Inhibitory zone diameters of *Aniba rosaeodora* essential oil, linalool and antibiotics on different bacteria isolated from marine environment after 24 h of treatment.

Compounds/Antibiotics	Bacteria Strain				
	Aeromonas caviae	*Aeromonas hydrophila*	*Enterococcus faecalis*	*Klebsiella pneumoniae*	*Providencia stuartii*
A. rosaeodora essential oil	22.33 ± 0.577	11.33 ± 0.577	21.33 ± 0.577	10.33 ± 0.577	9.33 ± 0.577
linalool	7.33 ± 0.577	11.66 ± 0.577	13.33 ± 0.577	11.66 ± 0.577	11.33 ± 0.577
pipemidic acid	22.33 ± 0.577	21.66 ± 0.577	18.66 ± 0.577	22.33±0.577	22.33 ± 0.577
ampicillin	0 ± 0.00	0 ± 0.00	15.66 ± 0.577	0 ± 0.00	0 ± 0.00
cefotaxime	24.67 ± 0.577	23.66 ± 0.577	17.66 ± 0.577	17.33 ± 0.577	25.33 ± 0.577
cefoxitin	0 ± 0.00	13.66 ± 0.577	22.66 ± 0.577	8.33 ± 0.577	15.33 ± 0.577
chloramphenicol	23.33 ± 1.155	20.66 ± 0.577	24.66 ± 0.577	13.66 ± 0.577	13.33 ± 0.577
erythromycin	0 ± 0.00	0 ± 0.00	25.66 ± 1.527	0 ± 0.00	0 ± 0.00
gentamycin	17.66 ± 0.577	14.66 ± 0.577	22.33 ± 0.577	14.33 ± 0.577	14.33 ± 0.577
lincomycin	0 ± 0.00	0 ± 0.00	29.66 ± 0.577	0 ± 0.00	0 ± 0.00
oxacillin	0 ± 0.00	0 ± 0.00	18.66 ± 0.577	0 ± 0.00	0 ± 0.00
sulfazotrin	21.66 ± 0.577	19.66 ± 0.577	21 ± 1.00	20.33 ± 0.577	21.33 ± 0.577
tetracycline	10.33 ± 0.577	22.66 ± 0.577	19.66 ± 0.577	0 ± 0.00	18.33 ± 0.577
vancomycin	0 ± 0.00	0 ± 0.00	20.66 ± 0.577	0 ± 0.00	0 ± 0.00

Data represents mean ± standard deviation of experiment realized in triplicate.

Table 3. Inhibitory zone diameters and minimum inhibitory concentration of *Aniba rosaeodora* essential oil on different bacterial cultures after 24 h of treatment.

Antimicrobial assay	Compounds	Bacteria Strain			
		Escherichia coli	*Staphylococcus aureus*	*Pseudomonas aeruginosa*	*Salmonella choleraesuis*
Inhibition zones (mm) by disc-diffusion	*A. rosaeodora*	15.0 ± 0.00	18.0 ± 0.00	13.0 ± 0.00	14.0 ± 0.00
	linalool	13.0 ± 0.00	15.0 ± 0.00	12.0 ± 0.00	12.0 ± 0.00
	gentamycin	14.0 ± 0.00	20.5 ± 0.70	17.0 ± 0.00	n.d.
	penicillin	n.d.	n.d.	n.d.	18.5 ± 0.70
MIC (μg/mL)	*A. rosaeodora*	350.0 ± 0.00	250.0 ± 0.00	450.0 ± 0.00	400.0 ± 0.00
	linalool	650.0 ± 0.00	550.0 ± 0.00	650.0 ± 0.00	650.0 ± 0.00
	amoxicillin	16.0 ± 0.00	8.0 ± 0.00	n.d.	n.d.
	gentamycin	n.d.	2.0 ± 0.00	n.d.	8.0 ± 0.00
	polymyxin B	n.d.	n.d.	16.0 ± 0.00	n.d.

MIC: minimum inhibitory concentration; n.d.: not determined. Data represents mean ± standard deviation of experiment realized in triplicate.

2.3. Antioxidant Activity of *A. rosaeodora* Essential Oil and Linalol

Aniba roseadora essential oil and linalool presented antioxidant activity concentration-dependent, as observed in the graph that relates *A. rosaeodora* essential oil and linalool concentration versus the percentage of inhibition of the 2,2-azinobis- (3-ethylbenzothiazoline-6-sulfonic acid) (ABTS) radical (Figure 3). The calculated EC_{50} was 15.46 μg/mL for *A. rosaeodora* essential oil and 6.78 μg/mL for linalool.

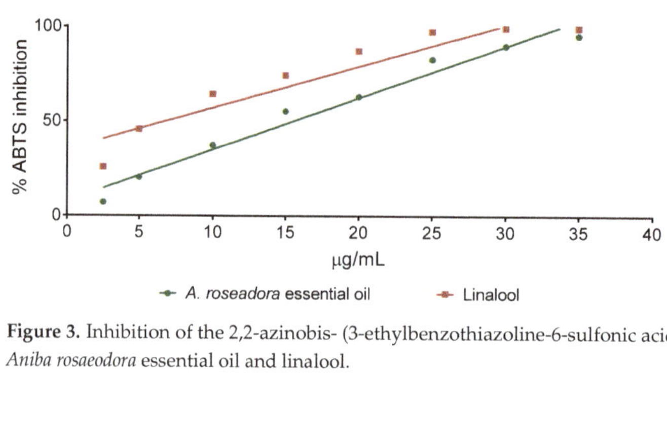

Figure 3. Inhibition of the 2,2-azinobis- (3-ethylbenzothiazoline-6-sulfonic acid) (ABTS) radical by *Aniba rosaeodora* essential oil and linalool.

2.4. *Cytotoxicity, Antitrypanosomal Activivy and Selectivity Index of A. rosaeodora Essential Oil and Linalool*

The activity of *A. rosaeodora* essential oil and linalool was evaluated against the epimastigote and intracellular amastigote forms of *Trypanosoma cruzi*, as well as its cytotoxic effect against mammal cells. Both compounds presented concentration-dependent inhibitory activity against epimastigote and intracellular amastigote forms of *T. cruzi* (Figure 4). The inhibitory concentration for 50% of parasites (IC_{50}) values for epimastigote forms was lower for *A. rosaeodora* essential oil than linalool. Analyzing the activity against different forms of the *T. cruzi*, *A. rosaeodora* essential oil exhibited IC_{50} value against epimastigote 6.0-fold higher in comparison to the IC_{50} against intracellular amastigotes forms. In contrast, linalool was 3.65-fold more effective against *T. cruzi* intracellular amastigote when compared to *A. rosaeodora* essential oil. Both compounds presented higher IC_{50} values when compared to benznidazole. Cytotoxicity assay revealed that *A. roseadora* essential oil and linalool not showed toxicity for BALB/c peritoneal macrophages even at the highest concentration analyzed (1000 μg/mL). Thus, linalool exhibited higher SI value than *A. roseadora* essential oil (Table 4).

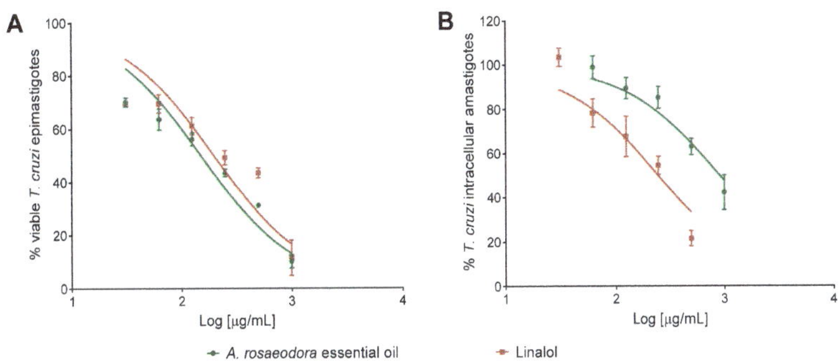

Figure 4. Activity of *Aniba rosaeodora* essential oil and linalool against *Trypanosoma cruzi* epimastigote (**A**) and intracellular amastigote forms (**B**) after 24 h of treatment. Data represent media ± standard deviation of three independent experiment realized in triplicate.

Table 4. BALB/c peritoneal macrophage cytotoxicity, trypanocidal activity and selectivity index of *Aniba rosaeodora* essential oil and linalool.

Essential oil/ Compounds	Citotoxicity CC_{50} (μg/mL)	*T. cruzi* IC_{50} (μg/mL)		SI
	Peritoneal Macrophage	Epimastigote	Intracellular Amastigote	
A. rosaeodora	>1000	150.5 ± 1.08	911.6 ± 1.15	>1.0
linalool	>1000	198.6 ± 1.12	249.6 ± 1.18	>4.0
benznidazole	162.0 ± 1.11	1.805 ± 1.13	0.4820 ± 1.17	336.0

IC_{50}: inhibitory concentration for 50% of parasites; CC_{50}: cytotoxic concentration for 50% of cells; SI: selectivity index, obtained from the ratio CC_{50}/IC_{50} intracellular amastigote. Data represents mean ± standard deviation of at least two independent experiments carried out in triplicate.

The parameters of infection analysis (Figure 5) showed that *A. rosaeodora* essential oil treatment displayed significant low number of amastigotes per 100 cells at 1000 μg/mL (p = 0.0001) and 500 μg/mL (p = 0.0011) (Figure 5A). Linalool showed a low number of amastigotes per 100 cells at 500 μg/mL (p = 0.0001), 250 μg/mL (p = 0.0014) and 125 μg/mL (p = 0.0290) (Figure 5B). On the other hand, the treatment with *A. rosaeodora* essential oil and linalool only presented a significant low mean number of amastigotes per infected cells at 1000 μg/mL (p = 0.0398, Figure 5C) and 500 μg/mL (p = 0.0229, Figure 5D), respectively. The alterations in intracellular amastigotes of *T. cruzi* after treatment with *A. rosaeodora* essential oil and linalool are represented in photomicrography images of Figure 5E.

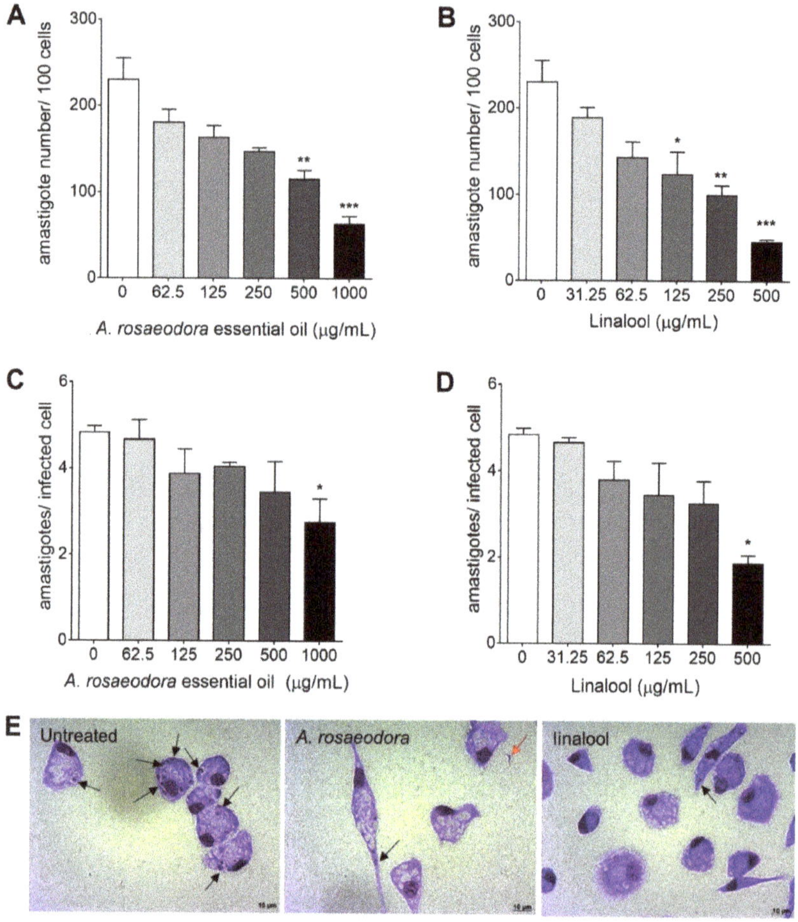

Figure 5. BALB/c peritoneal macrophages infected with *Trypanosoma cruzi* and treated for 24 h with *Aniba rosaeodora* essential oil or linalool. (**A–D**) Parameters of infection and (**E**) light microscopy after *A. rosaeodora* or linalool treatment at 1000 or 500 μg/mL respectively. Intracellular amastigotes inside macrophages (black arrows) and non-internalized parasite (red arrows). The images and data (mean ± standard deviation) represent two independent experiments performed in quadruplicate. * $p < 0.05$, ** $p < 0.01$ and *** $p < 0.001$ when compared with untreated infected cells by Kruskal–Wallis and Dunn's multiple comparison test. Giemsa, 40× objective.

2.5. Nitrite Quantification in T. Cruzi-Infected Peritoneal Macrophages Treated with A. rosaeodora Essential Oil and Linalool

The nitrite quantification in the supernatant of BALB/c peritoneal macrophages showed low nitrite levels in cells treated with *A. rosaeodora* essential oil (0.150 ± 0.220 μM $NaNO_2$, $p = 0.0259$) and linalool (0.175 ± 0.146 μM $NaNO_2$, $p = 0.0490$) when compared to untreated-unstimulated cells (1.129 ± 0.501 μM $NaNO_2$). In *T. cruzi*-stimulated cells, although nitrite levels after treatment with *A. rosaeodora* essential oil (0.952 ± 0.779 μM $NaNO_2$) and linalool (1.047 ± 0.702 μM $NaNO_2$) were lower than stimulated-untreated cells (1.347 ± 0.416 μM $NaNO_2$), the difference was not statistically significant for both compounds ($p = 0.945$ and $p > 0.999$, respectively) (Figure 6).

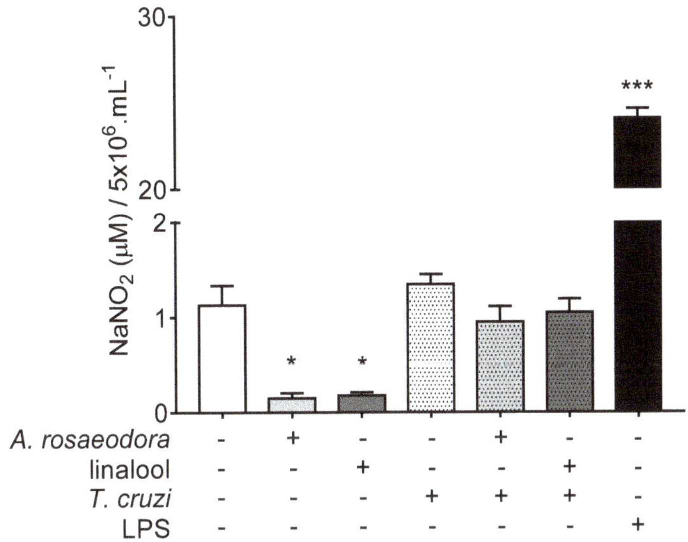

Figure 6. Nitrite quantification in the supernatant of the BALB/c peritoneal macrophage treated with *Aniba rosaeodora* essential oil (500 μg/mL) or linalool (125 μg /mL), and stimulated or not with *Trypanosoma cruzi*. Data represents mean ± standard deviation of experiment realized in sextuplicate; * $p < 0.05$, *** $p < 0.001$ when compared with untreated and unstimulated macrophages by Kruskal–Wallis and Dunn's multiple comparison test.

3. Discussion

Essential oil may change depending on the chemical nature of its constituents and can be modified by air, light, heat, water and various impurities of natural origin or from falsifications. The changes can be recognized both by changes in their organoleptic characteristics (aroma, color, taste, transparency, fluidity), as well as the values of their chemical and physical parameters. Thus the density, refractive index, solubility, color and appearance were analyzed and the physical characteristics of the essential oil were similar to the pattern described in previous studies of *A. rosaeodora* [14].

Studies has identified and quantified chemical compounds of *A. rosaeodora* essential oil, revealing that this species has chemotypes similar to essential oil extracted in Belém, state of Pará, Brazil, with linalool (84.8%) as the major compound, followed by α-terpineol (2.9%), geraniol (1.0%), benzyl benzoate (0.6%) and minimal amounts of monoterpene hydrocarbons and oxygenated sesquiterpenes (9.2%) [15]. The same was observed in the study of *A. rosaeodora* essential oil extracted in São Paulo, Brazil, where the presence of linalool (81.45%), trans-linalool oxide (1.19%), R-terpineol (1.09%) were observed [16]. Almeida et al. (2013) also reported that linalool is the main compound in essential oil obtained from wood, leaves and branches of the Brazilian rosewood [17].

The disk diffusion test carried out against standard strain bacteria and against bacteria isolated from a marine environment evidenced antibacterial activity, preliminarily. The mixed sensibility observed to the several antibiotics revealed the resistance pattern of marine environment bacteria. The sensibility observed to *A. rosaeodora* essential oil and linalool showed that both compounds have activity against marine environment bacteria used in this study. It is known that the bacterial cell wall influences in an important way on the action of certain antibiotics. The difference between Gram-positive and Gram-negative bacterial walls would be one of the responses to antibiotic resistance between two bacteria [14]. Bacterial resistance is best evidenced in environmental bacteria in the last few years. With the advent of modernization, an increasing amount of antibiotics was released into the environment along with the residues from domestic, industrial, agricultural and medical activities. This has ended up generating a selection of antibiotic-resistant bacteria or genes in the environment, which threatens the efficiency of antibiotics in fighting bacterial infections [18].

The MIC of *A. rosaeodora* essential oil resulted in concentrations lower than linalool. Holetz, et al. (2002) classifies samples that have MIC values below 100 μg/mL with good antibacterial activity; 100 to 500 μg/mL moderate; and 500 to 1000 μg/mL weak above 1000 μg/mL inactive [19]. Following this classification, *A. rosaeodora* essential oil showed moderate activity, while linalool displayed weak activity. The difference between the activity of both compounds may be related to the synergistic effect of the compounds present in the essential oil of *A. rosaeodora*. A synergistic interaction can be verified between the essential oils of *A. rosaeodora* and *Pelargonium graveolens* with gentamicin, and a very strong synergistic interaction against *Acinetobacter baumannii* ATCC 19606 (fractional inhibitory concentration/FIC index = 0.11) [20]. While research conducted with linalool showed low activity against Gram-positive and Gram-negative bacteria. Jabir et al. (2018) found that linalool loaded in gold nanoparticles modified with glutathione (LIN-GNPs) has effective antibacterial activity against Gram-positive bacteria, proving that LIN-GNPs acted on the bacterial cell membrane, in giving up and increasing cell wall permeability and stimulated reactive oxygen species (ROS) production that leads to bacterial nucleic acid damage [21].

Linalool is a compound widely used by the cosmetics industry [22] In the study by Herman et al. (2016) [10], a significant increase in antimicrobial efficacy was observed by the addition of linalool to essential oil, reducing its concentrations in products (cosmetics, medicine), making it possible to obtain its synergistic and additive effects. In addition, several studies have been reported on the commercial availability of oxidized linalool samples possibly causing allergic contact dermatitis [23–25]. Thus, essential oil from *A. rosaeodora* has potential applicability in edible and/or dermatological preparations.

Biological activity may be directly related to phenolic compounds as they are good electron donors and therefore have efficient antioxidant activity among secondary plant metabolites. These compounds are capable of control oxidative damage generated by reactive oxygen species or radicals [26]. We can also classify the antioxidant activity according to the excellent (IC_{50} < 15 μg/mL), good (15 μg/mL < IC_{50} < 50 μg/mL), medium (50 μg/mL < IC_{50} < 100 μg/mL), and weak activity ($IC_{50} \geq$ 100 μg/mL). *A. rosaeodora* essential oil antioxidant activity was considered good while the linalool was optimal, corroborating with previous studies that verified excellent antioxidant activity of *Aniba* species [26].

In traditional medicine, plant essential oils are known as a rich source of secondary metabolites with relevant biological activities, as an alternative in antiparasitic therapy [27,28]. The trypanocidal activity of essential oils of *Aniba* genus was described in the literature [14,29]. Currently, the drugs available for the treatment of Chagas disease are benznidazole and nifurtimox, which have limited efficacy, serious adverse effects and have been in use since the late 1960s [30]. Thus, in an attempt to search for new therapeutic alternatives for Chagas disease, we report the effect of *A. rosaeodora* essential oil and its main component linalool in the growth of epimastigote and intracellular amastigote forms of *T. cruzi*.

In the present study, *A. rosaeodora* essential oil showed activity against epimastigote forms. Literature data showed anti-*T. cruzi* activity in vitro in extracts and substances of different species of the *Aniba* genus of plants collected in the Amazon [29], with promising antileishmanial activity [14]. To understand whether linalool is responsible for the inhibitory activity, an analysis of linalool against epimastigote was performed. The results showed an inhibitory effect close to the values of *A. rosaeodora* essential oil. Therefore, it is worth inferring that the inhibitory effect of the essential oil occurs due to the high concentration of linalool, or due to a possible synergistic and/or additive effect of the constituents of the essential oil acting as trypanocidal agents [31].

Previous data demonstrated that the IC_{50}/24 h for linalool was 162.5 μg/mL for epimastigotes and 264 μg/mL for *T. cruzi* trypomastigotes (Y strain) [32], corroborating with data presented in this study. However, linalool had a potent trypanocidal effect against the trypomastigote form of *T. cruzi* (clone Dm28c) derived from cells, with IC_{50}/24 h of 306 ng/mL, indicating that different forms and/or origin and different strains may differ in their susceptibility to essential oil derivatives [33].

The search for new therapeutic drugs requires conditions that simulate the environment found by the parasite–cell interaction, therefore, the assay against intracellular amastigote forms of tripanosomatids may represent ideal conditions, with macrophages playing an important role in the evaluation of drug-mediated toxicity [34]. Thus, it was evaluated whether *A. rosaeodora* essential oil and linalool could inhibit *T. cruzi* intracellular amastigote. However, the inhibitory effect was observed only when infected cells were treated with linalool although in high concentration, while *A. rosaeodora* essential oil presented activity at an even higher concentration.

Piper aduncum essential oil (PaEO), with nerolidol (25.22%) and linalool (13.42%) as main constituents, effectively inhibits the intracellular survival/replication of T. cruzi amastigotes. PaEO at a concentration of 12.5 μg/mL decreased the rate of T. cruzi amastigote infection by 71.5%, with an IC_{50}/24 h of 9 μg/mL. As linalool showed trypanocidal activity, with IC_{50}/24 h of 306 ng/mL against trypomastigotes [33], it is possible to infer that activity against intracellular amastigote forms it is possibly due to linalool presence. In addition, previous data demonstrated that at low concentrations of purified linalool derived from the *Croton cajucara* essential oil, the number of parasites internalized in the macrophages decreased (treated before and after the interaction). On the other hand, no cytotoxic effects of essential oil and linalool were observed in peritoneal macrophages of Swiss mice and Vero cells [35]. As in our study, *A. rosaeodora* essential oil and linalool not exhibited cytotoxicity against peritoneal macrophages in the concentration range under analysis. As a result, linalool showed a select activity to the parasites when compared to mammalian cells [14].

Literature data with *L. infantum chagasi* determined that the post-interaction treatment with linalool has antiparasitic activity against intracellular amastigotes, inducing a decrease in the number of parasites within the macrophages [36]. In the same study, it was observed that linalool is capable of providing a drastic change in oxygen consumption, probably related to mitochondrial dysfunction. *P. aduncum* essential oil rich in linalool induced mitochondria dysfunction altering the mitochondrial membrane potential of the T. cruzi epimastigote [33]. Mitochondrial alterations as swelling and important changes in the organization of nuclear and kinetoplastic chromatins were observed by electron microscopy when *L. amazonensis* parasites were treated with *C. cajucara* essential oil [35]. Linalool may interfere with the integrity of protozoan mitochondria, however, further studies are needed to elucidate the mechanism involved in the trypanocidal activity observed in our study.

An indirect mechanism involved with antitrypanosomal activity is related with macrophage activation, particularly the nitric oxide (NO) induction. The NO-mediation directly kills *T. cruzi* in vitro [37]. Thus, we carried out an analysis of the nitrite quantification of *T. cruzi*-stimulated peritoneal macrophages treated with *A. rosaeodora* essential oil or linalool. However, a significant decrease in nitrite levels was observed in cells non-stimulated with *T. cruzi* and treated with *A. rosaeodora* essential oil or linalool. Reactive oxygen species decrease was also observed in cancer cells lines treated with *A. rosaeodora* essential oil, inhibiting apoptosis in these cells [13].

Otherwise, in *T. cruzi*-stimulated cells the treatment with *A. rosaeodora* essential oil or linalool did not significantly decrease nitrite levels. Linalool has a known anti-inflammatory activity [38] and inhibits NO formation in vitro [39], but interestingly, an in vitro experiment of macrophages treated with linalool (250 or 350 μg/mL) for 24 h before or after interactions with the *Leishmania infantum* was also not associated with any difference in NO production [36]. The inhibition of NO production observed in macrophages treated with *A.*

rosaeodora essential oil and linalool, although it is not associated with antitrypanosomal activity, is an interesting finding that should be better elucidated in further studies.

4. Materials and Methods

4.1. Reagents

Anhydrous sodium sulfate, ethanol, ethyl acetate, dimethyl sulfoxide (DMSO), eugenol, 2,2′-azino-bis(3-ethylbenzothiazoline-6-sulfonic acid) diammonium salt (ABTS), penicillin, streptomycin, N-benzyl-2-nitro-1H-imidazole-1-acetamide (Benznidazole), Brewer thioglycolate medium, RPMI 1640 medium, 3-(4,5-Dimethyl-2-thiazolyl)-2,5-diphenyl-2H-tetrazolium bromide (MTT), sulfanilamide, H_3PO_4, N-(1-naphthyl)ethylenediamine and sodium nitrite were purchased from Sigma, St Louis, MO, USA. Giemsa's azur-eosin-methylene blue, Brain Heart Infusion broth, Mueller-Hinton agar and Mueller-Hinton broth were purchased from MERK, Darmstadt, Germany. The other bacteria culture medium were purchased from BD, Becton Dickinson, Franklin Lakes, NJ, USA. API® 20 E system was purchased from bioMérieux, Durham, NC, USA. Fetal bovine serum (FBS) was purchased from Gibco, Gaithersburg, MD, USA.

4.2. Plant Material

Authentic samples of the *A. rosaeodora* species were obtained from three trees cultivated at the Adolpho Ducke Forest Reserve, Highway AM-010, km 26 (latitude −2.908185, longitude −59.975457), Manaus, Brazil. Leaves and thin branches were harvested with a trimmer from the treetops in the dry season, March 2017. The taxonomic identification was undertaken by the Herbarium of the Department of Botany of the Universidade Federal do Amazonas, registry number 5982. The leaves were selected and dried in an oven at 37 °C for 48 h and sprayed in an electric knife mill at the Food and Water Quality Control Laboratory of the Federal University of Maranhão.

4.3. Essential Oil Extraction

The extraction of the essential oil of *A. rosaeodora* was carried out with 100 g of dried leaves from thin branches diluted in water in the proportion of 1:10 by hydrodistillation using the Clevenger system for 3 h at 100 °C. The essential oil collected were dried with anhydrous sodium sulfate (Na_2SO_4) and the final volume found was used to determine the yield through the mass/volume ratio by measuring the density. Mass/volume ratios were calculated from the mass (g) of the initial vegetal material and the volume (mL) of essential oil obtained after extraction. The essential oil samples were kept at 25 °C and then weighed. For the verification of biological activity in vitro, the essential oil and the reference drugs were diluted in DMSO and subsequently made serial dilutions in an appropriate culture medium until reaching a final concentration below 1% DMSO.

4.4. Physical-Chemical Analysis of Essential Oil

Physical-chemical analyzes performed on *A. rosaeodora* essential oil were: density, measured with a glass pycnometer; refractive index, calculated with an ABBE 2WAJ refractometer (PCE Instruments, Southampton, UK); the color and appearance, that were visually verified by three different people; and the determination of solubility, carried out through the ratio of 1:1 of oil and 70% ethanol until its complete solubilization.

4.5. Gas Chromatography–Mass Spectrometry (GC–MS)

The standard used in the development of the analytical methodology was linalool. Standard solutions of monoterpenes were prepared by dilution in absolute ethyl alcohol and chloroform at different concentrations. The essential oil of *A. rosaeodora* was solubilized in ethyl acetate and was analyzed by a gas chromatograph Shimadzu QP 5000 (Shimadzu Corp., Kyoto, Japan), a column used with a capillary ZB-5 ms (5% phenyl arylene 95% dimethylpolysiloxane) coupled to 70 eV (40–500 Da) HP 5MS electronic impact detector with a transfer temperature of 280 °C. In the analysis, 0.3 µL of ethyl acetate and helium gas (99.99%) were injected at a temperature of 280 °C, using a split mode (1:10) with an initial temperature gradient of 40 to 300 °C.min^{-1}, with a chromatographic run that lasted 30 min.

4.6. Bacteria from Marine Enviroment

Bacteria strains isolated from the marine environment *Aeromonas caviae, Aeromonas hydrophila, Enterococcus faecalis, Klebsiella pneumoniae* and *Providencia stuartii* were gently provided by the Laboratory of Microbiology of the Water Quality Control Program at the Federal University of Maranhão. Water samples were aseptically collected from approximately 30 cm below the water surface of the Jansen lagoon, Maranhão Brazil (latitude −2.499629, longitude −44.301211). Then, the samples were transported to the Microbiology Laboratory of the Federal University of Maranhão in isothermal boxes containing ice to perform the identification. To *Aeromonas* isolation, successive decimal dilutions of water samples (10^{-1} to 10^{-7}) were prepared in alkaline peptone water (APA), with subsequent distribution of 1 mL aliquots in five series of five tubes containing tryptic soy broth (TSB Broth) and 0.1 mL in plates containing the selective medium, agar gelatin phosphate salt (GSP Agar) (duplicates), both with 20 µg/mL of ampicillin, an antibiotic used as an inhibitor of the accompanying microbiota of *Aeromonas*. Colonies suspected of being *Aeromonas* were seeded in tilted BD trypticase soy agar (TSA agar) tubes, followed by incubation at 28 °C for 24 h. After, the cultures on TSA agar were subjected to biochemical tests of oxidase, catalase, gas production from glucose for species identification, indole production, O/129 resistance, amino acid decarboxylation (test on triple sugar agar and iron—TSI agar), motility: nitrate reduction, esculine hydrolysis, Voges–Proskauer (VP) assay, carbohydrate fermentation and growth at 3% and 6% sodium chloride. To Enterobacteriaceae isolation and identification of *Klebsiella pneumoniae* and *Providencia stuartii* in the water samples, initially, 25 mL of each sample were homogenized in 225 mL of brain and heart infusion broth (BHI broth) and incubated in a bacteriological oven at 37 °C for three hours. After the incubation period, the entire inoculum was transferred to 250 mL of broth for *Escherichia coli* and incubated at 37 °C for 24 h. Isolation was performed using selective and differential media, methylene blue eosin agar (EMB agar) and MacConkey sorbitol agar (MCS agar). For the identification of the species, initially five colonies were selected from the selective culture media, small colonies with metallic green or black without gloss in EMB agar and those of intense pink color (positive sorbitol) and yellow (negative sorbitol) in MCS agar. Then, the colonies were isolated in tubes containing TSA agar inclined with subsequent incubation at 37 °C for 24 h. Biochemical identification was performed using conventional tests: indole, simmons citrate, methyl red, VP, malonate, carbohydrate fermentation—sorbitol, rhamnose, mannitol, arabinose, inositol and raffinose, decarboxylation of amino acids lysine and ornithine, motility and H_2S production in sulfide indole motility (SIM) agar [40] and by the API® 20 E system. For *Enterococcus* research, 9 mL of each sample were diluted in 90 mL of buffered peptide water and incubated for 24 h/35 °C. Subsequently dilutions (10^{-1} to 10^{-7}) and the highest dilution were plated on M-*Enterococcus* agar and incubated at 35 °C for 48 h. Brick red colonies were inoculated on TSI agar. Tubes that showed suggestive characteristics were analyzed by acid ramp, acid-base, H_2S (-), catalase, oxidase, 6% NaCl, glucose and esculin tests, and by the API® 20 E system.

4.7. Bacterial Strains and Culture Conditions

To perform the preliminary antimicrobial tests, the standard strains *Escherichia coli* (Migula) Castellani and Chalmers (ATCC® 25922™), *Staphylococcus aureus* subsp. *aureus* Rosenbach (ATCC® 12600™), *Pseudomonas aeruginosa* (Schroeter) Migula (ATCC® 27853™) and *Salmonella enterica* subsp. *enterica* (ex Kauffmann and Edwards) Le Minor and *Popoff serovar* Choleraesuis (ATCC® 12011™) were used. The tests were carried out at the Microbiology Laboratory of the Federal University of Maranhão. The strains were grown in BHI broth for 24 h at 37 °C and the inoculum was adjusted to a cell concentration of 10^8 colony forming unit (CFU)/mL following the MacFarland scale, recommended by the Clinical and Laboratory Standards Institute [41].

4.8. Antimicrobial Assays

During the preliminary test of diffusion in solid medium, 100 µL of inoculum of each bacterium sown on Mueller–Hinton agar plates were used, and on the agar surface, a paper disc impregnated with 50 µL of essential oil of *A. rosaeodora*, standard linalool or reference drugs were added; then the plates were incubated at 35 °C and after 24 h the inhibition zone was measured with a millimeter rule [42]. The MIC was also performed according to the broth dilution methodology performed in triplicate with the same bacteria used in the diffusion tests in solid medium [41]. Initially, serial dilutions were performed resulting in concentrations of 5–1000 µg/mL of *A. rosaeodora* essential oil, linalool or reference drugs and transferred to a test tube containing Mueller-Hinton broth. To each concentration, 100 µL of the microbial suspension containing 1.5×10^8 CFU/mL were added and subsequently incubated at 35 °C for 24 h. It was also reserved control of broth sterility and bacterial growth. After the incubation period, the MIC was determined, being defined as the lowest concentration that visibly inhibited bacterial growth (absence of visible turbidity). To confirm growth inhibition, the broth was subjected to the microbial seeding test of the inoculum on the surface of the plate-count agar.

4.9. Antioxidant Assay

Antioxidant activity was assessed using a reaction mixture of 2,2-azinobis- (3-ethylbenzothiazoline-6-sulfonic acid) (ABTS) at 3840 µg/mL with 88 µL of 37,840 µg/mL potassium persulfate solution left in the dark at room temperature for 16 h giving rise to the ABTS radical which was diluted in ethanol to obtain an absorbance of 0.7 to 734 nm. The results were obtained in a dark environment, in which 30 µL of each concentration of essential oil (200 to 15 µg/mL) and eugenol (90 to 5 µg/mL) was transferred in test tubes containing 3.0 mL of the cation radical ABTS and homogenized on a tube shaker, and after 6 min the absorbance of the reaction mixture was read on a spectrophotometer at a length of 734 nm [43]. The analyses were carried out in triplicates and the determination of the activity was demonstrated as percentage of inhibition (% I) of the ABTS radical cation according to the equation: % inhibition = (absorbance of the solution of the radical ABTS—absorbance of the sample)/(solution of ABTS absorbance radical) × 100 [44]. We also verified the efficient concentration or EC50% that represents the concentration necessary to sequester 50% of the ABTS root. The essential oil will be considered active when it has an EC50 < 500 µg/mL [45].

4.10. Parasites

Parasite cultures employed in this study were *Trypanosoma cruzi* (SC2005 strain). Trypomastigote forms were obtained from Vero cells infected and used to infect the macrophages. Epimatigote forms were originated from the suspension of cell culture trypomastigotes in 3 mL of liver infusion tryptose (LIT) medium supplemented with 10% fetal bovine serum (FBS), 100 U/mL of penicillin and 100 µg/mL of streptomycin), and incubated in an oven at 28 °C until complete differentiation of parasites.

4.11. Anti-Epimastigote Assay

Epimastigote forms of *T. cruzi*, from a 2- to 4-day-old culture were incubated for 24 h in the absence or in the presence of different concentrations (1000–15.625 µg/mL) of *A. rosaeodora* essential oil or linalool, obtained by serial dilutions (1:2), at a final volume of 100 µL per well. The controls were identified as blank (wells without parasites), untreated control (parasites and DMSO 1%) and reference drug (benznidazole). Incubation took place in a 96-wells plate, in a BOD incubator at 28 °C in LIT medium using a parasite concentration of 10^6 promastigotes/mL. After 24 h, with the aid of the Neubauer chamber and light microscopy [46], viability was evaluated by counting parasites and the results were used to calculate the IC_{50} (50% inhibition of parasite growth) following the formula: IC_{50} = (sample counting)/(control counting) ×100 [47].

4.12. Animals and Ethical Statements

BALB/c female mice from 4 to 6 weeks of age were purchased from the Institute of Science and Technology in Biomodels of the Institute of Science and Technology in Biomodels. All procedures were performed in accordance with the National Council for the Control of Animal Experimentation National Council for Animal Experimentation Control—CONCEA) and approved by the Ethics Committee on Animal Care and Utilization (CEUA/IOC—L018/2018).

4.13. Peritoneal Macrophage Collection and Culture

Peritoneal macrophages from BALB/c mice were collected after elicited with 3 mL 3% Brewer thioglycollate medium broth injection for 72 h, and maintained in RPMI 1640 medium supplemented with 10% FBS, 100 U/mL of penicillin and 100 µg/mL of streptomycin, overnight at 37 °C and 5% CO_2.

4.14. Cytotoxicity Assay

Peritoneal macrophages (5 × 10^5 cells/mL) were cultured in 96-well plates with different concentrations, obtained by serial dilutions (1:2), of *A. rosaeodora* essential oil or linalool (1000–7.8 µg/mL) or benznidazole (200–0.78 µg/mL) up to a final volume of 100 µL per well. The controls were categorized as blanks (wells with culture medium without cells), untreated control (cells and DMSO 1%) and reference drug (benznidazole). After 72 h, the cell viability was analyzed by the MTT colorimetric method [48]. Absorbance was measured in a spectrophotometer at 540 nm wavelength. The concentration inhibiting 50% of cell growth (CC_{50}) was calculated following the formula: CC_{50} = (sample absorbance-blank absorbance)/(control absorbance-blank absorbance) × 100 [49].

4.15. Activitiy Against Intracellular Amastigotes and Selectivity Index (SI)

BALB/c peritoneal macrophages cultured in 24-well plates (5 × 10^5 cells/well), with coverslips, were infected with trypomastigote forms of *T. cruzi*, obtained from cultured Vero cells, using the ratio of parasite/cell 10:1, at 37 °C and 5% CO_2 for 6 h. After incubation, well plates were washed with phosphate-buffered saline (PBS, pH 7.2) to remove the non-internalized parasites. The infected cells were treated with different concentrations of *A. rosaeodora* essential oil or linalool (1000–31.25 µg/mL), or benzonidazole (100–6.25 µg/mL) for 24 h. The amastigotes couting by analysis of light microscopy were carried out to determine the IC_{50} calculation. Selectivity index were obtained from the relationship of macrophage cytotoxicity and antiamastigote activity. Parameters of infection analysis were performed according to Teles et al. [50].

4.16. Nitrite Quantification

BALB/c peritoneal macrophages (5 × 10^6 cells/mL) was treated with *A. rosaeodora* essential oil (500 µg/mL) or linalool (250 µg/mL), and either stimulated or not stimulated with *T. cruzi* trypomastigotes (5 × 10^7 parasites/mL) for 48 h. Nitrite quantification of the supernatant of the cells was performed with Griess reagent. Briefly, 50 µL of culture supernatant and 50 µL of Griess reagent (25 µL of sulfanilamide 1% in 2.5% H_3PO_4 solution and 25 µL of N-(1-naphthyl)-ethylenediamine 0.1% solution) were added in 96-well plates. After incubation in a dark environment for 10 min, absorbance was obtained at 570 nm on the spectrophotometer. The nitrite values were obtained from the standard curve of sodium nitrite (100–1.5 µM) [51].

4.17. Statistical Analysis

The numerical results from at least two independent assays were expressed as mean ± standard deviation and the IC_{50} and CC_{50} determination were performed with the Graph-

Pad Prism 7.00 software package (GraphPad Software, San Diego, CA, USA). Kruskal-Wallis and Dunn's multiple comparison test was used to analyze the data and the difference at $p < 0.05$ was considered significant.

5. Conclusions

The essential oil of *A. rosaeodora* showed activity against all the strains tested, with a lower minimum inhibitory concentration being observed for *S. aureus*. An efficient antioxidant activity of the essential oil was evidenced by the ABTS radical discoloration technique, fully inhibiting the radical in relatively low concentrations. These results point to an important potential for use as an antimicrobial and antioxidant agent. The antitrypanosomal activity of *A. rosaeodora* essential oil and linalool were observed at high concentrations against epimatigote forms, and even higher against intracellular amastigotes of *T. cruzi*. Both *A. rosaeodora* essential oil and linalool reduced nitrite levels in unstimulated cells revealing a potential effect in NO production. These data revealed the pharmacological potential of the *A. rosaeodora* essential oil and linalool, which encourage further studies.

Author Contributions: Conceptualization, A.M.T., J.V.S.-S., A.N.M. and F.A.-S.; methodology, A.M.T., J.V.S.-S., J.M.P.F. and S.C.M.; formal analysis, A.M.T., J.V.S.-S., V.E.M.F., S.C.M. and F.A.-S.; resources, K.d.S.C., A.L.A.-S., A.N.M., V.E.M.F. and F.A.-S.; data curation, A.M.T., J.V.S.-S., A.N.M., S.C.M., V.E.M.F. and F.A.-S.; writing—original draft preparation, A.M.T., J.V.S.-S. and F.A.-S.; writing—review and editing, A.M.T., J.V.S.-S., K.d.S.C., A.L.A.-S., A.N.M., V.E.M.F., S.C.M. and F.A.-S.; visualization, J.M.P.F., V.E.M.F. and F.A.-S.; supervision, A.N.M., V.E.M.F. and F.A.-S.; project administration, K.d.S.C., A.L.A.-S., A.N.M., V.E.M.F. and F.A.-S.; funding acquisition, K.d.S.C., A.L.A.-S., A.N.M., V.E.M.F. and F.A.-S. All authors have read and agreed to the published version of the manuscript.

Funding: This research was funded by the Coordination for the Improvement of Higher Education Personnel (Coordenação de Aper feiçoamento de Pessoal de Nível Superior do Brazil—CAPES) [grant number Finance Code 001]; and the Carlos Chagas Filho Foundation for Research Support of the State of Rio de Janeiro (Fundação Carlos Chagas Filho de Amparo à Pesquisa do Estado do Rio de Janeiro—FAPERJ) [grant number E-26/010.001759/2019]. The APC was funded by Oswaldo Cruz Institute (Instituto Oswaldo Cruz—IOC). Fernando Almeida-Souza is postdoctoral researcher fellow of CAPES [grant number 88887.363006/2019-00]. Dra. Ana Lucia Abreu-Silva is research productivity fellow of National Scientific and Technological Development Council (Conselho Nacional de Desenvolvimento Científico e Tecnológico—CNPq) [grant number 309885/2017-5].

Institutional Review Board Statement: The experiments with animals were conducted in accordance with the guidelines for experimental procedures of the Conselho Nacional de Controle de Experimentação Animal (the National Council for the Control of Animal Experimentation) (CONCEA) and approved by the Comissão de Ética no Uso de Animais (the Animal Research Ethics Committee) of the Instituto Oswaldo Cruz (CEUA-IOC/FIOCRUZ), License n° L018/2018.

Informed Consent Statement: Not applicable.

Data Availability Statement: All datasets presented in this study are included in the article.

Conflicts of Interest: The authors declare no conflict of interest.

References

1. Chandra, H.; Bishnoi, P.; Yadav, A.; Patni, B.; Mishra, A.P.; Nautiyal, A.R. Antimicrobial Resistance and the Alternative Resources with Special Emphasis on Plant-Based Antimicrobials-A Review. *Plants* **2017**, *6*, 16. [CrossRef] [PubMed]
2. Huang, T.H.; Kota, B.P.; Razmovski, V.; Roufogalis, B.D. Herbal or natural medicines as modulators of peroxisome proliferator-activated receptors and related nuclear receptors for therapy of metabolic syndrome. *Basic Clin. Pharmacol. Toxicol.* **2005**, *96*. [CrossRef] [PubMed]
3. Orchard, A.; van Vuuren, S. Commercial Essential Oils as Potential Antimicrobials to Treat Skin Diseases. *Evid. Based Complementary Altern. Med. Ecam* **2017**, *2017*. [CrossRef]
4. Morais, M.C.; Souza, J.V.; da Silva Maia Bezerra Filho, C.; Dolabella, S.S.; Sousa, D.P. Trypanocidal Essential Oils: A Review. *Molecules* **2020**, *25*, 4568. [CrossRef]
5. Maia, J.G.S.; Andrade, E.H.A.; Couto, H.A.R.; Silva, A.C.M.; Marx, F.; Henke, C. Plant sources of amazon rosewood oil. *Química Nova* **2007**, *30*, 1906–1910. [CrossRef]

6. May, P.H.; Barata, L.E.S. Rosewood exploitation in the Brazilian Amazon: Options for sustainable production. *Econ. Bot.* **2004**, *58*, 257–265. [CrossRef]
7. Maia, J.G.S.; Zoghbi, M.G.B.; Andrade, E.H.A. Essential Oils of Aeollanthus suaveolens Matt. ex Spreng. *J. Essent. Oil Res.* **2003**, *15*, 86–87. [CrossRef]
8. Maia, J.G.S.; Andrade, E.H.A. Database of the Amazon aromatic plants and their essential oils. *Química Nova* **2009**, *32*, 595–622. [CrossRef]
9. Pattnaik, S.; Subramanyam, V.R.; Bapaji, M.; Kole, C.R. Antibacterial and antifungal activity of aromatic constituents of essential oils. *Microbios* **1997**, *89*, 39.
10. Herman, A.; Tambor, K.; Herman, A. Linalool Affects the Antimicrobial Efficacy of Essential Oils. *Curr. Microbiol.* **2016**, *72*, 165–172. [CrossRef]
11. Seol, G.H.; Kang, P.; Lee, H.S.; Seol, G.H. Antioxidant activity of linalool in patients with carpal tunnel syndrome. *BMC Neurol.* **2016**, 16. [CrossRef] [PubMed]
12. Bero, J.; Kpoviessi, S.; Quetin-Leclercq, J. Anti-Parasitic Activity of Essential Oils and their Active Constituents against Plasmodium, Trypanosoma and Leishmania. *Nov. Plant Bioresour.* **2014**, 455–469. [CrossRef]
13. Soeur, J.; Marrot, L.; Perez, P.; Iraqui, I.; Kienda, G.; Dardalhon, M.; Meunier, J.R.; Averbeck, D.; Huang, M.E. Selective cytotoxicity of *Aniba rosaeodora* essential oil towards epidermoid cancer cells through induction of apoptosis. *Mutat. Res.* **2011**, *718*. [CrossRef]
14. Da Silva, Y.C.; Silva, E.M.S.; Fernandes, N.S.; Lopes, N.L.; Orlandi, P.P.; Nakamura, C.V.; Costa, E.V.; da Veiga Junior, V.F. Antimicrobial substances from Amazonian *Aniba* (Lauraceae) species. *Nat. Prod. Res.* **2019**, 1–4. [CrossRef]
15. Sampaio Lde, F.; Maia, J.G.; de Parijos, A.M.; de Souza, R.Z.; Barata, L.E. Linalool from rosewood (*Aniba rosaeodora* Ducke) oil inhibits adenylate cyclase in the retina, contributing to understanding its biological activity. *Phytother. Res.* **2012**, *26*, 73–77. [CrossRef]
16. D'ACAMPORA Zellner, B.; Lo Presti, M.; Barata, L.E.; Dugo, P.; Dugo, G.; Mondello, L. Evaluation of leaf-derived extracts as an environmentally sustainable source of essential oils by using gas chromatography-mass spectrometry and enantioselective gas chromatography-olfactometry. *Anal. Chem.* **2006**, *78*. [CrossRef]
17. Almeida, M.R.; Fidelis, C.H.; Barata, L.E.; Poppi, R.J. Classification of Amazonian rosewood essential oil by Raman spectroscopy and PLS-DA with reliability estimation. *Talanta* **2013**, *117*, 305–311. [CrossRef]
18. Ye, M.; Sun, M.; Huang, D.; Zhang, Z.; Zhang, H.; Zhang, S.; Hu, F.; Jiang, X.; Jiao, W. A review of bacteriophage therapy for pathogenic bacteria inactivation in the soil environment. *Environ. Int.* **2019**, *129*. [CrossRef]
19. Holetz, F.B.; Pessini, G.L.; Sanches, N.R.; Cortez, D.A.; Nakamura, C.V.; Filho, B.P. Screening of some plants used in the Brazilian folk medicine for the treatment of infectious diseases. *Mem. Inst. Oswaldo Cruz.* **2002**, *97*, 1027–1031. [CrossRef]
20. Rosato, A.; Piarulli, M.; Corbo, F.; Muraglia, M.; Carone, A.; Vitali, M.E.; Vitali, C. In vitro synergistic antibacterial action of certain combinations of gentamicin and essential oils. *Curr. Med. Chem.* **2010**, *17*, 3289–3295. [CrossRef]
21. Jabir, M.S.; Taha, A.A.; Sahib, U.I. Linalool loaded on glutathione-modified gold nanoparticles: A drug delivery system for a successful antimicrobial therapy. *Artif. Cells Nanomed Biotechnol.* **2018**, *46*, 345–355. [CrossRef]
22. Vainstein, A.; Lewinsohn, E.; Pichersky, E.; Weiss, D. Floral fragrance. New inroads into an old commodity. *Plant Physiol.* **2001**, *127*, 1383–1389. [CrossRef] [PubMed]
23. Christensson, J.B.; Matura, M.; Gruvberger, B.; Bruze, M.; Karlberg, A.T. Linalool—A significant contact sensitizer after air exposure. *Contact Dermat.* **2010**, *62*. [CrossRef] [PubMed]
24. Bråred Christensson, J.; Andersen, K.E.; Bruze, M.; Johansen, J.D.; Garcia-Bravo, B.; Gimenez Arnau, A.; Goh, C.L.; Nixon, R.; White, I.R. Air-oxidized linalool: A frequent cause of fragrance contact allergy. *Contact Dermat.* **2012**, *67*. [CrossRef] [PubMed]
25. Audrain, H.; Kenward, C.; Lovell, C.R.; Green, C.; Ormerod, A.D.; Sansom, J.; Chowdhury, M.M.; Cooper, S.M.; Johnston, G.A.; Wilkinson, M.; et al. Allergy to oxidized limonene and linalool is frequent in the U.K. *Br. J. Dermatol.* **2014**, *171*. [CrossRef]
26. Martins, F.J.; Caneschi, C.A.; Vieira, J.L.; Barbosa, W.; Raposo, N.R. Antioxidant activity and potential photoprotective from amazon native flora extracts. *J. Photochem. Photobiol. B Biol.* **2016**, *161*. [CrossRef]
27. Oliveira de Souza, L.I.; Bezzera-Silva, P.C.; do Amaral Ferraz Navarro, D.M.; da Silva, A.G.; Dos Santos Correia, M.T.; da Silva, M.V.; de Figueiredo, R.C.B.Q. The chemical composition and trypanocidal activity of volatile oils from Brazilian Caatinga plants. *Biomed. Pharmacother.* **2017**, *96*. [CrossRef]
28. Dos Santos Sales, V.; Monteiro, Á.B.; Delmondes, G.A.; do Nascimento, E.P.; Sobreira Dantas Nóbrega de Figuêiredo, F.R.; de Souza Rodrigues, C.K.; Evangelista de Lacerda, J.F.; Fernandes, C.N.; Barbosa, M.O.; Brasil, A.X.; et al. Antiparasitic activity and essential oil chemical analysis of the piper tuberculatum jacq fruit. *Iran. J. Pharm. Res. IJPR* **2018**, *17*, 268.
29. Giongo, J.L.; Vaucher, R.A.; Da Silva, A.S.; Oliveira, C.B.; de Mattos, C.B.; Baldissera, M.D.; Sagrillo, M.R.; Monteiro, S.G.; Custódio, D.L.; Souza de Matos, M.; et al. Trypanocidal activity of the compounds present in Aniba canelilla oil against Trypanosoma evansi and its effects on viability of lymphocytes. *Microb. Pathog.* **2017**, *103*. [CrossRef]
30. Jackson, Y.; Wyssa, B.; Chappuis, F. Tolerance to Nifurtimox and Benznidazole in Adult Patients With Chronic Chagas' Disease. *J. Antimicrob. Chemother.* **2020**, *75*. [CrossRef]
31. Izumi, E.; Ueda-Nakamura, T.; Veiga, V.F.; Pinto, A.C.; Nakamura, C.V. Terpenes from Copaifera demonstrated in vitro antiparasitic and synergic activity. *J. Med. Chem.* **2012**, *55*. [CrossRef]

32. Santoro, G.F.; Cardoso, M.G.; Guimarães, L.G.; Mendonça, L.Z.; Soares, M.J. Trypanosoma cruzi: Activity of essential oils from Achillea millefolium L., Syzygium aromaticum L. and Ocimum basilicum L. on epimastigotes and trypomastigotes. *Exp. Parasitol.* **2007**, *116*, 283–290. [CrossRef]
33. Villamizar, L.H.; Cardoso, M.G.; Andrade, J.; Teixeira, M.L.; Soares, M.J. Linalool, a Piper aduncum essential oil component, has selective activity against Trypanosoma cruzi trypomastigote forms at 4 °C. *Memórias Do Inst. Oswaldo Cruz.* **2017**, *112*, 131–139. [CrossRef]
34. Suman Gupta, N. Visceral leishmaniasis: Experimental models for drug discovery. *Indian J. Med. Res.* **2011**, *133*, 27–39.
35. Do Socorro S Rosa, M.d.S.; Mendonça-Filho, R.R.; Bizzo, H.R.; de Almeida Rodrigues, I.; Soares, R.M.; Souto-Padrón, T.; Alviano, C.S.; Lopes, A.H. Antileishmanial activity of a linalool-rich essential oil from Croton cajucara. *Antimicrob. Agents Chemother.* **2003**, *47*. [CrossRef]
36. Dutra, F.L.; Oliveira, M.M.; Santos, R.S.; Silva, W.S.; Alviano, D.S.; Vieira, D.P.; Lopes, A.H. Effects of linalool and eugenol on the survival of *Leishmania* (L.) infantum chagasi within macrophages. *Acta Trop.* **2016**, *164*, 69–76. [CrossRef]
37. Vespa, G.N.; Cunha, F.Q.; Silva, J.S. Nitric oxide is involved in control of Trypanosoma cruzi-induced parasitemia and directly kills the parasite in vitro. *Infect. Immun.* **1994**, *62*. [CrossRef]
38. Kim, M.G.; Kim, S.M.; Min, J.H.; Kwon, O.K.; Park, M.H.; Park, J.W.; Ahn, H.I.; Hwang, J.Y.; Oh, S.R.; Lee, J.W.; et al. Anti-inflammatory effects of linalool on ovalbumin-induced pulmonary inflammation. *Int. Immunopharmacol.* **2019**, *74*. [CrossRef]
39. Peana, A.T.; Marzocco, S.; Popolo, A.; Pinto, A. (-)-Linalool inhibits in vitro NO formation: Probable involvement in the antinociceptive activity of this monoterpene compound. *Life Sci.* **2006**, *78*. [CrossRef]
40. Salfinger, Y.; Tortorello, M.L. *Compendium of Methods for the Microbiological Examination of Foods*; American Public Health Association (APHA): Washington, DC, USA, 1992.
41. CLSI. *Performance Standards for Antimicrobial disk and Dilution Susceptibility Tests for Bacteria Isolated from Animals*, 4th ed.; CLSI: Wayne, PA, USA, 2009; p. 99.
42. Bauer, A.W.; Kirby, W.M.; Sherris, J.C.; Turck, M. Antibiotic susceptibility testing by a standardized single disk method. *Am. J. Clin. Pathol.* **1966**, *45*, 493–496. [CrossRef]
43. Re, R.; Pellegrini, N.; Proteggente, A.; Pannala, A.; Yang, M.; Rice-Evans, C. Antioxidant activity applying an improved ABTS radical cation decolorization assay. *Free Radic. Biol. Med.* **1999**, *26*, 1231–1237. [CrossRef]
44. El Babili, F.; Bouajila, J.; Souchard, J.P.; Bertrand, C.; Bellvert, F.; Fouraste, I.; Moulis, C.; Valentin, A. Oregano: Chemical analysis and evaluation of its antimalarial, antioxidant, and cytotoxic activities. *J. Food Sci.* **2011**, *76*. [CrossRef] [PubMed]
45. Campos, M.G.; Webby, R.F.; Markham, K.R.; Mitchell, K.A.; Da Cunha, A.P. Age-induced diminution of free radical scav-enging capacity in bee pollens and the contribution of constituent flavonoids. *J. Agric. Food. Chem.* **2003**, *51*. [CrossRef] [PubMed]
46. Rottini, M.M.; Amaral, A.C.F.; Ferreira, J.L.P.; Oliveira, E.S.C.; Silva, J.R.A.; Taniwaki, N.N.; Dos Santos, A.R.; Almeida-Souza, F.; de Souza, C.S.F.; Calabrese, K.S. Endlicheria bracteolata (Meisn.) essential oil as a weapon against Leishmania amazonensis: In vitro assay. *Molecules* **2019**, *24*, 2525. [CrossRef]
47. Da Silva, V.D.; Almeida-Souza, F.; Teles, A.M.; Neto, P.A.; Mondego-Oliveira, R.; Mendes Filho, N.E.; Taniwaki, N.N.; Abreu-Silva, A.L.; Calabrese, K.d.S.; Mouchrek Filho, V.E. Chemical composition of Ocimum canum Sims. essential oil and the antimicrobial, antiprotozoal and ultrastructural alterations it induces in Leishmania amazonensis promastigotes. *Ind. Crop. Prod.* **2018**, *119*, 201–208. [CrossRef]
48. Mosmann, T. Rapid colorimetric assay for cellular growth and survival: Application to proliferation and cytotoxicity assays. *J. Immunol. Methods* **1983**, *65*. [CrossRef]
49. Oliveira, I.S.S.; Colares, A.V.; Cardoso, F.O.; Tellis, C.J.M.; Chagas, M.S.S.; Behrens, M.D.; Calabrese, K.S.; Almeida-Souza, F.; Abreu-Silva, A.L. Vernonia polysphaera baker: Anti-inflammatory activity in vivo and inhibitory effect in lps-stimulated RAW 264.7 cells. *PLoS ONE* **2019**, *14*, e0225275. [CrossRef]
50. Teles, A.M.; Rosa, T.; Mouchrek, A.N.; Abreu-Silva, A.L.; Calabrese, K.D.S.; Almeida-Souza, F. Cinnamomum zeylanicum, Origanum vulgare, and Curcuma longa essential oils: Chemical composition, antimicrobial and antileishmanial activity. *Evid Based Complement Alternat. Med.* **2019**, *2019*, 2421695. [CrossRef]
51. Almeida-Souza, F.; Silva, V.D.D.; Silva, G.X.; Taniwaki, N.N.; Hardoim, D.J.; Buarque, C.D.; Abreu-Silva, A.L.; Calabrese, K.D.S. 1,4-Disubstituted-1,2,3-Triazole compounds induce ultrastructural alterations in leishmania amazonensis promastigote: An in vitro antileishmanial and in silico pharmacokinetic study. *Int. J. Mol. Sci.* **2020**, *21*, 6839. [CrossRef]

antibiotics

MDPI

Article

Chemical Composition and Antimicrobial Activity of Selected Essential Oils against *Staphylococcus* spp. Isolated from Human Semen

Miroslava Kačániová [1,2,*], Margarita Terentjeva [3], Jana Štefániková [4], Jana Žiarovská [5], Tatsiana Savitskaya [6], Dmitrij Grinshpan [6], Przemysław Łukasz Kowalczewski [7], Nenad Vukovic [8] and Eva Tvrdá [9]

1 Department of Fruit Science, Viticulture and Enology, Faculty of Horticulture and Landscape Engineering, Slovak University of Agriculture, Tr. A. Hlinku 2, 94976 Nitra, Slovakia
2 Department of Bioenergetics, Food Analysis and Microbiology, Institute of Food Technology and Nutrition, University of Rzeszow, Cwiklinskiej 1, 35-601 Rzeszow, Poland
3 Institute of Food and Environmental Hygiene, Faculty of Veterinary Medicine, Latvia University of Life Sciences and Technologies, K. Helmaņa iela 8, LV-3004 Jelgava, Latvia; margarita.terentjeva@llu.lv
4 AgroBioTech Research Centre, Slovak University of Agriculture, Tr. A. Hlinku 2, 94976 Nitra, Slovakia; jana.stefanikova@uniag.sk
5 Department of Plant Genetics and Breeding, Faculty of Agrobiology and Food Resources, Slovak University of Agriculture, Tr. A. Hlinku 2, 949 76 Nitra, Slovakia; jana.ziarovska@uniag.sk
6 Research Institute for Physical Chemical Problems, Belarusian State University, Leningradskaya str. 14, 220030 Minsk, Belarus; savitskayaTA@bsu.by (T.S.); Grinshpan@bsu.by (D.G.)
7 Department of Food Technology of Plant Origin, Poznań University of Life Sciences, 31 Wojska Polskiego St., 60-624 Poznań, Poland; przemyslaw.kowalczewski@up.poznan.pl
8 Department of Chemistry, Faculty of Science, University of Kragujevac, P.O. Box 12, 34000 Kragujevac, Serbia; nvchem@yahoo.com
9 Department of Animal Physiology, Faculty of Biotechnology and Food Sciences, Slovak University of Agriculture, Tr. A. Hlinku 2, 949 76 Nitra, Slovakia; eva.tvrda@uniag.sk
* Correspondence: miroslava.kacaniova@gmail.com

Received: 10 October 2020; Accepted: 29 October 2020; Published: 31 October 2020

Abstract: *Staphylococcus* spp. is not only a commensal bacteria but also a major human pathogen that causes a wide range of clinical infections. Recent evidence suggests that *Staphylococcus* has the ability to colonize the reproductive system and to affect its structure and functions. The objective of this study was to determine the chemical properties and antibacterial effects of select essential oils (EOs): *Amyris balsamifera* L., *Boswellia carterii* Birdw., *Canarium luzonicum* (Blume) A. Gray, *Cinnamomum camphora* (L.) J. Presl., *Cinnamomum camphora* var. *linaloolifera* Y. Fuita, *Citrus x aurantium* L., *Gaultheria procumbens* L., *Litsea cubeba* (Lour.) Pers., *Melaleuca ericifolia* Smith., *Melaleuca leucadendra* L., *Pogostemon cablin* (Blanco) Benth., *Citrus limon* (L.) Osbeck, *Santalum album* L., and *Vetiveria zizanoides* (L.) Roberty against 50 *Staphylococcus* spp. cultures isolated from human semen, specifically *Staphylococcus aureus*, *S. capiti*, *S. epidermidis*, *S. haemoliticus*, and *S. hominis*. The disc diffusion and broth microdilution methods were used to assess the antimicrobial potential and to determine the minimum inhibitory concentration (MIC) of the selected EOs. The best anti-*Staphylococcus* activities were found with both methods for the essential oils of *C. luzonicum* (Blume) A. Gray, *A. balsamifera*, *C. camphora*, and *P. cabli*.

Keywords: *Staphylococcus* spp.; human semen; essential oils; antimicrobial activity; antimicrobial resistance

1. Introduction

As much as fifteen percent of infertility in males are a result of infections of the genitourinary tract [1]. Infections, both chronic and acute, lead to inflammation which compromises proper spermatogenic function [2–4]. This causes alterations in the sperm quality and quantity. Semen contamination occurs from microbiota present in the urinary tract or is transmitted via sexual intercourse [5].

Staphylococcus spp. has been frequently isolated from the reproductive system of men; furthermore, their ability to infect the male reproductive tract has been reported. *Staphylococcus* spp. may impair the secretory capacity of the epididymis, seminal vesicles, and prostate and may significantly affect sperm quality [6]. Essential oils (EOs) are a rich source of bioactive compounds, with some EOs exhibiting pronounced antimicrobial activity. Many plant parts, such as leaves, seeds, bark, resin, berries, flowers, roots, or fruits, contain EOs [7]. It has been shown that EOs of different plants and parts of the plant differ significantly in chemical composition and antimicrobial properties. Despite significant progress in the research of antimicrobial activity, extraction, and utilization of EOs, field studies on their application on opportunistic and pathogenic microbiota isolated from humans are needed [8]. Previous research suggests that the antimicrobial effect of EOs on human isolates could be used to prevent community- or hospital-acquired infections, which could become a suitable strategy to minimize the spread of antimicrobial resistance and to increase the efficiency of conservative treatment options [7–9].

The strongest antimicrobial activity of the *Juniperus communis* essential oil was found against *S. hominis* [10]. Salari et al. [11] used *Eucalyptus globulus* leaf extract to evaluate its activity on 56 isolates of *S. aureus*. The EOs extracted from all seven *Eucalyptus* spp. exhibited antibacterial activity against *S. aureus*. The best antimicrobial activity of *E. globulus* was found against *S. aureus* and *S. capiti*. In the meantime, *Cananga odorata* showed the best antimicrobial activity against *S. hominis* [10].

The objective of the present study was to investigate the chemical properties of selected essential oils and their antimicrobial effects against *Staphylococcus* spp. isolated from human semen.

2. Results and Discussion

2.1. Isolated Species of Staphylococci

In our study, 96 isolates were identified with mass spectrometry, with 50 isolates receiving a score higher than 2.00. The *Staphylococcus* spp. strains were *Staphylococcus aureus* (1 isolate), *S. capitis* (1 isolate), *S. epidermidis* (7 isolates), *S. haemoliticus* (26 isolates), and *S. hominis* (15 isolates) among the reliably identified isolates. The dendrogram of relatedness of mass spectra of *Staphylococcus* species is shown in Figure 1.

Two main branches with multiple subbranches can be seen in the constructed dendrogram. The diversity of spectra of all *Staphylococcus haemolyticus* were obtained as more narrow while the spectra of *Staphylococcus epidermis* were most diverse in comparison to all other *Staphylococcus* spp. that were analysed. *Staphylococcus capitis* and *Staphylococcus aureus* were assigned to be similar to the *Staphylococcus epidermis* group according to their protein profiles. A third compact group was created for the mass spectra of *Staphylococcus hominis* with two isolates that were related to other branches.

Infertility has become a commonly observed clinical diagnosis with infections of the genital tract being frequently identified in patients who undergo assisted reproductive therapy [12].

Infections of the genital tract are caused by microorganisms transmitted from the urinary tract or sexually transmitted as a result of sexual activity. Changes in the morphology and motility of spermatozoa as well as a reduced sperm viability have been identified as a result of the infection [13]. Up to 34.4% of semen samples were found to be contaminated with microorganisms, predominantly with *Staphylococcus* spp., *Enterococcus*, and *Escherichia coli* [14].

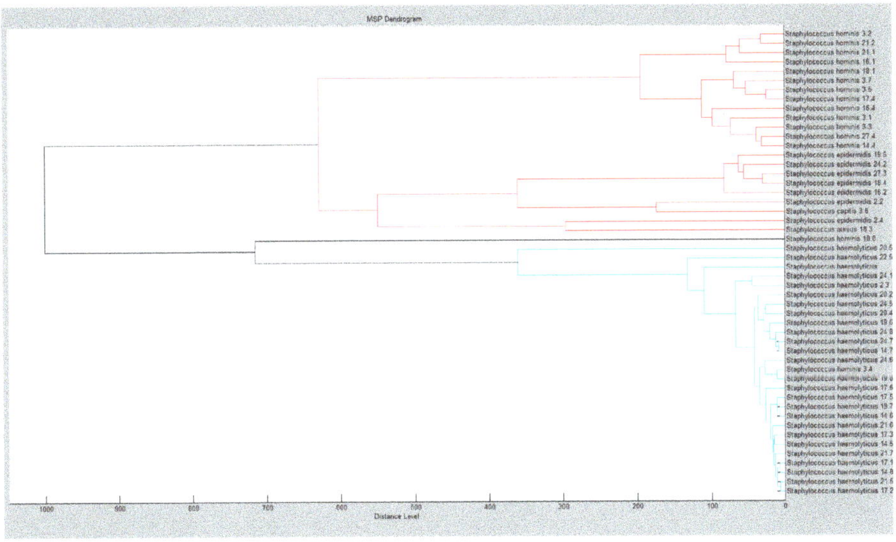

Figure 1. Dendogram of isolated *Staphylococcus* spp. from human semen constructed with a MALDI-TOF MS Biotyper.

2.2. Chemical Composition of Essential Oils

Different factors affecting the chemical composition of EOs. The most prominent endogenous factors are related to anatomical and physiological characteristics of the plants and to biosynthetic pathways of the volatiles, which might change depending on the plant tissue or season; however, it could also be influenced by DNA adaptation. On the other hand, exogenous factors might affect some of the genes responsible for volatiles formation, especially over a long period of time. Such changes may lead to ecotypes or chemotypes within the same plant species [15].

The chemical composition of *Amyris balsamifera* L. EO is shown in Table 1. The EO was obtained by steam distillation of crushed fresh wood. The presence of 15 chemical components with min 1% for each were identified. The compounds present in the highest amounts were valerianol (23.20%), guaiol (19.40%), and 10-epi-γ-eudesmol (14.80%). Different results were found in the study by Klouček et al. [16], where α-eudesmol (29.4%), β-eudesmol (10.4%), and valerianol (10.2%) were the main compounds of the amyris essential oil.

Table 1. Chemical composition of essential oil (%) *.

Essential Oil	Components	RT (min.)	Percentage of Components (n = 3)
Amyris balsamifera L.	Amorpha-4,11-diene	24.20	2.58
	β-cadinene	25.04	1.33
	Dihydroagarofuran	26.43	1.53
	β-dihydroagarofuran	26.56	1.05
	α-zingiberene	27.06	2.21
	Cedrene	28.72	4.91
	α-curcumen	28.89	2.44
	Nerolidol	35.64	1.57
	Elemol	39.25	9.62
	β-eudesmol	39.75	1.13
	γ-eudesmol	40.50	2.49
	Guaiol	40.94	19.40
	10-epi-γ-eudesmol	41.75	14.80
	Valerianol	43.16	23.20
	Drim-7-en-11-ol	50.65	1.84

Table 1. *Cont.*

Essential Oil	Components	RT (min.)	Percentage of Components (n = 3)
Boswelia carterii Birdw.	Sabinene	4.33	5.51
	3-Carene	5.05	1.39
	α-phellandrene	5.54	2.81
	β-myrcene	5.65	3.91
	D-limonene	6.68	26.40
	4-Thujanol	6.70	1.49
	prehnitene	9.25	8.65
	Copaene	17.96	1.59
	Bornyl acetate	21.6	1.00
	2-methylene-4,8,8-trimethyl-4-vinyl-bicyclo[5.2.0]nonane	21.93	7.83
	Farnesol	24.6	2.24
	trans-verbenol	25.65	1.55
	β-selinene	26.39	2.18
	γ-selinene	26.63	1.24
	γ-cadinene	28.02	1.67
	δ-cadinene	28.12	2.05
	Carveol	31.32	1.03
	Caryophyllene oxide	35.28	2.41
	tau-cadinol	41.74	2.13
Canarium luzonicum (Blume) A. Gray	β-Phellandrene	4.34	4.54
	α-Phellandrene	5.54	12.20
	D-limonene	6.68	36.40
	cis-Sabinene	6.90	3.06
	o-cymene	9.25	3.35
	α-terpinolen	9.75	1.59
	Terpinen-4-ol	22.71	1.15
	α-terpineol	26.41	3.83
	Elemol	39.25	16.70
	10-epi-γ-eudesmol	41.75	1.59
	Rosifoliol	43.27	1.08
	Elemicin	43.73	9.59
Cinnamomum camphora (L.) J. Presl.	Sabinene	4.33	6.07
	β-pinene	5.64	1.70
	D-limonene	6.68	25.90
	1,8-cineol	7.07	44.90
	γ-terpinene	8.35	1.43
	o-cymene	9.25	11.70
Cinnamomum caphora var. *linaloolifera* Y. Fuita	1,2-epoxylinalool	16.46	1.27
	Linalool	21.32	96.99
Citrus x aurantium L.	β-myrcene	5.65	2.32
	D-limonene	6.68	1.57
	1,8-cineole	7.07	2.70
	β-ocimene	8.75	2.39
	Linalyl acetate	21.14	63.4
	Caryophyllene	21.92	1.34
	α-terpineol	26.41	8.84
	Neryl acetate	27.51	3.77
	Geranyl acetate	28.64	6.02
	Geraniol	30.28	5.31
Gaultheria procumbens L.	Methyl salicylate	28.68	98.00
Litsea cubeba (Lour.) Pers.	Sabinene	4.33	2.20
	D-limonene	6.68	14.00
	1,8-cineole	7.07	1.62
	6-methyl-5-hepten-2-one	12.11	1.35
	Citronellal	17.86	1.00
	Linalool	21.32	1.82
	Caryophyllene	21.92	1.85
	(Z)-citral	25.56	31.00
	α-terpineol	26.41	1.06
	(E)-citral	27.51	35.2
	trans-geraniol	32.01	1.35
	nerolic acid	45.74	1.13

Table 1. *Cont.*

Essential Oil	Components	RT (min.)	Percentage of Components (n = 3)
Melaleuca leucadendron L.	D-limonene	6.68	8.12
	1,8-Cineole	7.07	49.20
	γ-terpinen	8.35	2.91
	o-cymene	9.25	3.16
	α-terpinolen	9.75	1.24
	Linalyl acetate	21.14	1.13
	Caryophyllene	21.92	5.65
	2,4-dihydroxy-2-methylpentane	24.12	4.11
	Z,Z,Z-1,5,9,9-tetramethyl-1,4,7,-cycloundecatriene	24.61	2.91
	α-terpineol acetate	26.07	1.84
	α-terpineol	26.41	9.92
	α-selinene	26.64	2.09
	Globulol	38.72	1.90
Melaleuca ericifolia Smith.	D-limonene	6.68	2.97
	1,8 cincol	7.07	23.10
	γ-terpinene	8.35	2.63
	o-cymene	9.25	3.64
	1,2-epoxylinalool	16.46	2.03
	Ethyl 2-(5-methyl-5-vinyltetrahydrofuran-2-yl)propan-2-yl carbonate	17.61	1.54
	β-linalool	21.24	36.70
	2-methylene-4,8,8-trimethyl-4-vinyl-bicyclo[5.2.0]nonane	21.93	1.17
	Alloaromadendrene	22.40	4.73
	Terpinen-4-ol	22.71	2.62
	Aromadendrene	23.75	1.86
	Leden	25.73	1.07
	α-terpineol	26.41	4.98
Pogostemon cabli (Blanco) Benth.	Patchoulene	17.51	3.28
	α-guaien	22.10	18.10
	α-patchoulene	23.23	6.09
	1,1,4a-Trimethyl-5,6-dimethylenedecahydronaphthalene	23.33	7.88
	γ-patchoulene	23.88	1.10
	Aciphyllene	26.11	3.53
	γ-guajene	26.69	18.20
	Patchouli alcohol	41.48	27.30
	Pogostol	42.74	2.65
	Dhelqangin	51.98	2.34
Citrus limon (L.) Osbeck	(E)-citral	27.51	2.44
	Sabinene	4.33	3.45
	β-myrcene	5.65	2.42
	D-limonene	6.68	67.10
	α-terpinolen	9.75	14.20
	α-bergamotene	21.89	1.33
	Neral	25.45	1.50
	β-bisabolene	27.22	1.30
Santalum album L.	α-santalene	21.31	5.03
	α-bergamotene	21.89	9.68
	α-santalal	41.14	2.54
	α-santalol	41.69	59.00
	β-santalol	50.36	9.02
	Lanceol	51.12	1.93
	(E)-nuciferol	51.87	1.74
	7-(5-hexynyl)-tricyclo[4.2.2.0(2,5)]dec-7-ene	52.26	1.38

Table 1. *Cont.*

Essential Oil	Components	RT (min.)	Percentage of Components (n = 3)
	3,3,5,6,8,8-Hexamethyltricyclo[5.1.0.0(2,4)]oct-5-ene	18.88	1.02
	Tricyclo[6.3.0.0(1,5)]undec-2-en-4-one, 5,9-dimethyl	20.76	2.00
	1,2,4,5-tetraethylbenzene	21.32	4.39
	α-muurolene	25.59	1.89
	α-cadinene	25.78	1.74
	Selina-5,11-diene	26.16	1.50
	δ-cadinene	26.64	1.57
	α-vetispirene	27.16	1.77
	β-vetispirene	27.42	2.03
	γ-vetivenene	30.19	3.30
	β-vetivenene	31.27	7.42
	Valencen	32.39	2.19
Vetiveria zizanoides (L.) Roberty	3,5,11-Eudesmatriene	34.91	1.65
	7,7-dichlorobicyclo[3.2.0]hept-2-en-6-one	41.06	1.22
	γ-himachalene	41.31	2.63
	Ziza-6(13)-en-12-al	41.99	1.02
	Khusiol	45.27	1.12
	β-guaiene	45.58	4.43
	Cyclocopacamphenol	46.03	1.66
	Zizanol	47.64	1.03
	(E)-isovalencenal	47.76	1.37
	Vetiselinenol	48.57	1.41
	Vetiverol	50.15	2.22
	Khusenol	50.68	5.24
	Vetiverone	51.09	3.02
	β-costol	52.11	3.52
	Khusenic acid	63.32	4.17

Note: * listed are the components that represented min. 1%. Values represent means of three replicate determinations (maximum relative standard deviation ± 5%).

The chemical composition of *Boswelia carterii* Birdw. EO is given in Table 1. The EO was obtained by steam distillation of hand-collected resin. Nineteen chemical components with min 1% were identified. D-limonene (26.40%) and prehnitene (prehnitol, 8.65%) were the main compounds, which is in agreement with Camarda et al. [17].

The chemical composition of *Canarium luzonicum* (Blume) A. Gray EO is shown in Table 1. The EO was collected by steam distillation of resin. The presence of 12 chemical components with min 1% was found. The main compounds were D-limonene (36.40%) and elemol (16.70%), similar to the report of Villanueva et al. [18].

The chemical composition of *Cinnamomum camphora* (L.) J. Presl. EO is provided in Table 1. The EO was obtained by redistillation of wood and branches by steam, so-called white fraction, which does not contain safrole. Six chemical components with min 1% were found. The dominant constituents were 1,8-cineol (eucalyptol, 44.90%), D-limonene (25.90%), and *o*-cymene (11.70%). A previous study on the EO from fruits in the Guizhou province reported D-camphor (26.10%), 1,8-cineole (19.90%), linalool (9.20%), α-terpineol (7.20%), and limonene (5.30%) [19]. The main constituents in the sample from Jiangxi were D-camphor (42.80%), 1,8-cineole (24.80%), α-terpineol (8.70%), and β-pinene (5.80%) [20].

The chemical composition of *Cinnamomum caphora* var. *linaloolifera* Y. Fuita EO is presented in Table 1. The EO was acquired by steam distillation of leaves. The main compound was linalool (96.99%). Linalool was found to be the major constituent of *C. caphora* var. *linaloolifera* leaf oil (95.00%), with no other compounds present at a level of more than 1% [21].

The chemical composition of *Citrus x aurantium* L. EO is given in Table 1. The EO was obtained by distillation of fresh leaves. The presence of 11 chemical components with min 1% was recorded. The main compounds were linalyl acetate (63.40%) and α-terpineol (*p*-menth-1-en-8-ol, 8.84%), with linalool and linalyl acetate in leaves and limonene being found in previous studies [22,23].

The chemical composition of *Gaultheria procumbens* L. EO is presented in Table 1. The EO was acquired by distillation of freshly fermented fresh leaves. Methyl salicylate (98.00%) was the main compound which is in agreement with a previous report [24]

The chemical composition of *Litsea cubeba* (Lour.) Pers. EO is shown in Table 1. The EO was obtained by distillation of fruits. The presence of 11 chemical components with min 1% was found: (E)-citral ((F)-geranial and (E)-neral, 35.20%), (Z)-citral ((Z)-neral, 31.00%), and D-limonene (14.00%). Our results are in agreement with Thielmann and Muranyi [25], who stated that citral and limonene were the major components of *L. cubeba* EO extracted from fruits.

The chemical composition of *Melaleuca leucadendron* L. EO is given in Table 1. The EO was obtained by steam distillation of young shoots and leaves. The presence of 11 chemical components with min 1% was recorded. The main compounds were 1,8-cineol (eucalyptol, 49.20%) and α-terpineol (9.92%), which is line with previously reported 1,8-cineole (44.8–60.2%), α-terpineol (5.93–12.5%), D-limonene (4.45–8.85%), and β-caryophyllene (3.78–7.64%) [26].

The chemical composition of *Melaleuca ericifolia* Smith. EO is provided in Table 1. The EO was collected by steam distillation of branches. The presence of 13 chemical components with min 1% was observed. The main compounds were *β*-linalool (linalyl alcohol, 36.70%) and 1,8-cineol (eucalyptol, 23.10%). The EO from the leaves of *M. leucadendra* from Vietnam were found to be rich in *α*-eudesmol (17.6–21.2%) and guaiol (10.9–12.5%), and linalool was present in smaller concentrations (4.9–5.1%) [27]. Other studies indicated that 1,8-cineole was the major compound of *M. leucadendron* oil [28–30].

The chemical composition of *Pogostemon cabli* (Blanco) Benth. EO is given in Table 1. The EO was obtained by distillation of fermented leaves with steam, followed by maturation of the EO over time. Ten chemical components were present at min 1%, including patchouli alcohol (27.30%), *γ*-guajene (*α*-bulnesene, 18.20%), and *α*-guaien (18.10%). The major components of the oil were reported to be acetophenone (51.00%), *β*-pinene (5.30%), (E)-nerolidol (5.40%), and patchouli alcohol (14.00%) [31].

The chemical composition of *Citrus limon* (L.) Osbeck EO is displayed in Table 1. The EO was acquired by cold pressing fresh fruit. The presence of 8 chemical components that represented min 1% was recorded. D-limonene (67.10%) and *p*-mentha-1,4(8)-diene (iso-terpinene and *α*-terpinolene, 14.20%) were the main compounds while limonene (55.40%), neral (10.40%), trans-verbenol (6.43%), and decanal (3.25%) were found to be the main components among 43 identified compounds in the EO of this fruit in India [32].

The chemical composition of *Santalum album* L. EO is given in Table 1. The EO was obtained by steam distillation of crushed wood. Twelve chemical components were identified with a min 1%. The main compounds were α-santalol (59.00%), *α*-bergamotene (9.68%), and *β*-santalol (9.02%). Among those, α- and β-santalol, which accounted for 19.60% and 16.00%, respectively, were identified in India, and cis-*α*-santalol was recorded in the EOs from Sri Lanka [33,34].

The chemical coposition of *Vetiveria zizanoides* (L.) Roberty EO is presented in Table 1. The EO was obtained by steam distillation of sun-dried roots. The analysis indicated the presence of 28 chemical components at min 1%. The main compounds were β-vetivenene (7.42%) and khusenol (5.24%). David et al., 2009, analyzed oils extracted with carbon dioxide-expanded ethanol and found valerenol (18.50%), valerenal (10.20%), and *β*-cadinene (6.23%) to be the most common compounds out of a total of 23 molecules identified. Interestingly, 48 more components were found in oils processed with conventional hydrodistillation [35].

2.3. Antibacterial Effect of Antimicrobials

In this study, 50 isolates of *Staphylococcus* spp. acquired from human semen were tested for antimicrobial resistance (Table 2) against chloramphenicol, tetracycline, tigecycline, and tobramycin, and the results were interpreted according to the European Committee on Antimicrobial Susceptibility Testing (EUCAST) guidelines [36]. In total, 37 (74%) isolates were resistant while 13 (26%) isolates were sensitive to chloramphenicol. All tested isolates were sensitive to tetracycline and tigecycline. Resistance to tobramycin was identified in the case of 32 isolates, while 10 were sensitive and 8 were intermediately resistant to tobramycin.

Table 2. Antimicrobial resistance of *Staphylococcus* spp.

Name of Bacteria/AMB	C	TET	TIG	TOB
Stapyloccocus aureus 18.3	S	S	S	S
Staphyloccous capitis 3.6	R	S	S	I
Staphylococcus epidermidis 2.2	R	S	S	R
Staphylococcus epidermidis 2.4	S	S	S	R
Staphylococcus epidermidis 16.2	S	S	S	R
Staphylococcus epidermidis 18.4	R	S	S	R
Staphylococcus epidermidis 19.5	S	S	S	R
Staphylococcus epidermidis 24.2	R	S	S	I
Staphylococcus epidermidis 27.3	R	S	S	R
Staphylococcus haemoliticus 2.3	R	S	S	R
Staphylococcus haemoliticus 14.5	R	S	S	R
Staphylococcus haemoliticus 14.6	R	S	S	R
Staphylococcus haemoliticus 14.7	S	S	S	R
Staphylococcus haemoliticus 14.8	R	S	S	R
Staphylococcus haemoliticus 17.1	R	S	S	R
Staphylococcus haemoliticus 17.2	R	S	S	R
Staphylococcus haemoliticus 17.3	R	S	S	R
Staphylococcus haemoliticus 17.5	S	S	S	R
Staphylococcus haemoliticus 17.6	R	S	S	S
Staphylococcus haemoliticus 19.7	R	S	S	R
Staphylococcus haemoliticus 19.8	S	S	S	R
Staphylococcus haemoliticus 20.2	S	S	S	R
Staphylococcus haemoliticus 20.4	R	S	S	I
Staphylococcus haemoliticus 20.5	R	S	S	R
Staphylococcus haemoliticus 21.5	R	S	S	S
Staphylococcus haemoliticus 21.6	S	S	S	I
Staphylococcus haemoliticus 21.7	R	S	S	R
Staphylococcus haemoliticus 22.5	R	S	S	R
Staphylococcus haemoliticus 24.1	R	S	S	S
Staphylococcus haemoliticus 24.2	R	S	S	R
Staphylococcus haemoliticus 24.5	R	S	S	S
Staphylococcus haemoliticus 24.6	S	S	S	I
Staphylococcus haemoliticus 24.7	R	S	S	R
Staphylococcus haemoliticus 24.8	R	S	S	R
Staphylococcus haemoliticus	R	S	S	S
Staphylococus hominis 3.1	S	S	S	R
Staphylococus hominis 3.2	R	S	S	R
Staphylococus hominis 3.3	R	S	S	I
Staphylococus hominis 3.4	R	S	S	R
Staphylococus hominis 3.5	S	S	S	R
Staphylococus hominis 3.7	R	S	S	S
Staphylococus hominis 14.4	R	S	S	R
Staphylococus hominis 16.1	R	S	S	S
Staphylococus hominis 16.4	S	S	S	R
Staphylococus hominis 17.4	R	S	S	R
Staphylococus hominis 18.1	R	S	S	R
Staphylococus hominis 18.8	R	S	S	I
Staphylococus hominis 21.1	R	S	S	S
Staphylococus hominis 21.2	R	S	S	I
Staphylococus hominis 27.4	R	S	S	S

Note: C—chloramphenicol, TET—tetracycline, TIG—tigecycline, TO—tobramycin, R—resistant, S—sensitive, I—intermediate.

Chloramphenicol is a broad spectrum antimicrobial which is active against gram-positive as well as gram-negative bacteria [37,38]. Because of chrolamphenicol toxicity and its application for life-treatening conditions, highly phenicol-resistant *S. aureus* strains of human origin have become a pressing area of scientific interest [39]. Resistance to tetracyclines is common as a result of their broad implementation in human and veterinary medicine. Furthermore, antimicrobial resistance to tetracycline has emerged in plants as well [40,41]. Resistance to tetracycline is encoded by genetic determinants and is fairly common in bacteria [42]. Tigecycline activity in vitro was observed against gram-positive and gram-negative microorganisms, such as *S. aureus*, *Enterococcus* spp., *S. pneumoniae*, *Haemophilus influenzae*, *Moraxella catarrhalis*, *Neisseria gonorrhoeae*, *N. peptostreptococci*, *Clostridium* spp., *Enterobacteriaceae*, and *Bacteroides* spp. [43,44]. It must be noted that differences in the antimicrobial resistance rates against gentamicin and tobramycin were found for *S. aureus* and *P. aeruginosa* across Europe [45].

2.4. Antimicrobial Assay

The antibacterial activities of 14 EOs against 50 *Staphylococcus* spp. isolates were determined with disc diffusion and broth dilution methods (Tables 3–6). The antimicrobial properties of the assessed oils exhibited broad variations.

Table 3. Antimicrobial activity of essential oils (EOs) with disc diffusion method in mm.

Isolate/EOs	1.	2.	3.	4.	5.	6.	7.
Stapyloccocus aureus 18.3	16.50 ± 1.32	11.00 ± 1.00	21.33 ± 1.15	7.67 ± 0.58	24.67 ± 0.58	14.67 ± 0.58	7.33 ± 0.58
Staphyloccous capitis 3.6	12.33 ± 0.58	8.67 ± 0.58	24.67 ± 0.58	4.67 ± 0.58	14.67 ± 0.58	7.67 ± 0.58	8.33 ± 0.58
Staphylococcus epidermidis 2.2	9.33 ± 0.58	12.67 ± 0.58	20.33 ± 0.58	9.00 ± 1.00	8.33 ± 0.58	15.33 ± 0.58	5.67 ± 0.58
Staphylococcus epidermidis 2.4	8.67 ± 0.58	11.67 ± 0.58	22.33 ± 0.58	7.67 ± 0.58	7.33 ± 0.58	15.00 ± 0.00	4.67 ± 0.58
Staphylococcus epidermidis 16.2	11.67 ± 0.58	12.67 ± 0.58	21.67 ± 0.58	4.67 ± 0.58	7.67 ± 0.58	16.33 ± 0.58	5.67 ± 0.58
Staphylococcus epidermidis 18.4	10.67 ± 0.58	13.33 ± 1.15	19.67 ± 0.58	5.33 ± 0.58	8.33 ± 0.58	15.67 ± 1.15	5.50 ± 0.50
Staphylococcus epidermidis 19.5	14.67 ± 0.58	12.67 ± 0.58	20.33 ± 0.58	4.67 ± 0.58	6.33 ± 1.15	16.67 ± 0.58	5.00 ± 1.00
Staphylococcus epidermidis 24.2	11.33 ± 1.15	11.67 ± 1.53	19.67 ± 0.58	6.33 ± 0.58	7.33 ± 1.15	16.00 ± 1.00	4.33 ± 0.58
Staphylococcus epidermidis 27.3	11.67 ± 1.53	11.67 ± 1.15	20.33 ± 1.15	5.67 ± 0.58	6.67 ± 0.58	17.33 ± 0.58	4.67 ± 0.58
Staphylococcus haemoliticus 2.3	8.67 ± 0.58	8.33 ± 0.58	20.67 ± 1.15	4.67 ± 0.58	4.67 ± 0.58	10.33 ± 0.58	5.67 ± 0.58
Staphylococcus haemoliticus 14.5	7.67 ± 0.58	8.67 ± 1.15	18.33 ± 0.58	6.33 ± 0.58	5.67 ± 0.58	8.67 ± 0.58	5.33 ± 1.15
Staphylococcus haemoliticus 14.6	11.00 ± 1.00	8.67 ± 1.15	19.67 ± 0.58	7.67 ± 0.58	4.67 ± 0.58	8.33 ± 0.58	6.33 ± 1.15
Staphylococcus haemoliticus 14.7	7.00 ± 1.00	7.67 ± 0.58	20.33 ± 0.58	8.33 ± 0.58	4.67 ± 1.15	7.67 ± 1.15	4.67 ± 0.58
Staphylococcus haemoliticus 14.8	11.00 ± 1.00	8.67 ± 0.58	18.67 ± 0.58	9.33 ± 1.15	5.33 ± 0.58	8.33 ± 0.58	6.00 ± 1.00
Staphylococcus haemoliticus 17.1	10.67 ± 1.15	10.67 ± 1.15	18.33 ± 0.58	6.33 ± 1.15	5.67 ± 0.58	6.67 ± 0.58	4.67 ± 0.58
Staphylococcus haemoliticus 17.2	9.67 ± 1.15	10.67 ± 0.58	17.33 ± 0.58	5.33 ± 0.58	6.67 ± 0.58	7.33 ± 0.58	5.67 ± 1.15
Staphylococcus haemoliticus 17.3	9.00 ± 1.00	8.67 ± 0.58	18.33 ± 0.58	5.33 ± 0.58	6.33 ± 1.15	8.33 ± 0.58	5.33 ± 0.58
Staphylococcus haemoliticus 17.5	13.33 ± 1.15	8.67 ± 0.58	17.67 ± 1.15	6.33 ± 0.58	7.33 ± 0.58	6.33 ± 1.15	5.67 ± 0.58
Staphylococcus haemoliticus 17.6	11.33 ± 1.15	9.33 ± 0.58	20.33 ± 0.58	5.67 ± 1.15	6.33 ± 0.58	5.67 ± 0.58	5.00 ± 1.00
Staphylococcus haemoliticus 19.7	14.33 ± 0.58	7.67 ± 0.58	18.33 ± 1.15	6.00 ± 1.73	7.67 ± 0.58	7.33 ± 0.58	4.67 ± 0.58
Staphylococcus haemoliticus 19.8	13.67 ± 1.53	10.67 ± 0.58	18.67 ± 1.15	5.67 ± 1.15	7.33 ± 0.58	8.67 ± 0.58	6.33 ± 0.58
Staphylococcus haemoliticus 20.2	9.00 ± 1.00	10.67 ± 1.15	20.33 ± 1.15	5.67 ± 0.58	5.67 ± 0.58	7.33 ± 0.58	4.67 ± 1.15
Staphylococcus haemoliticus 20.4	10.33 ± 0.58	8.67 ± 0.58	19.33 ± 1.15	5.00 ± 1.00	6.33 ± 0.58	7.67 ± 1.15	5.67 ± 1.15
Staphylococcus haemoliticus 20.5	13.00 ± 1.00	9.00 ± 1.00	18.33 ± 1.15	7.33 ± 0.58	5.33 ± 0.58	7.67 ± 0.58	6.00 ± 1.00
Staphylococcus haemoliticus 21.5	14.33 ± 0.58	8.33 ± 0.58	20.33 ± 0.58	7.67 ± 0.58	6.00 ± 1.00	8.33 ± 0.58	5.83 ± 0.29
Staphylococcus haemoliticus 21.6	9.33 ± 0.58	9.67 ± 1.15	18.67 ± 0.58	7.33 ± 0.58	6.67 ± 1.15	7.67 ± 0.58	6.17 ± 0.29
Staphylococcus haemoliticus 21.7	8.33 ± 0.58	10.33 ± 0.58	17.67 ± 0.58	6.67 ± 0.58	4.67 ± 1.15	6.33 ± 0.58	5.17 ± 0.29
Staphylococcus haemoliticus 22.5	7.33 ± 0.58	8.33 ± 0.58	17.33 ± 1.15	5.67 ± 0.58	4.67 ± 0.58	6.33 ± 1.15	4.33 ± 0.58
Staphylococcus haemoliticus 24.1	11.67 ± 0.58	10.33 ± 0.58	18.67 ± 0.58	6.33 ± 1.15	5.33 ± 1.53	6.67 ± 1.15	5.67 ± 1.15
Staphylococcus haemoliticus 24.2	10.33 ± 0.58	10.67 ± 0.58	19.33 ± 1.15	8.67 ± 1.15	5.67 ± 1.15	6.67 ± 0.58	4.67 ± 1.15
Staphylococcus haemoliticus 24.5	12.17 ± 0.29	9.67 ± 0.58	18.67 ± 0.58	7.67 ± 0.58	6.67 ± 0.58	6.00 ± 1.00	5.33 ± 0.58
Staphylococcus haemoliticus 24.6	10.67 ± 0.58	8.33 ± 0.58	17.67 ± 0.58	8.33 ± 1.15	5.67 ± 0.58	5.67 ± 0.58	4.33 ± 0.58
Staphylococcus haemoliticus 24.7	12.00 ± 2.00	8.33 ± 0.58	19.33 ± 1.15	8.00 ± 1.00	6.33 ± 1.15	4.67 ± 0.58	4.67 ± 0.58
Staphylococcus haemoliticus 24.8	10.33 ± 0.58	8.33 ± 0.58	18.33 ± 0.58	7.67 ± 0.58	6.33 ± 0.58	5.33 ± 0.58	5.83 ± 0.76
Staphylococcus haemoliticus	11.33 ± 0.58	12.67 ± 0.58	20.33 ± 0.58	8.33 ± 0.58	7.67 ± 0.58	6.33 ± 0.58	4.67 ± 0.58
Staphylococus hominis 3.1	11.33 ± 0.58	10.33 ± 0.58	21.67 ± 0.58	10.67 ± 0.58	18.33 ± 0.58	8.00 ± 1.73	3.33 ± 1.15
Staphylococus hominis 3.2	9.33 ± 0.58	11.33 ± 1.15	22.00 ± 1.73	9.33 ± 0.58	19.67 ± 0.58	8.33 ± 0.58	3.67 ± 0.58
Staphylococus hominis 3.3	7.67 ± 0.58	10.00 ± 1.73	21.33 ± 0.58	8.33 ± 0.58	19.33 ± 0.58	8.00 ± 1.73	4.33 ± 0.58
Staphylococus hominis 3.4	11.67 ± 0.58	8.33 ± 0.58	21.33 ± 0.58	9.00 ± 1.00	18.33 ± 0.58	7.33 ± 1.15	4.67 ± 0.58
Staphylococus hominis 3.5	9.67 ± 0.58	9.67 ± 0.58	21.67 ± 0.58	8.33 ± 0.58	20.33 ± 0.58	8.67 ± 0.58	4.33 ± 0.58
Staphylococus hominis 3.7	10.33 ± 1.15	10.33 ± 1.15	18.67 ± 0.58	8.67 ± 0.58	21.67 ± 0.58	8.00 ± 1.00	4.33 ± 0.58
Staphylococus hominis 14.4	10.67 ± 1.15	9.33 ± 1.15	20.33 ± 0.58	7.33 ± 0.58	18.67 ± 1.15	7.33 ± 1.15	3.67 ± 0.58
Staphylococus hominis 16.1	10.67 ± 1.53	8.67 ± 0.58	18.67 ± 0.58	7.67 ± 1.15	17.67 ± 1.15	7.33 ± 0.58	3.67 ± 1.15
Staphylococus hominis 16.4	9.33 ± 0.58	8.33 ± 0.58	20.33 ± 0.58	7.33 ± 1.15	19.67 ± 0.58	9.33 ± 0.58	3.67 ± 0.58
Staphylococus hominis 17.4	11.33 ± 1.15	9.00 ± 1.00	21.67 ± 0.58	8.33 ± 0.58	19.33 ± 0.58	7.67 ± 0.58	5.33 ± 0.58
Staphylococus hominis 18.1	12.67 ± 1.15	10.67 ± 0.58	20.67 ± 0.58	8.67 ± 0.58	18.33 ± 0.58	8.67 ± 0.58	5.67 ± 0.58
Staphylococus hominis 18.8	9.00 ± 1.00	11.67 ± 0.58	21.67 ± 0.58	8.67 ± 1.15	17.33 ± 0.58	7.67 ± 0.58	4.33 ± 0.58
Staphylococus hominis 21.1	11.33 ± 0.58	11.67 ± 0.58	20.50 ± 0.50	8.55 ± 0.50	18.33 ± 0.58	8.33 ± 0.58	4.67 ± 0.58
Staphylococus hominis 21.2	11.00 ± 1.00	11.00 ± 1.00	21.33 ± 0.58	7.67 ± 0.58	19.33 ± 1.15	8.67 ± 1.15	4.00 ± 0.00
Staphylococus hominis 27.4	10.67 ± 1.15	9.33 ± 0.58	22.33 ± 0.58	7.67 ± 0.58	19.33 ± 1.53	7.67 ± 0.58	4.33 ± 1.15

Note: 1—Amyris balsamifera L., 2—Boswelia carterii Birdw., 3—Canarium luzonicum (Blume) A. Gray, 4—Cinnamomum camphora (L.) J. Presl., 5—Cinnamomum camphora var. linaloolifera Y. Fuita, 6—Citrus x aurantium L., 7—Gaultheria procumbens L.

Table 4. Antimicrobial activity of EOs with disc diffusion method in mm.

Isolate/EOs	8.	9.	10.	11.	12.	13.	14.
Stapyloccocus aureus 18.3	23.33 ± 0.58	4.67 ± 0.58	14.67 ± 0.58	9.00 ± 1.00	10.67 ± 1.15	5.33 ± 0.58	11.33 ± 0.58
Staphyloccous capitis 3.6	25.33 ± 0.58	3.67 ± 1.15	8.67 ± 0.58	8.33 ± 0.58	12.67 ± 1.15	7.33 ± 0.58	12.67 ± 0.58
Staphylococcus epidermidis 2.2	14.66 ± 0.58	4.33 ± 0.58	6.67 ± 0.58	7.33 ± 0.58	8.33 ± 0.58	4.33 ± 0.58	6.67 ± 1.15
Staphylococcus epidermidis 2.4	12.67 ± 0.58	4.00 ± 1.00	5.67 ± 0.58	6.33 ± 0.58	8.67 ± 1.15	4.33 ± 0.58	6.67 ± 1.15
Staphylococcus epidermidis 16.2	14.33 ± 0.58	3.67 ± 0.58	6.33 ± 1.15	6.00 ± 1.00	9.00 ± 1.00	5.00 ± 1.00	6.67 ± 0.58
Staphylococcus epidermidis 18.4	14.50 ± 0.87	4.67 ± 0.58	7.33 ± 0.58	5.67 ± 1.15	9.67 ± 1.15	5.33 ± 0.58	7.00 ± 1.73
Staphylococcus epidermidis 19.5	11.67 ± 0.58	3.67 ± 0.58	6.33 ± 0.58	5.67 ± 0.58	8.33 ± 0.58	4.33 ± 1.15	7.67 ± 0.58
Staphylococcus epidermidis 24.2	12.67 ± 1.53	3.33 ± 0.58	7.33 ± 0.58	6.67 ± 0.58	9.33 ± 0.58	5.33 ± 0.58	8.33 ± 0.58
Staphylococcus epidermidis 27.3	11.33 ± 1.15	4.67 ± 0.58	5.67 ± 0.58	5.33 ± 0.58	8.67 ± 0.58	4.67 ± 0.58	8.33 ± 0.58
Staphylococcus haemoliticus 2.3	12.33 ± 0.58	4.67 ± 1.15	5.33 ± 0.58	10.67 ± 1.15	10.33 ± 0.58	5.33 ± 0.58	8.67 ± 0.58
Staphylococcus haemoliticus 14.5	9.33 ± 0.58	3.67 ± 1.15	5.33 ± 0.58	10.33 ± 0.58	9.00 ± 1.00	4.00 ± 1.00	7.33 ± 0.58
Staphylococcus haemoliticus 14.6	9.33 ± 1.15	3.33 ± 0.58	6.33 ± 0.58	11.00 ± 1.00	8.33 ± 0.58	4.67 ± 0.58	7.00 ± 1.73
Staphylococcus haemoliticus 14.7	9.67 ± 1.15	4.67 ± 1.15	5.67 ± 1.15	11.33 ± 1.15	9.33 ± 1.15	5.00 ± 1.00	7.33 ± 0.58
Staphylococcus haemoliticus 14.8	9.33 ± 1.53	4.67 ± 0.58	5.67 ± 0.58	10.33 ± 2.08	9.67 ± 1.15	4.67 ± 0.58	6.33 ± 0.58
Staphylococcus haemoliticus 17.1	10.67 ± 1.15	3.67 ± 0.58	5.67 ± 1.53	12.33 ± 0.58	8.33 ± 1.53	4.67 ± 1.15	7.67 ± 1.53
Staphylococcus haemoliticus 17.2	10.00 ± 1.73	3.67 ± 0.58	5.33 ± 0.58	10.67 ± 0.58	10.00 ± 2.00	5.00 ± 0.00	6.33 ± 0.58
Staphylococcus haemoliticus 17.3	10.33 ± 1.53	4.33 ± 0.58	5.33 ± 1.15	11.67 ± 0.58	9.67 ± 1.53	5.50 ± 0.50	8.33 ± 0.58
Staphylococcus haemoliticus 17.5	8.00 ± 1.00	4.67 ± 0.58	6.33 ± 0.58	12.67 ± 1.15	8.33 ± 0.58	4.67 ± 0.58	8.33 ± 1.53
Staphylococcus haemoliticus 17.6	6.67 ± 1.53	4.33 ± 0.58	5.67 ± 0.58	11.33 ± 1.15	9.00 ± 1.00	5.17 ± 0.76	8.67 ± 0.58
Staphylococcus haemoliticus 19.7	6.33 ± 1.15	3.33 ± 1.15	6.33 ± 0.58	9.33 ± 1.15	7.67 ± 0.58	5.67 ± 1.15	7.67 ± 0.58
Staphylococcus haemoliticus 19.8	7.33 ± 0.58	4.33 ± 1.15	5.67 ± 0.58	10.67 ± 0.58	8.33 ± 0.58	5.33 ± 0.58	6.67 ± 1.15
Staphylococcus haemoliticus 20.2	6.67 ± 1.15	3.67 ± 1.15	6.33 ± 1.15	10.33 ± 2.08	9.33 ± 1.15	3.67 ± 0.58	6.67 ± 0.58
Staphylococcus haemoliticus 20.4	8.33 ± 1.15	4.00 ± 0.00	6.00 ± 1.73	10.67 ± 1.15	10.67 ± 1.53	6.00 ± 1.00	5.33 ± 0.58
Staphylococcus haemoliticus 20.5	8.67 ± 0.58	3.67 ± 1.15	5.33 ± 0.58	12.33 ± 0.58	10.67 ± 0.58	4.67 ± 0.58	5.67 ± 0.58
Staphylococcus haemoliticus 21.5	8.33 ± 0.58	3.83 ± 0.76	6.67 ± 1.53	11.67 ± 0.58	10.67 ± 1.15	5.33 ± 1.15	4.66 ± 1.15
Staphylococcus haemoliticus 21.6	9.33 ± 0.58	4.33 ± 0.58	8.67 ± 0.58	11.33 ± 1.15	9.00 ± 1.00	4.67 ± 0.58	6.33 ± 1.15
Staphylococcus haemoliticus 21.7	8.00 ± 1.73	4.33 ± 0.58	8.67 ± 1.15	11.33 ± 0.58	10.67 ± 1.15	5.33 ± 0.58	6.67 ± 1.15
Staphylococcus haemoliticus 22.5	7.33 ± 0.58	3.67 ± 0.58	8.33 ± 1.15	10.67 ± 1.15	9.00 ± 1.00	5.00 ± 1.73	6.33 ± 0.58
Staphylococcus haemoliticus 24.1	10.00 ± 1.00	4.33 ± 0.58	8.00 ± 1.00	10.67 ± 0.58	9.00 ± 1.73	5.00 ± 0.87	7.00 ± 1.73
Staphylococcus haemoliticus 24.2	10.33 ± 1.53	4.67 ± 0.58	8.67 ± 1.15	10.67 ± 1.15	9.00 ± 1.00	5.33 ± 0.58	6.33 ± 0.58
Staphylococcus haemoliticus 24.5	9.67 ± 1.15	4.67 ± 1.15	8.33 ± 1.15	12.67 ± 1.15	10.67 ± 1.15	5.67 ± 1.15	7.33 ± 0.58
Staphylococcus haemoliticus 24.6	7.67 ± 0.58	4.67 ± 0.58	9.67 ± 0.58	10.67 ± 0.58	9.33 ± 1.15	5.33 ± 0.58	6.67 ± 1.15
Staphylococcus haemoliticus 24.7	8.00 ± 1.00	5.67 ± 0.58	8.33 ± 0.58	12.33 ± 0.58	9.33 ± 1.53	4.33 ± 0.58	6.67 ± 1.53
Staphylococcus haemoliticus 24.8	8.33 ± 0.58	4.33 ± 0.58	9.33 ± 0.58	10.67 ± 0.58	8.67 ± 1.15	4.67 ± 0.58	5.67 ± 0.58
Staphylococcus haemoliticus	6.67 ± 0.58	4.67 ± 0.58	8.67 ± 1.15	12.67 ± 1.15	9.67 ± 0.58	5.33 ± 0.58	5.33 ± 0.58
Staphylococus hominis 3.1	21.33 ± 1.15	6.67 ± 0.58	15.33 ± 0.58	8.67 ± 1.15	8.33 ± 0.58	7.67 ± 0.58	10.67 ± 1.15
Staphylococus hominis 3.2	20.67 ± 0.58	5.67 ± 0.58	14.33 ± 0.58	8.33 ± 0.58	7.33 ± 0.58	6.67 ± 0.58	11.33 ± 0.58
Staphylococus hominis 3.3	19.33 ± 0.58	6.33 ± 1.15	15.33 ± 0.58	10.33 ± 0.58	8.33 ± 0.58	7.67 ± 0.58	11.67 ± 0.58
Staphylococus hominis 3.4	21.00 ± 1.73	6.67 ± 1.15	14.33 ± 0.58	10.67 ± 1.15	7.67 ± 0.58	8.67 ± 0.58	11.67 ± 1.53
Staphylococus hominis 3.5	21.67 ± 1.15	5.67 ± 0.58	13.33 ± 0.58	9.67 ± 0.58	7.67 ± 1.15	7.67 ± 0.58	11.67 ± 0.58
Staphylococus hominis 3.7	20.67 ± 1.15	5.33 ± 1.53	14.33 ± 0.58	12.33 ± 0.58	8.33 ± 0.58	7.67 ± 0.58	12.33 ± 0.58
Staphylococus hominis 14.4	21.67 ± 0.58	4.67 ± 0.58	13.67 ± 1.53	10.33 ± 0.58	7.67 ± 0.58	8.33 ± 0.58	12.67 ± 1.15
Staphylococus hominis 16.1	22.67 ± 1.15	5.67 ± 0.58	11.33 ± 1.15	12.67 ± 1.15	6.67 ± 1.15	6.67 ± 1.15	11.33 ± 1.15
Staphylococus hominis 16.4	20.33 ± 1.15	6.67 ± 0.58	14.33 ± 0.58	10.33 ± 1.15	6.33 ± 0.58	7.33 ± 0.58	10.33 ± 0.58
Staphylococus hominis 17.4	22.33 ± 0.58	6.33 ± 0.58	12.67 ± 1.15	11.33 ± 1.15	5.33 ± 1.15	8.67 ± 0.58	10.67 ± 0.58
Staphylococus hominis 18.1	22.67 ± 0.58	6.00 ± 1.00	11.33 ± 1.15	9.33 ± 0.58	6.33 ± 1.15	7.33 ± 0.58	9.33 ± 0.58
Staphylococus hominis 18.8	19.33 ± 0.58	7.00 ± 1.73	11.67 ± 1.15	11.67 ± 1.53	5.67 ± 1.15	8.67 ± 0.58	11.33 ± 0.58
Staphylococus hominis 21.1	18.67 ± 0.58	6.33 ± 0.58	12.67 ± 0.50	13.67 ± 1.53	7.33 ± 1.15	6.33 ± 1.53	11.67 ± 0.58
Staphylococus hominis 21.2	18.33 ± 0.58	7.67 ± 0.58	12.00 ± 1.73	11.33 ± 1.53	6.33 ± 0.58	7.00 ± 1.73	12.33 ± 0.58
Staphylococus hominis 27.4	19.00 ± 1.73	7.67 ± 1.15	11.67 ± 0.58	12.67 ± 0.58	7.33 ± 0.58	8.33 ± 0.58	12.33 ± 0.58

Note: 8—Litsea cubeba (Lour.) Pers., 9—Melaleuca leucadendron L., 10—Melaleuca ericifolia Smith., 11—Pogostemon cabli (Blanco) Benth., 12—Citrus limon (L.) Osbeck, 13—Santalum album L., 14—Vetiveria zizanoides (L.) Roberty.

Table 5. Antimicrobial activity of EO detected with minimal inhibitory concentration in μL/mL.

Microorganism/EOs	1.	2.	3.	4.	5.	6.	7.
Stapyloccocus aureus 18.3	1.56	1.56	0.39	25.00	0.39	3.12	12.50
Staphyloccous capitis 3.6	1.56	3.12	0.78	25.00	1.56	12.50	12.50
Staphylococcus epidermidis 2.2	3.12	1.56	0.78	12.50	3.12	3.12	25.00
Staphylococcus epidermidis 2.4	3.12	1.56	0.78	25.00	3.12	3.12	25.00
Staphylococcus epidermidis 16.2	3.12	1.56	0.39	25.00	3.12	3.12	25.00
Staphylococcus epidermidis 18.4	3.12	1.56	1.56	25.00	3.12	3.12	25.00
Staphylococcus epidermidis 19.5	1.56	1.56	0.78	25.00	3.12	3.12	25.00
Staphylococcus epidermidis 24.2	3.12	1.56	1.56	25.00	3.12	3.12	25.00
Staphylococcus epidermidis 27.3	3.12	1.56	0.78	25.00	3.12	3.12	25.00
Staphylococcus haemoliticus 2.3	6.25	6.25	0.78	25.00	12.50	6.25	25.00
Staphylococcus haemoliticus 14.5	6.25	6.25	1.56	25.00	12.50	12.50	25.00

Table 5. *Cont.*

Microorganism/EOs	1.	2.	3.	4.	5.	6.	7.
Staphylococcus haemoliticus 14.6	3.12	6.25	1.56	25.00	12.50	12.50	25.00
Staphylococcus haemoliticus 14.7	12.50	6.25	0.78	3.12	12.50	12.50	25.00
Staphylococcus haemoliticus 14.8	3.12	6.25	1.56	3.12	12.50	12.50	25.00
Staphylococcus haemoliticus 17.1	3.12	3.12	1.56	25.00	12.50	12.50	25.00
Staphylococcus haemoliticus 17.2	3.12	3.12	1.56	25.00	12.50	12.50	25.00
Staphylococcus haemoliticus 17.3	3.12	6.25	1.56	25.00	12.50	12.50	25.00
Staphylococcus haemoliticus 17.5	1.56	6.25	1.56	25.00	6.25	25.00	25.00
Staphylococcus haemoliticus 17.6	3.12	6.25	0.78	25.00	6.25	25.00	25.00
Staphylococcus haemoliticus 19.7	1.56	6.25	1.56	25.00	6.25	25.00	25.00
Staphylococcus haemoliticus 19.8	1.56	3.12	1.56	25.00	6.25	12.50	12.50
Staphylococcus haemoliticus 20.2	6.25	3.12	0.78	25.00	12.50	12.50	25.00
Staphylococcus haemoliticus 20.4	6.25	6.25	1.56	25.00	12.50	12.50	25.00
Staphylococcus haemoliticus 20.5	1.56	6.25	1.56	25.00	25.00	12.50	12.50
Staphylococcus haemoliticus 21.5	1.56	6.25	0.78	25.00	12.50	12.50	25.00
Staphylococcus haemoliticus 21.6	3.12	6.25	1.56	25.00	12.50	12.50	12.50
Staphylococcus haemoliticus 21.7	6.25	3.12	1.56	25.00	25.00	12.50	25.00
Staphylococcus haemoliticus 22.5	6.25	6.25	1.56	25.00	25.00	12.50	25.00
Staphylococcus haemoliticus 24.1	3.12	1.56	1.56	25.00	12.50	12.50	25.00
Staphylococcus haemoliticus 24.2	3.12	1.56	1.56	25.00	12.50	12.50	25.00
Staphylococcus haemoliticus 24.5	1.56	3.12	1.56	25.00	12.50	12.50	25.00
Staphylococcus haemoliticus 24.6	1.56	3.12	1.56	25.00	12.50	25.00	25.00
Staphylococcus haemoliticus 24.7	1.56	3.12	1.56	25.00	12.50	25.00	25.00
Staphylococcus haemoliticus 24.8	1.56	3.12	1.56	25.00	12.50	25.00	25.00
Staphylococcus haemoliticus	1.56	1.56	0.78	25.00	12.50	12.50	25.00
Staphylococus hominis 3.1	1.56	1.56	0.78	12.50	3.12	12.50	25.00
Staphylococus hominis 3.2	3.12	1.56	0.78	12.50	3.12	12.50	25.00
Staphylococus hominis 3.3	6.25	1.56	0.78	12.50	3.12	12.50	25.00
Staphylococus hominis 3.4	3.12	3.12	0.78	12.50	3.12	12.50	25.00
Staphylococus hominis 3.5	3.12	3.12	0.78	12.50	1.56	12.50	25.00
Staphylococus hominis 3.7	3.12	1.56	1.56	12.50	0.78	12.50	25.00
Staphylococus hominis 14.4	3.12	3.12	0.78	12.50	1.56	12.50	25.00
Staphylococus hominis 16.1	3.12	3.12	1.56	6.25	1.56	12.50	25.00
Staphylococus hominis 16.4	3.12	3.12	0.78	12.50	3.12	12.50	25.00
Staphylococus hominis 17.4	1.56	3.12	0.78	25.00	3.12	12.50	25.00
Staphylococus hominis 18.1	1.56	3.12	0.78	25.00	3.12	12.50	25.00
Staphylococus hominis 18.8	3.12	3.12	0.78	12.50	1.56	12.50	25.00
Staphylococus hominis 21.1	3.12	3.12	0.78	12.50	1.56	12.50	25.00
Staphylococus hominis 21.2	3.12	3.12	0.78	25.00	1.56	12.50	25.00
Staphylococus hominis 27.4	3.12	6.25	0.39	25.00	3.12	12.50	25.00

Note: 1—Amyris balsamifera L., 2—Boswelia carterii Birdw., 3—Canarium luzonicum (Blume) A. Gray, 4—Cinnamomum camphora (L.) J. Presl., 5—Cinnamomum camphora var. linaloolifera Y. Fuita, 6—Citrus x aurantium L., 7—Gaultheria procumbens L.

Table 6. Minimal inhibitory concentration of EOs in µL/mL.

Name of bacteria/EOs	8.	9.	10.	11.	12.	13.	14.
Stapyloccocus aureus 18.3	0.39	25.00	3.12	3.12	3.12	12.50	3.12
Staphyloccous capitis 3.6	0.39	25.00	6.25	3.12	3.12	6.25	3.12
Staphylococcus epidermidis 2.2	1.56	25.00	12.50	6.25	6.25	12.50	25.00
Staphylococcus epidermidis 2.4	1.56	25.00	12.50	6.25	6.25	12.50	25.00
Staphylococcus epidermidis 16.2	1.56	25.00	12.50	6.25	6.25	12.50	25.00
Staphylococcus epidermidis 18.4	1.56	25.00	12.50	6.25	6.25	12.50	25.00
Staphylococcus epidermidis 19.5	1.56	25.00	12.50	6.25	6.25	12.50	12.50
Staphylococcus epidermidis 24.2	1.56	25.00	12.50	6.25	6.25	12.50	12.50
Staphylococcus epidermidis 27.3	1.56	25.00	12.50	6.25	6.25	12.50	12.50
Staphylococcus haemoliticus 2.3	1.56	25.00	12.50	3.12	3.12	12.50	12.50
Staphylococcus haemoliticus 14.5	3.12	25.00	12.50	3.12	3.12	12.50	12.50
Staphylococcus haemoliticus 14.6	3.12	25.00	12.50	3.12	3.12	12.50	12.50
Staphylococcus haemoliticus 14.7	3.12	25.00	12.50	3.12	3.12	12.50	12.50
Staphylococcus haemoliticus 14.8	3.12	25.00	12.50	3.12	3.12	12.50	12.50

Table 6. *Cont.*

Name of bacteria/EOs	8.	9.	10.	11.	12.	13.	14.
Staphylococcus haemoliticus 17.1	3.12	25.00	12.50	3.12	3.12	12.50	12.50
Staphylococcus haemoliticus 17.2	3.12	25.00	12.50	3.12	3.12	12.50	12.50
Staphylococcus haemoliticus 17.3	3.12	25.00	12.50	3.12	3.12	12.50	12.50
Staphylococcus haemoliticus 17.5	6.25	25.00	12.50	3.12	3.12	25.00	12.50
Staphylococcus haemoliticus 17.6	6.25	25.00	12.50	3.12	3.12	25.00	12.50
Staphylococcus haemoliticus 19.7	6.25	25.00	12.50	3.12	6.25	25.00	12.50
Staphylococcus haemoliticus 19.8	6.25	25.00	12.50	3.12	6.25	25.00	12.50
Staphylococcus haemoliticus 20.2	6.25	25.00	12.50	3.12	6.25	25.00	12.50
Staphylococcus haemoliticus 20.4	6.25	25.00	12.50	3.12	3.12	25.00	25.00
Staphylococcus haemoliticus 20.5	6.25	25.00	12.50	3.12	3.12	25.00	25.00
Staphylococcus haemoliticus 21.5	6.25	25.00	12.50	3.12	3.12	25.00	25.00
Staphylococcus haemoliticus 21.6	6.25	25.00	6.25	3.12	3.12	25.00	25.00
Staphylococcus haemoliticus 21.7	6.25	25.00	6.25	3.12	3.12	25.00	25.00
Staphylococcus haemoliticus 22.5	6.25	25.00	6.25	3.12	3.12	25.00	25.00
Staphylococcus haemoliticus 24.1	3.12	25.00	6.25	3.12	3.12	25.00	25.00
Staphylococcus haemoliticus 24.2	3.12	25.00	6.25	3.12	3.12	25.00	25.00
Staphylococcus haemoliticus 24.5	3.12	25.00	6.25	3.12	3.12	25.00	25.00
Staphylococcus haemoliticus 24.6	6.25	25.00	6.25	3.12	3.12	25.00	25.00
Staphylococcus haemoliticus 24.7	6.25	25.00	6.25	3.12	3.12	25.00	25.00
Staphylococcus haemoliticus 24.8	6.25	25.00	6.25	3.12	3.12	25.00	25.00
Staphylococcus haemoliticus	6.25	25.00	6.25	3.12	3.12	25.00	25.00
Staphylococus hominis 3.1	0.78	12.50	3.12	6.25	3.12	6.25	3.12
Staphylococus hominis 3.2	0.78	12.50	3.12	6.25	6.25	6.25	3.12
Staphylococus hominis 3.3	0.78	12.50	3.12	3.12	3.12	6.25	3.12
Staphylococus hominis 3.4	0.78	12.50	3.12	3.12	3.12	6.25	3.12
Staphylococus hominis 3.5	0.78	25.00	3.12	3.12	3.12	6.25	3.12
Staphylococus hominis 3.7	0.78	25.00	3.12	3.12	3.12	6.25	3.12
Staphylococus hominis 14.4	0.78	25.00	3.12	3.12	3.12	6.25	3.12
Staphylococus hominis 16.1	0.78	25.00	3.12	3.12	3.12	6.25	3.12
Staphylococus hominis 16.4	0.78	12.50	3.12	3.12	3.12	6.25	3.12
Staphylococus hominis 17.4	0.78	12.50	3.12	3.12	3.12	6.25	3.12
Staphylococus hominis 18.1	0.78	12.50	3.12	6.25	3.12	6.25	3.12
Staphylococus hominis 18.8	0.78	12.50	3.12	3.12	3.12	6.25	3.12
Staphylococus hominis 21.1	0.78	12.50	3.12	3.12	6.25	6.25	3.12
Staphylococus hominis 21.2	0.78	12.50	3.12	3.12	6.25	6.25	3.12
Staphylococus hominis 27.4	0.78	12.50	3.12	3.12	6.25	6.25	3.12

Note: 8—Litsea cubeba (Lour.) Pers., 9—Melaleuca leucadendron L., 10—Melaleuca ericifolia Smith., 11—Pogostemon cabli (Blanco) Benth., 12—Citrus limon (L.) Osbeck, 13—Santalum album L., 14—Vetiveria zizanoides (L.) Roberty.

The best antimicrobial activity of *A. balsamifera* L. was found against *S. aureus* (16.50 ± 1.32 mm). *B. carterrii* Birdw. revealed the best antimicrobial effect against *S. epidermidis* (13.33 ± 1.15 mm). *C. luzonicum* (Blume) A. Gray showed the best antimicrobial activity against *S. capitis* (24.67 ± 0.58 mm), and *C. camphora* (L.) J. Presl. was found to be most effective against *S. hominis* (10.67 ± 0.58 mm). The best antimicrobial activity of *C. camphora* var. *linaloolifera* Y. Fuita was recorded against *S. aureus* (24.67 ± 0.58 mm), and *C. x aurantium* L. exhibited the highest antimicrobial properties against *S. epidermidis* (17.33 ± 0.58 mm). The EO of *G. procumbens* L. was most effective against *S. capitis* (8.33 ± 0.58 mm).

The best antimicrobial activity of *L. cubeba* (Lour.) Pers was found against *S. capitis* (25.33 ± 0.58 mm). *M. leucadendron* L. showed the best antimicrobial effect against *S. hominis* (7.67 ± 0.58 and 7.67 ± 1.15 mm, respectively). *M. ericifolia* Smith. was highly effective against *S. hominis* (15.33 ± 0.58 mm), while *P. cabli* (Blanco) Benth. exhibited the highest antimicrobial potential against *S. haemoliticus* and *S. hominis* (12.67 mm). The best antimicrobial activity of *C. limon* (L.) Osbeck was found against *S. capitis* (12.67 ± 1.15 mm), and *S. album* L. was highly efficient against *S. hominis* (8.67 ± 0.58 mm). The most effective antimicrobial activity of *V. zizanoides* (L.) Roberty EO was recorded against *S. capitis* and *S. hominis* (12.67 mm).

For the analysed EOs, significant differences in their activity were observed against *Staphylococcus* spp. (Table 7). The most pronounced activity was recorded for *C.luzonicum* (Blume) A. Gray, *A. Balsamifera* L., *C. camphora* var. *linaloolifera*, and *P. cabli* (Blanco) Benth. EOs.

Table 7. Differences between the analysed essential oils in their activity against *Staphylococcus* spp.

Essential Oil	Average/Standard Deviation/Statistical Difference
A. balsamifera L.	$10.82 \pm 2.09\ ^{a}$
B. carterii Birdw.	$9.95 \pm 1.64\ ^{b,a}$
C. luzonicum (Blume) A. Gray	$19.84 \pm 1.94\ ^{c,a,b}$
C. camphora (L.) J. Presl.	$7.18 \pm 1.60\ ^{d,a,b,c}$
C. camphora var. *linaloolifera* Y. Fuita	$10.65 \pm 6.30\ ^{e,c,d}$
C. x aurantium L.	$8.84 \pm 3.33\ ^{f,a,b,c,d,e}$
G. procumbens L.	$5.07 \pm 1.13\ ^{g,a,b,c,d,e,f}$
L. cubeba (Lour.) Pers.	$13.54 \pm 5.90\ ^{h,a,b,c,d,e,g}$
M. leucadendron L.	$4.82 \pm 1.34\ ^{i,a,b,c,d,e,f,h}$
M. ericifolia Smith.	$8.93 \pm 3.40\ ^{j,a,b,c,d,e,g,h,i}$
P. cabli (Blanco) Benth.	$10.29 \pm 2.20\ ^{k,c,d,g,h,i,j}$
C. limon (L.) Osbeck	$8.71 \pm 1.66\ ^{l,a,b,c,d,e,g,h,i,k}$
S. album L.	$5.81 \pm 1.53\ ^{m,a,b,c,d,e,g,h,i,j,k,l}$
V. zizanoides (L.) Roberty	$8.47 \pm 2.49\ ^{n,a,b,c,d,e,g,h,i,k,m}$

Note: Individual letters (a–n) in upper case indicate the statistical difference. $p \leq 0.05$.

No significant differences were found against *A. balsamifera* L. vs. *P. cabli* (Blanco) Benth.; *G. procumbens* L. vs. *M. leucadendron* L.; *B. carterii* Birdw. vs. *P. cabli* (Blanco) Benth.; *M. ericifolia* Smith. vs. *V. zizanoides* (L.) Roberty; *B. carterii* Birdw. vs. *C. camphora* var. *linaloolifera* Y. Fuita; *C. x aurantium* L. vs. *V. zizanoides* (L.) Roberty; *C. limon* (L.) Osbeck vs. *V. zizanoides* (L.) Roberty; *M. ericifolia* Smith. vs. *C. limon* (L.) Osbeck; *C. camphora* var. *linaloolifera* Y. Fuita vs. *P. cabli* (Blanco) Benth.; *C. x aurantium* L. vs. *C. limon* (L.) Osbeck; *A. balsamifera* L. vs. *C. camphora* var. *linaloolifera* Y. Fuita; and *C. x aurantium* L. vs. *M. ericifolia* Smith (Figure 2).

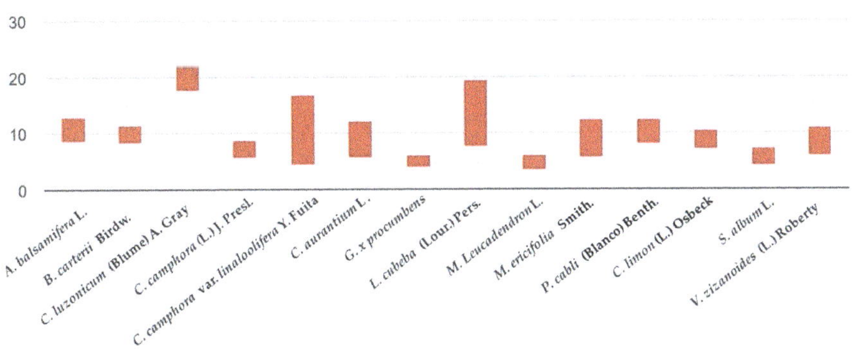

Figure 2. Mean (mm) and standard deviation for analysed essential oils in their activity against *Staphylococcus* spp.

In this study, the EO of *A. balsamifera* L. showed the best antimicrobial activity with the disc diffusion test against *S. aureus* with an inhibition zone of 16.50 mm. Minimum inhibitory concentration (MIC) values obtained with the broth microdilution method were 1.59 μL/mL against *S. aureus*, *S. capitis*, one strain of *S. epidermidis*, 10 strains of *S. haemoliticus*, and three strains of *S. hominis*. *A. balsamifera* was reported to possess antimicrobial activity against gram-positive and gram-negative bacteria, including *Staphylococcus aureus*, *Salmonella paratyphi*, *Escherichia coli*, *Klebsiella pneumoniae*, and microscopic fungi [46].

B. carterii Birdw. EO was found to be the most effective against one strain of *S. epidermidis* (13.33 mm) tested with the disc diffusion method. With the microdilution method, MIC = 1.59 μL/mL was found against *S. aureus*, all strains of *S. epidermidis*, three strains of *S. haemoliticus*, and three strains of *S. hominis*. The antimicrobial activity of EOs of *B. carteri*, *B. neglecta*, *B. sacra*, *B. thurifera*, and *B. frereana* varied from moderate to poor against *S. aureus* (ATCC 12600) [47].

The EO of *C. luzonicum* (Blume) A. Gray exhibited the best antimicrobial activity against *S. capitis* (24.67 mm) with the disc diffusion method. Using the broth microdilitution method, MIC = 0.39 μL/mL was recorded against *S. aureus*, one strain of *S. epidermidis*, and one strain of *S. hominis*. *C. luzonicum* was reported to show antifungal activity without expressing toxicity or other negative side effects [48].

C. camphora (L.) J. Presl. EO revealed the best antimicrobial activity against *S. homins* with an inhibition zone of 10.67 mm with the disc diffusion test and MIC = 3.12 μL/mL against two strains of *S. haemoliticus*. *C. camphora* var. *linaloolifera* Y. Fuita showed the best antimicrobial activity against *S. aureus*, with an inhibition zone of 24.67 mm with the disc diffusion method and MIC = 0.39 μL/mL against *S. aureus*. The EO of *C. camphora* was found to possess antifungal activity against *A. niger* (MIC = 20 μg/mL) and exhibited an inhibitory effect against *B. cereus* and *S. aureus* [49]. Previously identified antimicrobial properties of the EOs of *C. camphora* were in agreement with our results [50–54].

The EO of *C. x aurantium* was the most active against one strain of *S. epidermidis* with the disc diffusion method (inhibition zone of 17.33 mm). With the broth microdilution method, MIC = 3.12 μL/mL was found against *S. aureus* and all strains of *S. epidermidis*. *C. aurantium* was found to inhibit *B. subtilis* and *P. crustosum* [55]. A study on the antimicrobial activity of the *C. aurantium* EO against pathogenic bacteria (*Staphylococcus aureus*, *Salmonella* sp., *Pseudomonas aeruginosa*, *Bacillus subtilis*, and *Escherichia coli)* revealed that gram-positive bacteria were more susceptible than gram-negative bacteria [56].

G. procumbens L. EO exhibited the strongest antimicrobial activity against one strain of *S. aureus* with the disc diffusion test (7.33 mm). An MIC value of 12.50 μL/mL was found for *S. aureus*, *S. capitis*, and one strain of *S. haemoliticus*, determined with the broth microdilution method. Hammer et al. [57] reported a higher activity of *G. procumbens* EO against reference strains of gram-negative bacteria (*Acinetobacter baumanii*, *Aeromonas sobria*, *Escherichia coli*, *Klebsiella pneumoniae*, *Salmonella typhi*, and *Serratia marcescens*) observed in comparison to gram-positive microorganisms (*Staphylococcus aureus* and *Enterococcus faecalis*). A higher resistance of gram-positive bacteria against *G. procumbens* EO was shown by Nikolic et al. [24], who studied the bacteriostatic and bactericidal activity of the oil against microbial isolates.

L. cubeba (Lour.) Pers. EO exhibited the best antimicrobial activity against *S. capitis* with the disc diffusion test (25.33 mm) and an MIC of 0.39 μL/mL against *S. aureus* and *S. capitis* with the broth microdilitution test. The antibacterial activity of *L. cubeba* EO against food-borne pathogens has been reported as well [58–60]. A notably high antimicrobial activity was found against methicyllin-resistant *Staphylococcus aureus* (MRSA) [61,62].

The EOs of *M. ericifolia* Smith showed the strongest antimicrobial activity against *S. aureus* with respect to *S. hominis*. *Melaleuca* EOs have been reported to possess antibacterial activity against common food-borne pathogens [63] and were suggested for the eradication of MRSA in hospitals [64]. Even a concentration of 5% *M. alternifolia* was active against pathogenic bacteria of skin, and a potential application of *M. alternifolia* oil for wound treatment was suggested as well [65–67]. Furthermore, antimicrobial, antifungal, antiviral, and antioxidant properties were described in *M. ericifolia* [26]. Leaf extracts acquired from this plant exhibited antimicrobial activity against gram-positive and gram-negative bacteria, including *S. aureus* [68].

The EO of *Pogestemon cabli* was the most effective against two strains of *S. haemoliticus* and *S. homins* (inhibition zone of 12.67 mm) using the disc diffusion method. The recorded MIC values against two strains of *S. aureus*, *S. capitis*, all strains of *S. haemoliticus*, and all but two strains of *S. hominis* were 3.12 μL/mL. The EO from *P. cabli* was found to be more active against gram-positive than gram-negative bacteria, with the largest inhibition zone (35 mm with 20 μL of oil) and the lowest MIC (250 μg/mL) and minimum bactericidal concentration (MBC) (750 μg/mL) found against *Bacillus cereus*. A moderate antifungal activity was recorded against *Candida albicans* in comparison to *Saccharomyces cerevisiae* (16- vs. 14-mm zone diameters with 20 µL of oil). The lowest MIC and minimal fungicidal concentration(MFC) (both were 750 µg/mL) were found for *Candida albicans* [69].

The EO of *C. limon* (L.) Osbeck was found to be the most effective against one strain of *S. capitis*, with an inhibition zone of 12.67 mm with the disc diffusion test. The broth microdilution method showed MICs of 3.12 µL/mL against *S. aureus*, *S. capitis*, as well as several strains of *S. haemoliticus* and *S. hominis*. The antimicrobial activity of EOs from *C. limon* was recorded against *S. aureus*, *E. coli*, and *B. subtilis* [70], with inhibitory effects against gram-positive bacteria [71]. Hydro-distillated EOs from *C. limon* were reported to be more active due to a high content of limonene [72].

The EOs of *S. album* L. exhibited the highest antimicrobial activity against one strain of *S. hominis* (8.67 mm). An MIC of 6.25 µL/mL was detected against *S. capitis* and all strains of *S. hominis*. A previously reported MIC for *S. album* ranged between 0.078 and 5 µg/mL [73], and an antimicrobial activity against *Staphylococcus aureus* and *Klebsiella pneumoniae* was described as well [74].

The EO of *V. zizanioides* (L.) Roberty showed the highest activity against *S. capitis* and one strain of *S. homins* with an inhibition zone of 12.67 mm using the disc diffusion test. With the broth microdilution tests, the MIC was 3.12 µL/mL against *S. aureus*, *S. capitis*, and all strains *S. hominis*. Gupta et al. [75] found a higher antimicrobial activity of the EO against gram-positive in comparison to gram-negative bacteria. Antifungal and antimicrobial activity against *Candida albicans* as well as wildtype and drug-resistant strains of *M. smegmatis* and drug-resistant strains of *E. coli* have been previously reported [76].

3. Materials and Methods

3.1. Essential Oil Samples

The following essential oils were used in the present study (Table 8): *Amyris balsamifera* L., *Boswellia carterii* Birdw., *Canarium luzonicum* (Blume) A. Gray, *Cinnamomum camphora* (L.) J. Presl., *Cinnamomum camphora* var. *linaloolifera* Y. Fuita, *Citrus x aurantium* L., *Gaultheria procumbens* L., *Litsea cubeba* (Lour.) Pers., *Melaleuca ericifolia* Smith., *Melaleuca leucadendra* L., *Pogostemon cablin* (Blanco) Benth., *Citrus limon* (L.) Osbeck, *Santalum album* L., and *Vetiveria zizanoides* (L.) Roberty. All EOs were produced in Slovakia (Hanus a.s., Nitra) and used in original packaging. All tested oils were stored in the dark at 4 °C.

Table 8. List of essential oils.

Botanical Species	Common Name	Family	Part
Amyris balsamifera L.	amyris	Rutaceae	crushed wood
Boswellia carterii Birdw.	frankincense	Burseraceae	resin
Canarium luzonicum (Blume) A. Gray	elemi	Burseraceae	resin
Cinnamomum camphora (L.) J. Presl.	camphor three bark	Lauraceae	wood, branches
Cinnamomum camphora var. *linaloolifera* Y. Fuita	ho leaf	Lauraceae	leaves
Citrus x aurantium L.	petitgrain	Rutaceae	leaves
Gaultheria procumbens L.	wintergreen	Ericaceae	leaves
Litsea cubeba (Lour.) Pers.	litsea cubeba fruit	Lauraceae	fruits
Melaleuca leucadendron L.	cajeput	Myrtaceae	shoots of leaves
Melaleuca ericifolia Smith.	rosalina	Myrtaceae	branches
Pogostemon cabli (Blanco) Benth.	patchouli	Lamiaceae	fermented leaves
Citrus limon (L.) Osbeck	lemon	Rutaceae	fruits
Santalum album L.	sandalwood	Santalaceae	crushed wood
Vetiveria zizanoides (L.) Roberty	vetiver	Poaceae	dried roots

3.2. Chemical Composition of EOs

Gas chromatographic-mass spectrometric analysis (GC Agilent 7890B and MS Agilent 5977A, Agilent Technologies Inc., Santa Clara, CA, USA) of the EOs was performed as described by Kačániová et al. [77] with a slightly modified version. Prior to the analysis, EO samples were diluted in hexane (HPLC ≥ 97%, Sigma Aldrich GmbH, Darmstad, Germany) to a concentration of 10 µL/mL. One microliter of diluted sample was injected into the inlet (250 °C) operated in split mode

1:10. The separation was achieved using a HP-5ms capillary column (30 m × 0.25 mm × 0.25 μm film; Agilent Technologies). The oven temperature program was set to 50 °C for the first 5 min and subsequently increased to 240 °C at the rate of 3 °C/min, where it was kept constant for 2 min. Helium was used as a carrier gas at constant flow (1.2 mL/min). The mass detector parameters were as follows: ionization energy of the filament—70 eV, transfer line temperature—250 °C, MS source temperature—230 °C, and quadrupole temperature—150 °C. The mass spectrometer was programmed under electron impact (EI) in a full scan mode at *m*/*z* 40–350 with a scanning rate of 2.4 scans/s. The identification of compounds was carried out by comparing mass spectra (over 80% match) with a commercial database NIST® 2017 and the Wiley library for retention times of reference standards (D-limonene, β-myrcene, and γ-terpinene; Sigma-Aldrich GmbH) to compare data on occurrence in EOs with the literature. The relative content of the identified compounds was calculated by dividing the individual peak area by the total area of all peaks. Peaks under 1% were not counted. Each sample was measured in triplicate.

3.3. Microorganisms

Semen samples were obtained from 27 males following 2 days of sexual abstinence. The specimens were taken by masturbation into a sterile wide mouth container. The samples were liquefied at 37 °C for 30 min. All experiments were performed within 1 h after sampling. Only ejaculates showing normal semen parameters (concentration > 20 × 10^6/mL, motility > 40%, viability > 40%, and morphology > 4%) and free from leukocytes were used. The experiments were approved by the Ethic Committee at the Specialized Hospital Sv. Svodar Zobor, protocol no. 030809/2015. Tryptone Soya agar (TSA, Merck, Darmstadt, Germany) and Blood agar (BA, Merck, Darmstadt, Germany) were inoculated with the semen samples, and after incubation (24 h, 37 °C), individual colonies were selected for further confirmation with MALDI-TOF MS Biotyper (Brucker Daltonics, Bremen, Germany) [78]. The isolates were maintained in Mueller Hinton Agar (MHA, Merck, Darmstadt, Germany) and cultured 24 h before the experiment to reach a concentration of 10^5 cfu/mL.

3.4. Antimicrobial Susceptibility Testing

The antimicrobial susceptibility test was performed with the disc diffusion method against (10 mcg) chloramphenicol, tetracycline, tigecycline, and tobramycin. The discs were obtained from Oxoid (Basingstoke, UK). The results were interpreted according to EUCAST [36].

3.5. Disc Diffusion Method

A suspension of the tested culture (0.1 mL of 10^5 cells/mL) was spread onto Mueller Hinton Agar (MHA, Oxoid, Basingstoke, UK). Filter paper discs (6 mm) were impregnated with 15 μL of the EO and placed on the inoculated plates. The agars were incubated at 4 °C for 2 h and subsequently placed into an incubator at 37 °C for 24 h. The diameters of the inhibition zones were measured in mm. All the tests were performed in triplicate [79]. The results were evaluated as follows (disk diameter included): ≥15 mm was strongly inhibitory, <15–10 mm was moderately/mildly inhibitory, and <10 mm was not inhibitory [78–82].

3.6. Determination of Minimum Inhibitory Concentration

The broth microdilution assay was used for determination of the minimal inhibition concentration (MIC) according to the Clinical and Laboratory Standards Institute [83]. All tests were performed in Mueller Hinton Broth (MHB, Oxoid, Basingstoke, UK). The bacterial strains were cultured overnight at 37 °C in MHA. The tested strains were suspended in MHB to give a final density of 10^6 cfu/mL confirmed by viable counts. The EO solution was prepared in dimethyl sulphoxide (DMSO, Penta, Prague, Czech Republic). An amount of 50 μL of MHB was added to each 96-well micro-titer plate, and 100 μL of MHB was added to the 10th well for sterility control. For the growth control, MHB with 5% DMSO was added to the 9th well. Fifty microliters of EOs initially dissolved in 5% DMSO were added into the first well. A serial 2-fold dilution was performed by transferring 50 μL of the suspension to the subsequent wells up to the 8th well; bacterial inoculum of 0.5 McFarland was diluted in the ratio of 1:100 and added into the 1st–8th wells in order to acheive the final concentration of 5×10^5 cfu/mL. Bacterial cell viability and MIC values were determined by observing the turbidity. The lowest concentrations of the EOs with clear suspension were considered as the MIC values. The test was performed in triplicate alongside cefoxitin (30 mcg), used as a positive control.

3.7. Statistical Analysis

The basic variation (disc diffusion method) in statistical values from obtained data were calculated with Statgraphic, Tukey HSD test. Mean, standard deviation, minimum, maximum, coefficient of variation, and frequency of size of inhibition zones were calculated for the antimicrobial activity of essential oils.

4. Conclusions

In this study, 50 different strains of *Staphylococcus* spp. isolated from human semen were tested for susceptibility against 14 different essential oils alongside determination of their chemical composition. The antimicrobial resistance of the tested isolates was evaluated as well. The antimicrobial resistance of *Staphylococcus* spp. against chloramphenicol and tobramycin was found, while all isolates were sensitive to tetracycline and tigecycline. *C. luzonicum* (Blume) A. Gray exhibited a strong inhibitory effect; *A. balsamifera* L., *C. camphora* var. *linaloolifera* Y. Fuita, *L. cubeba* (Lour.) Pers., and *P. cabli* (Blanco) Benth. possessed a moderately inhibitory effect; and *B. carterii* Birdw., *C. camphora* (L.) J. Presl., *C. aurantium* L., *G. procumbens* L., *M. leucadendron* L., *M. ericifolia* Smith., *C. limon* (L.) Osbeck, *S. album* L., and *V. zizanoides* (L.) Roberty revealed no inhibitory activity on *Staphylococcus* spp. isolated from human ejaculates. As such, we may suggest the use of the selected essential oils against *Staphylococcus* spp. contamination of human semen samples.

Author Contributions: Conceptualization, M.K. and E.T.; data curation, M.K., M.T., J.Ž., J.Š., P.Ł.K., and N.V.; investigation, M.K., M.T., J.Ž., T.S., and D.G.; methodology, M.K.; supervision, M.K.; writing—original draft, M.K., M.T., E.T., J.Š., J.Ž., T.S., D.G., P.Ł.K., and N.V. All authors have carefully revised and approved the final version of the manuscript.

Funding: This work was supported by grants APVV SK-BY-RD-19-0014 and BRFBR № X20SLKG-003 "The formulation of novel compositions and properties study of the polysaccharide-based edible films and coatings with antimicrobial and antioxidant plant additives".

Acknowledgments: The paper was supported by the Slovak Research and Development Agency project no. APVV-15-0544.

Conflicts of Interest: The authors declare no conflict of interest.

References

1. Pellati, D.; Mylonakis, I.; Bertoloni, G.; Fiore, C.; Andrisani, A.; Ambrosini, G.; Armanini, D. Genital tract infections and infertility. *Eur. J. Obstet. Gynecol. Reprod. Biol.* **2008**, *140*, 3–11. [CrossRef]
2. Henkel, R.; Schill, W.-B. Sperm separation in patients with urogenital infections. *Andrologia* **2009**, *30*, 91–97. [CrossRef] [PubMed]
3. Urata, K.; Narahara, H.; Tanaka, Y.; Egashira, T.; Takayama, F.; Miyakawa, I. Effect of endotoxin-induced reactive oxygen species on sperm motility. *Fertil. Steril.* **2001**, *76*, 163–166. [CrossRef]
4. Sanocka-Maciejewska, D.; Ciupińska, M.; Kurpisz, M. Bacterial infection and semen quality. *J. Reprod. Immunol.* **2005**, *67*, 51–56. [CrossRef] [PubMed]
5. Purvis, K.; Christiansen, E. Infection in the male reproductive tract. Impact, diagnosis and treatment in relation to male infertility. *Int. J. Androl.* **1993**, *16*, 1–13. [CrossRef]
6. Marconi, M.; Pilatz, A.; Wagenlehner, F.; Diemer, T.; Weidner, W. Impact of infection on the secretory capacity of the male accessory glands. *Int. Braz. J. Urol.* **2009**, *35*, 299–309. [CrossRef] [PubMed]

7. Greive, K.A.; Barnes, T.M. The efficacy of Australian essential oils for the treatment of head lice infestation in children: A randomised controlled trial. *Australas. J. Dermatol.* **2018**, *59*, e99–e105. [CrossRef]
8. Brochot, A.; Guilbot, A.; Haddioui, L.; Roques, C. Antibacterial, antifungal, and antiviral effects of three essential oil blends. *Microbiologyopen* **2017**, *6*, e00459. [CrossRef]
9. Man, A.; Gâz, A.Ş.; Mare, A.D.; Berţa, L. Effects of low-molecular weight alcohols on bacterial viability. *Rev. Rom. Med. Lab.* **2017**, *25*, 335–343. [CrossRef]
10. Kačániová, M.; Ivanišová, E.; Ondrejovič, M.; Chmelová, D.; Terentjeva, M.; Kunová, S.; Tvrdá, E. BIological activity of essential oils against *Staphylococcus* spp. isolated from human semen. *J. Microbiol. Biotechnol. Food Sci.* **2019**, *9*, 404–408. [CrossRef]
11. Salari, M.H.; Amine, G.; Shirazi, M.H.; Hafezi, R.; Mohammadypour, M. Antibacterial effects of *Eucalyptus globulus* leaf extract on pathogenic bacteria isolated from specimens of patients with respiratory tract disorders. *Clin. Microbiol. Inf.* **2006**, *12*, 194–196. [CrossRef] [PubMed]
12. Fraczek, M.; Kurpisz, M. Mechanisms of the harmful effects of bacterial semen infection on ejaculated human spermatozoa: Potential inflammatory markers in semen. *Folia Histochem. Cytobiol.* **2015**, *53*, 201–217. [CrossRef] [PubMed]
13. Wallach, E.E.; Wolff, H. The biologic significance of white blood cells in semen. *Fertil. Steril.* **1995**, *63*, 1143–1157. [CrossRef]
14. Khalili, M.; Sharifi-Yazdi, M. The effect of bacterial infection on the quality of human's spermatozoa. *Iran. J. Public Health* **2001**, *35*, 62–67.
15. Barra, A. Factors affecting chemical variability of essential oils: A review of recent developments. *Nat. Prod. Commun.* **2009**, *4*. [CrossRef]
16. Kloucek, P.; Smid, J.; Flesar, J.; Havlik, J.; Titera, D.; Rada, V.; Drabek, O.; Kokoska, L. In vitro inhibitory activity of essential oil vapors against *Ascosphaera apis*. *Nat. Prod. Commun.* **2012**, *7*. [CrossRef]
17. Camarda, L.; Dayton, T.; Di Stefano, V.; Pitonzo, R.; Schillaci, D. Chemical composition and antimicrobial activity of some oleogum resin essential oils from *Boswellia* spp. (Burseraceae). *Ann. Chim.* **2007**, *97*, 837–844. [CrossRef]
18. Villanueva, M.A.; Torres, R.C.; Başer, K.H.C.; Özek, T.; Kürkçüoğlu, M. The composition of Manila elemi oil. *Flavour Fragr. J.* **1993**, *8*, 35–37. [CrossRef]
19. Liang, G.Y.; Qiu, D.W.; Wei, H.F.; Li, H.Y.; Zhao, S.; He, Z.Y.; Liu, N. The essential oil of the fruits from *Cinnamomum camphora*. *J. Guiyang Coll. Tradit. Chin. Med.* **1994**, *16*, 59–60.
20. Li, S.L.; Luo, Y. The analysis of essential oil from camphor tree fruits in Jinggangshan. *J. Jiangxi Univ. Tradit. Chin. Med.* **2005**, *17*, 36–37.
21. Frizzo, C.D.; Santos, A.C.; Paroul, N.; Serafini, L.A.; Dellacassa, E.; Lorenzo, D.; Moyna, P. Essential oils of camphor tree (*Cinnamomum camphora* nees & eberm) cultivated in Southern Brazil. *Braz. Arch. Biol. Technol.* **2000**, *43*, 313–316. [CrossRef]
22. Macleod, A.J.; Macleod, G.; Subramanian, G. Volatile aroma constituents of orange. *Phytochemistry* **1988**, *27*, 2185–2188. [CrossRef]
23. Kabouche, Z.; Benkiki, N.; Seguin, E.; Bruneau, C. A new dicoumarinyl ether and two rare furocoumarins from *Ruta montana*. *Fitoterapia* **2003**, *74*, 194–196. [CrossRef]
24. Nikolić, M.; Marković, T.; Mojović, M.; Pejin, B.; Savić, A.; Perić, T.; Marković, D.; Stević, T.; Soković, M. Chemical composition and biological activity of *Gaultheria procumbens* L. essential oil. *Ind. Crops Prod.* **2013**, *49*, 561–567. [CrossRef]
25. Thielmann, J.; Muranyi, P. Review on the chemical composition of *Litsea cubeba* essential oils and the bioactivity of its major constituents citral and limonene. *J. Essent. Oil Res.* **2019**, *31*, 361–378. [CrossRef]
26. Pujiarti, R.; Ohtani, Y.; Ichiura, H. Physicochemical properties and chemical compositions of *Melaleuca leucadendron* leaf oils taken from the plantations in Java, Indonesia. *J. Wood Sci.* **2011**, *57*, 446–451. [CrossRef]
27. An, N.T.G.; Huong, L.T.; Satyal, P.; Tai, T.A.; Dai, D.N.; Hung, N.H.; Ngoc, N.T.B.; Setzer, W.N. Mosquito larvicidal activity, antimicrobial activity, and chemical compositions of essential oils from four species of myrtaceae from Central Vietnam. *Plants* **2020**, *9*, 544. [CrossRef]
28. Silva, C.J.; Barbosa, L.C.A.; Maltha, C.R.A.; Pinheiro, A.L.; Ismail, F.M.D. Comparative study of the essential oils of seven *Melaleuca* (Myrtaceae) species grown in Brazil. *Flavour Fragr. J.* **2007**, *22*, 474–478. [CrossRef]
29. Farag, R.S.; Shalaby, A.S.; El-Baroty, G.A.; Ibrahim, N.A.; Ali, M.A.; Hassan, E.M. Chemical and biological evaluation of the essential oils of different *Melaleuca* species. *Phyther. Res.* **2004**, *18*, 30–35. [CrossRef]

30. Sakasegawa, M.; Hori, K.; Yatagai, M. Composition and antitermite activities of essential oils from *Melaleuca* species. *J. Wood Sci.* **2003**, *49*, 181–187. [CrossRef]
31. Murugan, R.; Mallavarapu, G.R.; Padmashree, K.V.; Rao, R.R.; Livingstone, C. Volatile oil composition of *Pogostemon heyneanus* and comparison of its composition with patchouli oil. *Nat. Prod. Commun.* **2010**, *5*. [CrossRef]
32. Paw, M.; Begum, T.; Gogoi, R.; Pandey, S.K.; Lal, M. Chemical composition of *Citrus limon* L. burmf peel essential oil from North East India. *J. Essent. Oil Bear. Plants* **2020**, *23*, 337–344. [CrossRef]
33. Subasinghe, U.; Gamage, M.; Hettiarachchi, D.S. Essential oil content and composition of Indian sandalwood (*Santalum album*) in Sri Lanka. *J. For. Res.* **2013**, *24*, 127–130. [CrossRef]
34. Teixeira da Silva, J.A.; Kher, M.M.; Soner, D.; Page, T.; Zhang, X.; Nataraj, M.; Ma, G. Sandalwood: Basic biology, tissue culture, and genetic transformation. *Planta* **2016**, *243*, 847–887. [CrossRef]
35. David, A.; Wang, F.; Sun, X.; Li, H.; Lin, J.; Li, P.; Deng, G. Chemical composition, antioxidant, and antimicrobial activities of *Vetiveria zizanioides* (L.) nash essential oil extracted by carbon dioxide expanded ethanol. *Molecules* **2019**, *24*, 1897. [CrossRef]
36. EUCAST. The European Committee on Antimicrobial Susceptibility Testing. Available online: http://www.eucast.org/fileadmin/src/media/PDFs/EUCAST_files/Breakpoint_tables/v_9.0_Breakpoint_Tables.pdf (accessed on 13 September 2020).
37. Schwarz, S.; Kehrenberg, C.; Doublet, B.; Cloeckaert, A. Molecular basis of bacterial resistance to chloramphenicol and florfenicol. *FEMS Microbiol. Rev.* **2004**, *28*, 519–542. [CrossRef]
38. Schwarz, S.; White, D.G. Phenicol resistance. In *Frontiers in Antimicrobial Resistance: A Tribute to Stuart B. Levy*; White, D.G., Alekshun, M.N., McDermott, P.F., Eds.; ASM Press: Washington, DC, USA, 2005; pp. 124–147.
39. Witte, W. Biomedicine: Medical consequences of antibiotic use in agriculture. *Science* **1998**, *279*, 996–997. [CrossRef]
40. Falkiner, F. The consequences of antibiotic use in horticulture. *J. Antimicrob. Chemother.* **1998**, *41*, 429–431. [CrossRef]
41. Levy, S.B. *The Antibiotic Paradox*; Springer: Boston, MA, USA, 1992; ISBN 978-0-306-44331-2.
42. Roberts, M.C. Tetracycline resistance determinants: Mechanisms of action, regulation of expression, genetic mobility, and distribution. *FEMS Microbiol. Rev.* **1996**, *19*, 1–24. [CrossRef]
43. Biedenbach, D.J.; Beach, M.L.; Jones, R.N. In vitro antimicrobial activity of GAR-936 tested against antibiotic-resistant gram positive blood stream infection isolates and strains producing extended-spectrum β-lactamases. *Diagn. Microbiol. Infect. Dis.* **2001**, *40*, 173–177. [CrossRef]
44. Edlund, C.; Nord, C.E. In-vitro susceptibility of anaerobic bacteria to gar-936, a new glycylcycline. *Clin. Microbiol. Infect.* **2000**, *6*, 159–163. [CrossRef] [PubMed]
45. Sanfilippo, C.M.; Morrissey, I.; Janes, R.; Morris, T.W. Surveillance of the activity of aminoglycosides and fluoroquinolones against ophthalmic pathogens from Europe in 2010–2011. *Curr. Eye Res.* **2015**, 1–9. [CrossRef] [PubMed]
46. Dahiya, P.; Manglik, A. Evaluation of antibacterial, antifungal and antioxidant potential of essential oil from *Amyris balsamifera* against multi drug resistant clinical isolates. *Asian J. Pharm. Clin. Res.* **2013**, *6*, 57–60.
47. Van Vuuren, S.F.; Kamatou, G.P.P.; Viljoen, A.M. Volatile composition and antimicrobial activity of twenty commercial frankincense essential oil samples. *S. Afr. J. Bot.* **2010**, *76*, 686–691. [CrossRef]
48. Screening of the antifungal activity of essential oils against human and plant pathogenic filamentous fungi. *Flora Mediterr.* **2019**, *29*. [CrossRef]
49. Satyal, P.; Paudel, P.; Poudel, A.; Dosoky, N.S.; Pokharel, K.K.; Setzer, W.N. Bioactivities and compositional analyses of *Cinnamomum* essential oils from Nepal: *C. camphora, C. tamala,* and *C. glaucescens*. *Nat. Prod. Commun.* **2013**, *8*, 1777–1784. [CrossRef]
50. Mishra, A.K.; Dwivedi, S.K.; Kishore, N.; Dubey, N.K. Fungistatic properties of essential oil of *Cinnamomum camphora*. *Int. J. Pharmacogn.* **1991**, *29*, 259–262. [CrossRef]
51. Liu, C.H.; Mishra, A.K.; He, B.; Tan, R.X. Composition and antifungal activity of essential oils from *Artemisia princeps* and *Cinnamomum camphora*. *Int. Pest Control* **2001**, *43*, 72–74.
52. Singh, P.; Srivastava, B.; Kumar, A.; Dubey, N.K. Fungal contamination of raw materials of some herbal drugs and recommendation of *Cinnamomum camphora* oil as herbal fungitoxicant. *Microb. Ecol.* **2008**, *56*, 555–560. [CrossRef]

53. Ho, C.-L.; Eugene, I.; Wang, C.; Su, Y.-C. Essential oil compositions and bioactivities of the various parts of *Cinnamomum camphora* Sieb. var. linaloolifera Fujuta. *For. Res. Q.* **2009**, *31*, 77–95.
54. Rangari, P. Screening of antibacterial sensitivty of essential oils of camphor and cinnamon. *Shodh Anusandhan Samachar* **2011**, *1*, 22–26.
55. Kačániová, M.; Terentjeva, M.; Galovičová, L.; Ivanišová, E.; Štefániková, J.; Valková, V.; Borotová, P.; Kowalczewski, P.Ł.; Kunová, S.; Felšöciová, S.; et al. Biological activity and antibiofilm molecular profile of *Citrus aurantium* essential oil and its application in a food model. *Molecules* **2020**, *25*, 3956. [CrossRef] [PubMed]
56. Teneva, D.; Denkova-Kostova, R.; Goranov, B.; Hristova-Ivanova, Y.; Slavchev, A.; Denkova, Z.; Kostov, G. Chemical composition, antioxidant activity and antimicrobial activity of essential oil from *Citrus aurantium* L. zest against some pathogenic microorganisms. *Z. Nat. C* **2019**, *74*, 105–111. [CrossRef]
57. Hammer, K.A.; Carson, C.F.; Riley, T.V. Antimicrobial activity of essential oils and other plant extracts. *J. Appl. Microbiol.* **1999**, *86*, 985–990. [CrossRef] [PubMed]
58. Li, W.-R.; Shi, Q.-S.; Liang, Q.; Xie, X.-B.; Huang, X.-M.; Chen, Y.-B. Antibacterial activity and kinetics of *Litsea cubeba* oil on *Escherichia coli*. *PLoS ONE* **2014**, *9*, e110983. [CrossRef]
59. Liu, T.-T.; Yang, T.-S. Antimicrobial impact of the components of essential oil of *Litsea cubeba* from Taiwan and antimicrobial activity of the oil in food systems. *Int. J. Food Microbiol.* **2012**, *156*, 68–75. [CrossRef] [PubMed]
60. Yang, Y.; Jiang, J.; Qimei, L.; Yan, X.; Zhao, J.; Yuan, H.; Qin, Z.; Wang, M. The fungicidal terpenoids and essential oil from *Litsea cubeba* in Tibet. *Molecules* **2010**, *15*, 7075–7082. [CrossRef]
61. Cui, H.; Li, W.; Li, C.; Vittayapadung, S.; Lin, L. Liposome containing cinnamon oil with antibacterial activity against methicillin-resistant Staphylococcus aureus biofilm. *Biofouling* **2016**, *32*, 215–225. [CrossRef]
62. Hu, W.; Li, C.; Dai, J.; Cui, H.; Lin, L. Antibacterial activity and mechanism of *Litsea cubeba* essential oil against methicillin-resistant *Staphylococcus aureus* (MRSA). *Ind. Crops Prod.* **2019**, *130*, 34–41. [CrossRef]
63. Siddique, S.; Parveen, Z.; Firdaus-e-Bareen; Mazhar, S. Chemical composition, antibacterial and antioxidant activities of essential oils from leaves of three *Melaleuca* species of Pakistani flora. *Arab. J. Chem.* **2020**, *13*, 67–74. [CrossRef]
64. Dryden, M.; Dailly, S.; Crouch, M. A randomized, controlled trial of tea tree topical preparations versus a standard topical regimen for the clearance of MRSA colonization. *J. Hosp. Infect.* **2004**, *56*, 283–286. [CrossRef] [PubMed]
65. Messager, S.; Hammer, K.A.; Carson, C.F.; Riley, T.V. Effectiveness of hand-cleansing formulations containing tea tree oil assessed ex vivo on human skin and in vivo with volunteers using European standard EN 1499. *J. Hosp. Infect.* **2005**, *59*, 220–228. [CrossRef]
66. Edmondson, M.; Newall, N.; Carville, K.; Smith, J.; Riley, T.V.; Carson, C.F. Uncontrolled, open-label, pilot study of tea tree (*Melaleuca alternifolia*) oil solution in the decolonisation of methicillin-resistant Staphylococcus aureus positive wounds and its influence on wound healing. *Int. Wound J.* **2011**, *8*, 375–384. [CrossRef] [PubMed]
67. Blackwood, B.; Thompson, G.; McMullan, R.; Stevenson, M.; Riley, T.V.; Alderdice, F.A.; Trinder, T.J.; Lavery, G.G.; McAuley, D.F. Tea tree oil (5%) body wash versus standard care (Johnson's Baby Softwash) to prevent colonization with methicillin-resistant *Staphylococcus aureus* in critically ill adults: A randomized controlled trial. *J. Antimicrob. Chemother.* **2013**, *68*, 1193–1199. [CrossRef]
68. Hussein, S.A.M.; Hashim, A.N.M.; El-Sharawy, R.T.; Seliem, M.A.; Linscheid, M.; Lindequist, U.; Nawwar, M.A.M. Ericifolin: An eugenol 5-O-galloylglucoside and other phenolics from *Melaleuca ericifolia*. *Phytochemistry* **2007**, *68*, 1464–1470. [CrossRef]
69. Das, P.; Dutta, S.; Begum, J.; Anwar, M.N. Antibacterial and antifungal activity analysis of essential oil of *Pogostemon cablin* (Blanco) benth. *Bangladesh J. Microbiol.* **2016**, *30*, 7–10. [CrossRef]
70. Saeb, S.; Amin, M.; Seyfi Gooybari, R.; Aghel, N. Evaluation of antibacterial activities of *Citrus limon*, *Citrus reticulata*, and *Citrus grandis* against pathogenic bacteria. *Int. J. Enteric Pathog.* **2016**, *4*, 11–15. [CrossRef]
71. Dhiman, A.; Ahmad, S.; Nanda, A.; Narasimhan, B. In vitro antimicrobial status of methanolic extract of *Citrus sinensis* Linn. fruit peel. *Chron. Young Sci.* **2012**, *3*, 204. [CrossRef]
72. Hamdan, D.; Ashour, M.L.; Mulyaningsih, S.; El-Shazly, A.; Wink, M. Chemical composition of the essential oils of variegated pink-fleshed lemon (Citrus × limon L. Burm. f.) and their anti-inflammatory and antimicrobial activities. *Z. Nat. C* **2013**, *68*, 0275. [CrossRef]

73. Misra, B.B.; Dey, S. Comparative phytochemical analysis and antibacterial efficacy of in vitro and in vivo extracts from East Indian sandalwood tree (*Santalum album* L.). *Lett. Appl. Microbiol.* **2012**, *55*, 476–486. [CrossRef]
74. Jirovetz, L.; Buchbauer, G.; Denkova, Z.; Stoyanova, A.; Murgov, I.; Gearon, V.; Birkbeck, S.; Schmidt, E.; Geissler, M. Comparative study on the antimicrobial activities of different sandalwood essential oils of various origin. *Flavour Fragr. J.* **2006**, *21*, 465–468. [CrossRef]
75. Gupta, R.; Sharma, K.K.; Afzal, M.; Damanhouri, Z.A.; Ali, B.; Kaur, R.; Kazmi, I.; Anwar, F. Anticonvulsant activity of ethanol extracts of *Vetiveria zizanioides* roots in experimental mice. *Pharm. Biol.* **2013**, *51*, 1521–1524. [CrossRef] [PubMed]
76. Champagnat, P.; Sidibé, L.; Forestier, C.; Carnat, A.; Chalchat, J.-C.; Lamaison, J.-L. Antimicrobial activity of essential oils from *Vetiveria nigritana* and *Vetiveria zizanioides* roots. *J. Essent. Oil Bear. Plants* **2007**, *10*, 519–524. [CrossRef]
77. Kačániová, M.; Galovičová, L.; Ivanišová, E.; Vukovic, N.L.; Štefániková, J.; Valková, V.; Borotová, P.; Žiarovská, J.; Terentjeva, M.; Felšöciová, S.; et al. Antioxidant, antimicrobial and antibiofilm activity of coriander (*Coriandrum sativum* L.) essential oil for its application in foods. *Foods* **2020**, *9*, 282. [CrossRef] [PubMed]
78. Kačániová, M.; Kunová, S.; Sabo, J.; Ivanišová, E.; Žiarovská, J.; Felšöciová, S.; Fatrcová-Šramková, K.; Terentjeva, M. Isolation and identification of lactic acid bacteria in wine production by MALDI-TOF MS biotyper. *Acta Hortic. Regiotect.* **2020**, *23*, 21–24. [CrossRef]
79. Zheljazkov, V.D.; Sikora, V.; Semerdjieva, I.B.; Kačániová, M.; Astatkie, T.; Dincheva, I. Grinding and fractionation during distillation alter hemp essential oil profile and its antimicrobial activity. *Molecules* **2020**, *25*, 3943. [CrossRef]
80. Xiangwei, Z.; Xiaodong, W.; Peng, N.; Yang, Z.; JiaKuan, C. Chemical composition and antimicrobial activity of the essential oil of *Sagittaria trifolia*. *Chem. Nat. Compd.* **2006**, *42*, 520–522. [CrossRef]
81. Rota, M.C.; Herrera, A.; Martínez, R.M.; Sotomayor, J.A.; Jordán, M.J. Antimicrobial activity and chemical composition of *Thymus vulgaris*, *Thymus zygis* and *Thymus hyemalis* essential oils. *Food Control* **2008**, *19*, 681–687. [CrossRef]
82. Lv, F.; Liang, H.; Yuan, Q.; Li, C. In vitro antimicrobial effects and mechanism of action of selected plant essential oil combinations against four food-related microorganisms. *Food Res. Int.* **2011**, *44*, 3057–3064. [CrossRef]
83. National Committee for Clinical Laboratory Standards. *Performance Standards for Antimicrobial Susceptibility Testing*; 9th International Supplement, M100-S9; NCCLS: Wayne, PA, USA, 1999.

antibiotics MDPI

Article

Origanum vulgare Essential Oil vs. a Commercial Mixture of Essential Oils: In Vitro Effectiveness on *Salmonella* spp. from Poultry and Swine Intensive Livestock

Maura Di Vito [1,2,†], Margherita Cacaci [2,3,†], Lorenzo Barbanti [1], Cecilia Martini [2], Maurizio Sanguinetti [2,3,*], Stefania Benvenuti [4], Giovanni Tosi [5], Laura Fiorentini [5], Maurizio Scozzoli [6], Francesca Bugli [2,3,†] and Paola Mattarelli [1,†]

1 Dipartimento di Scienze e Tecnologie Agro-Alimentari, Università di Bologna, Viale G. Fanin 42, 40127 Bologna, Italy; wdivit@gmail.com (M.D.V.); lorenzo.barbanti@unibo.it (L.B.); paola.mattarelli@unibo.it (P.M.)

2 Dipartimento di Scienze Biotecnologiche di Base, Cliniche Intensivologiche e Perioperatorie, Università Cattolica del Sacro Cuore, Largo A. Gemelli 8, 00168 Rome, Italy; margherita.c86@gmail.com (M.C.); ceciliamartini84@gmail.com (C.M.); francesca.bugli@unicatt.it (F.B.)

3 Dipartimento di Scienze di Laboratorio e Infettivologiche, Fondazione Policlinico Universitario A. Gemelli IRCCS, Largo A. Gemelli 8, 00168 Rome, Italy

4 Dipartimento di Scienze della Vita, Università di Modena e Reggio Emilia, Via G. Campi 103, 41125 Modena, Italy; stefania.benvenuti@unimore.it

5 Istituto Zooprofilattico Sperimentale della Lombardia e dell'Emilia-Romagna "Bruno Ubertini", Sede Territoriale di Forlì, Via Don E. Servadei 3E/3F, 47122 Forlì, Italy; giovanni.tosi@izsler.it (G.T.); laura.fiorentini@izsler.it (L.F.)

6 Società Italiana per la Ricerca sugli Oli Essenziali (SIROE), Viale Regina Elena 299, 00161 Rome, Italy; mscozzoli@gmail.com

* Correspondence: maurizio.sanguinetti@unicatt.it; Tel.: +39-063-015-4218; Fax: +39-063-051-152

† These authors contributed equally to this work.

Received: 9 October 2020; Accepted: 28 October 2020; Published: 31 October 2020

Abstract: *Salmonella* spp. represent a public health concern for humans and animals due to the increase of antibiotic resistances. In this scenario, the use of essential oils (EOs) could be a valid tool against *Salmonella* contamination of meat. This work compares the in vitro effectiveness of an Italian mixture of feed additives based on EOs (GR-OLI) with EO of *Origanum vulgare* L., recently admitted by European Food Safety Authority (EFSA) for animal use. Twenty-nine *Salmonella* serotypes isolated from poultry and pig farms were used to assess GR-OLI and *O. vulgare* EO antimicrobial propeties. *O. vulgare* EO was active on the disaggregation of mature biofilm, while GR-OLI was capable of inhibiting biofilm formation and disaggregating preformed biofilm. Furthermore, GR-OLI inhibited bacterial adhesion to Caco-2 cells in a dose-dependent manner. Both products showed inhibition of bacterial growth at all time points tested. Finally, the synergistic action of GR-OLI with commonly used antibiotics against resistant strains was investigated. In conclusion, the mixture could be used both to reduce the meat contamination of *Salmonella* spp. before slaughter, and in synergy with low doses of ciprofloxacin against resistant strains. Although EOs as feed additives are already used in animal husbandry, no scientific study has ever highlighted their real antimicrobial potential.

Keywords: *Salmonella*; *Origanum vulgare*; ciprofloxacin; poultry farms; pig farms

1. Introduction

The serotypes of *Salmonella* spp, a pathogenic genus of Enterobacteriaceae, are responsible for animal infections from a sub-clinical to severe level, as well as typhoid fever and severe diarrhea in humans [1]. These Gram-negative bacteria are one of the major causes of concern for the veterinary industry. In particular, chickens and pigs are known as *Salmonella*'s major vehicles. Great efforts are aimed at controlling the *Salmonella* spp. colonization in pig and chicken reservoirs since these animals, if infected, generally host the bacterium asymptomatically in the tonsils, intestine, and lymphoid tissue associated with the intestine [2]. According to the community summary report on trends and sources of zoonoses and zoonotic agents and foodborne outbreaks in the European Union [3], the three most commonly reported zoonoses in Europe are the foodborne enteric diseases campylobacteriosis, salmonellosis, and yersiniosis. Recent studies have identified *Salmonella* spp. multi-resistance to the most common antibiotics used in livestock [4,5]. For this reason, the European Food Safety Authority (EFSA) considers *Salmonella* serotypes isolated from livestock as a danger to public health [6], and the European Union banned the use of antibiotics in animal food production as growth promoters [7]. Subsequently, control measures are aimed at reducing the prevalence of *Salmonella* spp. in livestock, especially in chickens and pigs, and the search for valid alternatives to the use of antibiotics has been stimulated. Probiotics, prebiotics, acidifiers, plant extracts, and nutraceuticals are new alternatives to antibiotics that are widely investigated by researchers [8,9]. Among them, essential oils (EOs) have a prominent place [9].

According to the International Organization for Standardization, essential oils are volatile products obtained from "natural raw material of plant origin by steam distillation, by mechanical processes from the epicarp of citrus fruits, or by dry distillation, after separation of the aqueous phase—if any—by physical processes" [10].

To date, many studies have evaluated the in vitro and in vivo efficacy of EOs on livestock microbial strains, including Salmonella spp. Several studies have analyzed the effects of EOs or their preparations on live animals from pig and poultry farms [11–13], on strains isolated from farms, on reference strains [14], and even on meat after slaughtering [15–18], in order to reduce the infection of Salmonella spp. or increase the shelf-life of the products intended for human consumption. However, not all EOs have been shown to be suitable for in vivo use, as the minimum inhibitory concentration (MIC) of most EOs is much higher than the acceptable dose levels in animal industry in terms of cost-effectiveness and feed palatability [9].

In November 2019, the EFSA published a report expressing the Agency's Endorsement of the safety and efficacy of *Origanum vulgare* L. EO use in feed of all animal species [19]. For each animal species, this report indicates the allowed dosages of the EO expressed in mg/kg live weight. The doses of 22 mg/kg for fattening chickens, 33 mg/kg for laying hens, 30 mg/kg for fattening turkeys, 40 mg/kg for piglets, and 48 mg/kg for fattening pigs are established as safe for both humans and animals, and they are not expected to pose a risk for the environment. According to European Commission Regulation No 1334/20083, *O. vulgare* EO can be used as a flavouring additive in all animal feed, without additional evaluation and approval [4]. The antimicrobial action of *O. vulgare* EO on both bacteria and fungi has been documented in several studies. Generally, the antimicrobial action of this EO is more effective in fungi than it is in bacteria. More specifically, it is more effective against Gram-positive bacteria than Gram-negative bacteria [20]. The EO of *O. vulgare* L. acts on fungal cells by thinning the morphology of the hyphae, and inducing an oxidative stress until cell lysis [21,22]. In bacteria, the main target of the EO active chemicals is the cellular phospholipid bilayer [23]. In particular, the two major chemicals of the EO (thymol and carvacrol) disrupt the outer membrane, alter the proton gradient, and inhibit the production of ATP of Gram-negative cells (including *Salmonella* spp cells) [24,25]. Carvacrol is the principal active compound of the *O. vulgare* phytocomplex, and it belongs to phenols that could exert toxic effects.

The aim of this study was to compare the in vitro antimicrobial effectiveness of the *O. vulgare* EO vs. a mixture of feed additives, namely GR-OLI, characterised by a lower concentration of carvacrol and already approved for use in animal feed, on 29 strains of *Salmonella* spp. isolated from poultry and pig farms.

2. Results

2.1. Compositional Analysis

The constituents identified from *O. vulgare* EO and GR-OLI, their linear retention indices (LRIs), and their percent composition are reported in Table 1. Specifically, the EO of *O. vulgare* is composed of 66.98% carvacrol, while the GR-OLI mixture is characterized by four major compounds with concentrations greater than 10% (limonene 15.32%, carvacrol 12.50%, 1–8 cineol 11.95%, and p-cymene 10.62%).

Table 1. Chemical composition and LRIs of natural compounds.

		%	
Components	**LRI**	***O. vulgare***	**GR-OLI**
α−thujene	927	0.82	0.13
α−pinene	934	0.98	2.48
camphene	948	0.13	0.24
sabinene	974	n.d.	0.38
β−pinene	976	0.15	2.20
octen-3-ol	980	0.46	n.d.
2-octanone	988	0.22	n.d.
β-myrcene	993	1.51	0.90
3-octanol	998	0.05	n.d.
α−phellandrene	1005	0.18	n.d.
α− terpinene	1017	1.06	1.24
p-cymene	1026	7.04	10.62
limonene	1029	0.47	15.32
1,8-cineol	1034	n.d.	11.95
cis-ocimene	1040	n.d.	1.36
trans ocimene	1049	0.05	0.22
γ−terpinene	1060	5.50	3.80
trans sabinene hydrate	1067	0.18	n.d.
cis linalool oxide	1073	n.d.	0.11
terpinolene	1089	0.15	0.63
linalool	1102	1.48	8.73
fenchol	1115	n.d.	0.21
camphor	1146	n.d.	0.45
borneol	1167	0.19	0.44
terpinen-4-ol	1179	0.53	7.12
p-cimen-8-ol	1188	n.d.	0.10
α-terpineol	1193	n.d.	2.75
linalyl acetate	1263	n.d.	5.03
thymol	1296	2.78	5.98
carvacrol	1315	66.98	12.50
neryl acetate	1369	n.d.	0.14
geranyl acetate	1387	n.d.	0.21
β−caryophyllene	1427	1.64	0.65
α-trans bergamotene	1442	n.d.	0.08
aromadendrene	1446	n.d.	0.24
β-farnesene	1462	n.d.	0.27
γ-cadinene	1512	n.d.	0.09
caryophyllene oxide	1594	0.14	n.d.
α-bisabolool	1694	n.d.	0.39

2.2. Antibiograms

All *Salmonella* spp. strains were tested for their sensitivity against antibiotics commonly used in human medicine. As indicated in Table 2, all *S.* Typhimurium strains, with the exception of the monophasic variant, were sensitive to most antibiotics except amoxicillin/clavulanic acid and ciprofloxacin, to which about 31% (S24, S31, S32 S34) and 54% (S3, S7, S12, S17, S18, S21 S32) of strains were resistant, respectively. Only the S17 and S32 strains were also resistant to gentamicin, while S24 and S31 were also resistant to trimethoprim/sulfamethoxazole. All monophasic *S.* Typhimurium strains were resistant to amoxicillin/clavulanic acid; S19 and S28 strains were resistant to ciprofloxacin, and only S29 to trimethoprim/sulfamethoxazole. Whereas, S19 and S29 strains were susceptible by increased exposure sensitivity to piperacillin/tazobactam and ceftazidime, respectively. The S13 strain was sensitive to all antibiotics tested. *S.* Infantis strains showed various multi-resistances. Specifically, no strain was resistant to piperacillin/tazobactam, ertapenem, imipenem, meropenem, amikacin, and gentamicin, while all strains were resistant to ciprofloxacin. In total, 75% of the tested strains (i.e., all excluding S10, S26, and S42) were resistant to trimethoprim/sulfamethoxazole, 67% (all except S4, S10, S40, and S42) to amoxicillin/clavulanic acid, 58% (all minus S4, S25, S38, S39, and S42) to cefotaxime, and 50% (all apart S4, S10, S25, S38, S39 and S42) to cefepime. Increased exposure sensitivities were shown by the S10 strain vs. cefepime, and by 58% of the strains vs. ceftazidime (i.e, all strains except S4, S25, S38, S39, and S42).

2.3. Broth Microdilution Susceptibility Testing

The broth microdilution susceptibility test was performed to study the in vitro sensitivity of *Salmonella* spp. vs. *O. vulgare* EO and GR-OLI. As shown in Table 1, the two natural products have the same activity against both *S.* Typhimurium and *S.* Infantis. In particular, the *O. vulgare* EO and the GR-OLI have a respective minimal inhibitory concentration of the 90% of strains (MIC90) equal to 2% *v/v* (20 μL/mL) and 16% *v/v* (equal to 40 μL of EOs content /mL), for both *S.* Typhimurium and *S.* Infantis.

2.4. Biofilm Formation Assay

The ability of low concentrations of *O. vulgare* EO and GR-OLI to inhibit bacterial biofilm formation or disaggregate a preformed biofilm was evaluated. As shown in Figure 1, the two concentrations tested of GR-OLI were capable of inhibiting biofilm formation or disaggregating any preformed biofilm for *S.* Typhimurium. Both concentrations showed activity in the inhibition of the biofilm formation of *S.* Infantis, but only the higher concentration was active in disaggregating the preformed biofilm. *O. vulgare* EO was active only in the disaggregation of mature biofilm of both serotypes, but no inhibiting activity was exerted on biofilm formation.

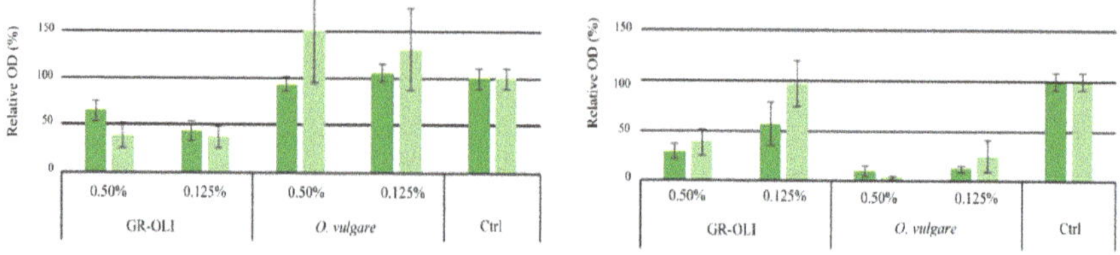

Figure 1. Control-related ratios of the OD values measured for *S.* Typhimurium and *S.* Infantis biofilm inhibition and biofilm disaggregation by GR-OLI and *O. vulgare* EO. Vertical bars indicate ± standard errors.

Table 2. Sensitivity of strains to antibiotics and natural products.

D.	Origin	Sample Source	SP.	AMC	TZP	CTX	CAZ	FEP	ETP	IPM	MEM	AMK	GEN	CIP	SXT	GR-OLI	*O. vulgare*
3	Swine	Feces	S.T.	S	S	S	S	S	S	S	S	S	S	R	S	4	0.25
7	Swine	Feces	S.T.	S	S	S	S	S	S	S	S	S	S	R	S	8	2
12	Swine	Feces	S.T.	S	S	S	S	S	S	S	S	S	S	R	S	8	0.5
13	Swine	Feces	S.T.	S	S	S	S	S	S	S	S	S	S	S	S	4	0.5
17	Swine	Feces	S.T.	S	S	S	S	S	S	S	S	S	R	R	S	16	2
18	Swine	Feces	S.T.	S	S	S	S	S	S	S	S	S	S	R	S	16	2
21	Swine	Feces	S.T.	S	S	S	S	S	S	S	S	S	S	R	S	8	2
24	Swine	Feces	S.T.	R	S	S	S	S	S	S	S	S	S	S	R	16	1
30	Swine	Feces	S.T.	S	S	S	S	S	S	S	S	S	S	S	S	16	0.5
31	Swine	Feces	S.T.	R	S	S	S	S	S	S	S	S	S	S	R	8	1
32	Swine	Feces	S.T.	R	S	S	S	S	S	S	S	S	R	R	S	8	1
33	Swine	Feces	S.T.	S	S	S	S	S	S	S	S	S	S	S	S	8	1
34	Swine	Feces	S.T.	R	S	S	S	S	S	S	S	S	S	S	S	2	<0.25
19	Swine	Feces	m.S.T.	R	I	S	S	S	S	S	S	S	S	R	S	4	1
27	Swine	Feces	m.S.T.	R	S	S	S	S	S	S	S	S	S	S	S	4	1
28	Swine	Feces	m.S.T.	R	S	S	S	S	S	S	S	S	S	R	S	16	1
29	Swine	Feces	m.S.T.	R	S	S	I	S	S	S	S	S	S	S	R	0.5	<0.25
4	Chicken	Boot swabs	S.I.	S	S	S	S	S	S	S	S	S	S	R	R	8	<0.25
10	Chicken	Boot swabs	S.I.	S	S	R	I	I	S	S	S	S	S	R	S	8	1
25	Chicken	Boot swabs	S.I.	R	S	S	S	S	S	S	S	S	S	R	R	16	1
26	Chicken	Boot swabs	S.I.	R	S	R	I	R	S	S	S	S	S	R	S	16	1
35	Chicken	Boot swabs	S.I.	R	S	R	I	R	S	S	S	S	S	R	R	4	0.5
36	Chicken	Boot swabs	S.I.	R	S	R	I	R	S	S	S	S	S	R	R	>16	1
37	Chicken	Boot swabs	S.I.	R	S	R	I	R	S	S	S	S	S	R	R	16	<0.25
38	Chicken	Boot swabs	S.I.	R	S	S	S	S	S	S	S	S	S	R	R	16	1
39	Chicken	Boot swabs	S.I.	R	S	S	S	S	S	S	S	S	S	R	R	16	1
40	Chicken	Boot swabs	S.I.	S	S	R	I	R	S	S	S	S	S	R	R	4	0.5
41	Chicken	Boot swabs	S.I.	R	S	R	I	R	S	S	S	S	S	R	R	>16	1
42	Chicken	Boot swabs	S.I.	S	S	S	S	S	S	S	S	S	S	R	S	16	<0.25
															MIC90	**16**	**2**

Note: D = Designation, SP = Species, AMC = Amicacin/Clavulonic Acid, TZP = Piperacillin/Tazobactam, CTX = Cefotaxime, CAZ = Ceftazidime, FEP = Cefepime, ETP = Ertapenem, IPM = Imiprenem, MEM = Meropenem, AMK = Amikacin, GEN = Gentamicin, CIP = Ciprofloxacin, SXT = Trimethoprim/Sulfamethoxazole, S = Sensitivity, I = Increased exposure sensitivity, R = Resistance, S.T. = *S.* Typhimurium, m.S.T. = monophasic *S.* Typhimurium, S.I. = *S.* Infantis.

2.5. Cell Adhesion Assay

Since GR-OLI was effective not only in the disaggregation of the preformed biofilm but also on the inhibition of biofilm formation, we decided to investigate the interference of this natural mixture on bacterial adhesion to target cells. An adhesion assay was performed using eight *Salmonella* spp. strains. In particular, two *Salmonella* spp. strains sensitive to almost all antibiotics (one sensible *S.* Typhimurium and one *S.* Infantis resistant only to ciprofloxacin), and six multi-resistant strains (two *S.* Typhimurium strains, two monophasic *S.* Typhimurium, and two *S.* Infantis strains) were tested. Figure 2 shows the control-related ratios of the colony-forming unit (CFU) count obtained after the adhesion of the bacteria strains, pre-treated with GR-OLI and untreated, to the Caco-2 monolayer. The ratios refer to bacteria that are still able to adhere to Caco-2 cells after treatment with GR-OLI. Data show that *S.* Typhimurium strains were inhibited in a proportion to the GR-OLI concentration. Particularly, the inhibition was significant at the lower concentration for both monophasic and non-monophasic *S.* Typhimurium (bars of standard error not attaining the 100% level of the control), while the higher concentration was effective only on non-monophasic *S.* Typhimurium. In reverse, the only significant inhibition of *S.* Infantis adhesion was obtained at the minimal concentration tested.

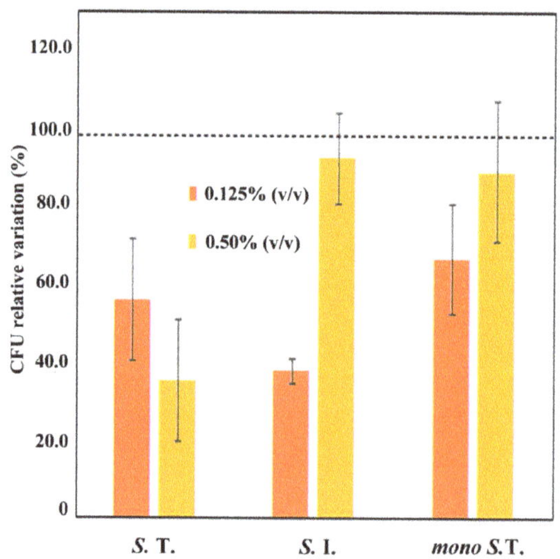

Figure 2. Control-related ratios of the CFU count recovered from GR-OLI-treated bacteria strains (S.T. = *S.* Typhimurium, S. I. = *S.* Infantis, mono S. T. = monophasic *S.* Typhimurium) after the adhesion to the Caco-2 monolayer. Vertical bars indicate ± standard errors.

2.6. Growth Curves

The same strains selected for the cell adhesion assay were treated with GR-OLI and *O. vulgare* EO to evaluate the impact of the two compounds on the bacterial growth curve. Growth curves showed that both products, when tested at the first sub-MIC concentrations, were able to inhibit the growth of most of the tested *Salmonella* strains. Furthermore, the inhibition determined by *O. vulgare* EO on *S.* Typhimurium strains was significantly stronger than that of the GR-OLI at 15 and 20 h (Table 3). Whereas, if compared to the control (Ctrl), both products showed an equal and significant inhibition of *S.* Infantis growth at all the time points tested.

Table 3. Relative growth of *S.* Typhimurium and *S.* Infantis at 10, 15, and 20 h after treatment with *O. vulgare* EO and GR-OLI vs. untreated control (=100).

Strain	*S.* Typhimurium			*S.* Infantis		
Product	10 h	15 h	20 h	10 h	15 h	20 h
Ctrl	100 a	100 a	100 a	100 a	100 a	100 a
O. vulgare EO	57 b	63 c	66 b	64 b	67 b	67 b
GR-OLI	65 b	82 b	99 a	66 b	64 b	64 b

Note. In each *Salmonella* species and time after treatment, different letters indicate significantly different means (LSD test at $p < 0.05$). Ctrl = Control.

2.7. Checkerboard Titration Method

The four strains (S26, S35, S36, S37) equally resistant to amoxicillin/clavulanic acid, cefotaxime, and ciprofloxacin were chosen to test the capability of GR-OLI to synergically reactivate the sensitivity to one or more of these drugs. Table 4 shows the values of the minimal bactericidal concentration (MBC) of each single compound, those of the GR-OLI/antibiotics interaction, and the relative values of the fractional bactericidal concentration index (FBCI). Due to the turbidity of the wells containing both the dilutions of the GR-OLI and those of the antibiotic, it was not possible to perform the OD450 to obtain the MIC values and the relative fractional bactericidal concentration (FIC) index. Ciprofloxacin is the only antibiotic showing synergy with the GR-OLI for all the samples analyzed, while amoxicillin/clavulanic and cefotaxime had a variable interaction depending on the strain analyzed.

Table 4. Average MBC values of amoxicillin/clavulanic acid, cefotaxime, and ciprofloxacin alone and combined with GR-OLI, and relative FBCI values.

Strain	MBC_{AMC}	MBC_{CTX}	MBC_{CIP}	MBC_{GR}	$MBC_{AMC/GR}$	$MBC_{CTX/GR}$	$MBC_{CIP/GR}$	$FBCI_{AMC/GR}$	$FBCI_{CTX/GR}$	$FBCI_{CIP/GR}$
26	32	>128	0.5	12.5	0.25/1.56	0.5/1.56	0.01/3.125	0.133 (s)	0.270 (s)	0.128 (s)
35	64	128	0.5	12.5	0.125/1.56	0.25/3.12	0.03/6.25	0.127 (s)	0.530 (a)	0.321 (s)
36	64	128	0.5	12.5	0.25/6.25	0.5/1.56	0.005/6.25	0.504 (a)	0.505 (a)	0.128 (s)
37	64	>128	0.5	12.5	0.125/12.5	0.25/3.12	0.03/3.125	1.002 (i)	0.280 (s)	0.251 (s)

Note. AMC = Amoxicillin/clavulanic acid, CTX = cefotaxime, CIP = ciprofloxacin, (s) = synergy, (a) = additivity, (i) = indifference.

3. Discussion

Salmonella is the second human bacterial zoonosis delivered especially by chicken and pork meats, but also milk, eggs, and seafood. The WHO estimates 550 million people (including 220 million children under the age of 5 years) fall ill each year due to diarrhoeal diseases due to unsafe food [26]. In 2018, 91,857 confirmed cases of salmonellosis in humans were reported with an EU notification (EFSA, 2019), while the U.S. Center for Disease Control and Prevention (CDC) estimates that *Salmonella* bacteria cause about 1.35 million infections per year in the United States, of which only 41,930 in 2011 were laboratory confirmed [27,28]. Along with the world population increase, the consumption of meat is also increasing. Foley et al. [19] reported that since the early 1900s, the consumption of chicken in the U.S. has increased about sixfold, while pork consumption by about 20%. Whereas, the European Union data show that in 2018, Europe increased its chicken meat production by a quarter, and 70% of this production was in six member states: Poland (16.8%), the United Kingdom (12.9%), France (11.4%), Spain (10.7%), Germany (10.4%), and Italy (8.5%) [18]. An upward trend, although less steep than in the case of poultry meat, was recorded for pork meat whose consumption, in Europe, has increased by about 3.5% per person in 10 years [19]. In order to meet consumer demands, unavoidable changes in animal production were necessary. The introduction of intensive animal husbandry practices has on the one hand increased the exposure of consumers to zoonosis, and, on the other hand, has probably modified the characteristics of *Salmonella* spp. colonization in farms by selecting strains resistant to antibiotics. In animals, *Salmonella* infection can cause fever, diarrhea, prostration, and mortality. Most of the animals that survive this infection remain asymptomatic carriers, posing a threat to human health as, during slaughtering, their carcasses can contaminate others [20,21]. Within *Salmonella* serotypes, *S.* Typhimurium, *S.* Enteritidis, *S.* Heidelberg, *S.* Montevideo, and *S.* Infantis are among the major pig and poultry serotypes most frequently associated with human infections [1]. Strains of *Salmonella* spp. with antimicrobial drug resistance acquired in the animal host are now widespread in all countries [22]. Resistance to ciprofloxacin, which belongs to the group of fluoroquinolones and was, until the last decade, the treatment of choice, and to cephalosporins is increasingly being documented [22–25]. Therefore, the WHO listed resistant *Salmonella* spp. among priority pathogens for which new antibiotics were urgently needed, and several countries have established *Salmonella* surveillance and control programmes. Our data agree with the above-reported concern, because the presence of widespread resistance to ciprofloxacin is confirmed by the circumstance that 21 of the 29 analyzed strains (72.4%) were resistant to fluoroquinolone, and highlight the resistance or a reduced sensitivity to cephalosporins (cefotaxime, ceftazidime, and cefepime), especially in the *S.* Infantis serotype. Amoxicillin/clavulanic acid is another drug showing decreased efficacy especially against *S.* Infantis and monophasic *S.* Typhimurium strains. Although the combination of amoxicillin with clavulanic acid overcomes the intrinsic resistance of beta-lactamase-producing strains, and therefore makes it one of the main antimicrobial substances in swine medicine for the treatment and control of infections, the fair percentage of resistance (55.2% of strains) supports the choice of the European Medicine Agency [26] to classify this association in category C. This category includes antibiotics that are approved for use in livestock and pet animals, but which must be used with caution, only when there are few or no alternatives belonging to category D [26,27]. Natural substances represent a valid resource in the search for alternatives to current antibiotics. Thanks to their high antimicrobial potential, EOs are widely studied to counteract the development of antibiotic resistance and respond to the growing demand of consumers for antibiotic-free foods [12,28]. As noted in the introduction, the *O. vulgare* EO was found to be active against a broad spectrum of microorganisms. The antimicrobial activity is essentially mediated by the main chemicals carvacrol and thymol, which, because of their amphipathic nature, interact with the bacterial and fungal cell membrane. In particular, carvacrol is able to accumulate in the cell membrane of *Salmonella* spp and other bacteria strains, where it can bind to hydrogen by altering the cell membrane potential and inducing a conformational and metabolic modification (decrease of ATP production) up to the time of cell death [20]. This antimicrobial activity of the *O. vulgare* EO on bacterial and fungal membranes is common to many EOs caracterised by the same amphypathic chemical compounds. Despite their strong antimicrobial action, the use of EOs in farms is limited by their poor water solubility. This characteristic makes it necessary to convey them with suitable surfactants or through biotechnological processes. The Italian product GR-OLI is a water-soluble mixture of EOs emulsified in an inert carrier additive, which is regularly authorized as additive for use in animal feed. This mixture has been compared with the activity of the *O. vulgare* OE that recently received a positive opinion from the EFSA for use in animal production. The chemical analysis of both products shows that the *O. vulgare* OE and GR-OLI have respectively three (carvacrol, p-cymene, and γ-terpinene) and eight (limonene, carvacrol, 1-8 cineol, p-cimene, linalool, terpinen-4-ol, and thymol) chemicals with a concentration >5%. Furthermore, if compared to the *O. vulgare* EO, the GR-OLI has a lower concentration of carvacrol and a higher concentration of the other terpenic molecules with known antimicrobial action. If, on the one hand, the antimicrobial action of carvacrol is well known [29,30], on the other hand, this phenolic compound is acknowledged to be potentially toxic, depending on the concentration of use [31]. For this reason, a preliminary in vitro comparison between the antimicrobial properties of *O. vulgare* EO and this commercial aromatic mixture was needed. Data show that the MIC90 of the *O. vulgare* EO is slightly lower than that of GR-OLI against the different *Salmonella* strains tested, and that the sub-MIC of *O. vulgare* EO inhibits over time the *S.* Typhimurium growth more effectively than GR-OLI. However, while the *O. vulgare* EO is only capable of disaggregating a formed biofilm, GR-OLI is simultaneously capable of inhibiting the formation of the biofilm and disaggregating the formed one at minimal concentrations potentially compatible with animal palatability. The ability to prevent the early stages of bacterial adhesion to intestinal cells is critical for the establishment of chronic colonization in animals, which are the reservoir for acute events. In this regard, data obtained from the cell adhesion assay confirmed that GR-OLI, at very low concentrations, is actually able to inhibit bacterial adhesion to the intestinal cell line Caco-2. Inhibition occurs in different ways depending on the serotype. Specifically, the monophasic *S.* Typhimurium and *S.* Infantis strains showing the greatest resistance to antibiotics were sensitive only to the higher concentration tested, while the other strains tested were sensitive to both concentrations. These activities could be useful also with animals carrying *Salmonella* spp. asymptomatically. In these animals, it is important to inhibit both the adhesion and the formation of the biofilm to prevent contamination of the carcasses at the time of slaughtering. Furthermore, data obtained from the checkerboard test indicate that GR-OLI has synergistic action with ciprofloxacin at concentrations much lower than MIC. This data identifies a possible new resource in the fight against antibiotic resistances, as it indicates the possibility of reactivating the sensitivity to ciprofloxacin with low doses of natural compounds mixed with commercial antibiotics. Moreover, given the heterogeneity of the phytocomplex of each EO, the use of concentrations lower than MIC is not currently correlated with the development of resistance [32]. This makes the use of sub-MIC of the EOs mixtures safer against the development of potential resistances.

4. Materials and Methods

4.1. Natural Substances, Antibiotics, and Reagents

O. vulgare L. EO and GR-OLI (by APA-CT, Forlì, Italy), a confidential solution (under patent processing) containing the 25% *v/v* of nine EOs (*Eucalyptus globulus*, *Satureja hortensis*, *Citrus aurantium* var. *dulcis*, *Thymus vulgaris*, *Melaleuca alternifolia*, *Citrus limon*, *Lavandula hybrida*, *Melaleuca cajeputi*, *Thymus capitatus*) dispersed in a surfactant (Glyceryl polyethyleneglycol ricinoleate cod. E484), admitted in animal feed, were tested against *Salmonella* spp. No preservatives or other substances were added to the mixture.

Amoxicillin/clavulanic acid, cefotaxime, and ciprofloxacin (Sigma Aldrich, St. Louis, MO, USA) were used to test their interaction with GR-OLI. C8-C40 n-alkanes mixture, p-cymene, limonene, 1,8-cineol, thymol, carvacrol, and n-hexane were purchased from Sigma-Aldrich (Milan, Italy) and used as standards. All reference standards used for GC analysis, chromatographic-grade organic solvents, and reagents were purchased from Sigma-Aldrich (Milan, Italy).

4.2. Bacterial Strains and Growth Media

To study the effectiveness of the natural products, 29 isolates of *Salmonella enterica* subsp. *enterica* (specifically, 17 *S. enterica* subsp. *enterica* serovar Typhimurium, of which 4 monophasic, and 12 *S.* Infantis). *Salmonella* spp. strains were isolated, during 2017, from swine and poultry intensive farms with no epidemiological correlation and provided by Istituto Zooprofilattico of Forlì (Italy). *Salmonella* spp. strains were isolated from environmental samples (faeces, boot swabs) as part of monitoring plans for the reduction of the most important public health-related *Salmonella* serovars (*S.* Typhimurium, including monophasic variants, *S.* Enteritidis, *S.* Infantis, *S.* Virchow, and *S.* Hadar) in poultry and swine farms. The detection of *Salmonella* spp. was carried out using a culture method according to Amendment 1: Annex D of EN/ISO 6579:2002 [29]. Based on this method, colonies of presumptive *Salmonella* were subcultured and their identiy was confirmed by means of biochemical tests. The pure colonies showing typical biochemical reactions for Salmonella were also tested for the presence of *Salmonella* somatic antigens (O-antigens) and flagellar antigens (H-antigens) by slide-agglutination using polyvalent antisera (BD Difco™—Becton, Dickinson and Company, Franklin Lakes, NJ, USA). Serotyping of *Salmonella* spp. strains was carried out using a slide-agglutination test following the White-Kauffmann-Le Minor scheme according to the part 3 of ISO/TR 6579-3:2014 [30]. For this purpose, a colony from a pure culture of each *Salmonella* spp. strain was cultured on nutrient agar and incubated at 37 °C ± 1 °C overnight. After the incubation, each strain was investigated for auto-agglutination by the slide-agglutination test using a 3.5% solution of sodium chloride. Once auto-agglutination was excluded, each strain was submitted to the agglutination test for serotyping the most important public health-related *Salmonella* serovars: *S.* Typhimurium (including monophasic variants), *S.* Enteritidis, *S.* Infantis, *S.* Virchow, and *S.* Hadar. For this purpose, the following somatic antisera (O-antisera) were used: O:4, O:5, O:6, O:7, O:8, O:9, and O:46 (BD Difco™—Becton, Dickinson and Company, Franklin Lakes, NJ, USA); after agglutination with the O-antisera, the agglutination with flagellar antisera (H-antisera) was performed using the following flagellar H-antisera: H:i, H:2, H:g, H:m, H:q, H:s, H:t, H:r, H:5, H:z_{10}, and H:x (BD Difco™—Becton, Dickinson and Company, Franklin Lakes, NJ, USA). For biphasic H-antigens strains (e.g., *S.* Typhimurium), if one H-phase was negative, a phase inversion was carried out using the Sven Gard method according to the part 3 of ISO/TR 6579-3:2014. Based on their antigenic formula, the *Salmonella* spp. strains were identified according to the White-Kauffmann-Le Minor scheme [31]. The antigenic formula of the *Salmonella* spp. strains used in this study is summarized in Table 5. Muller Hinton medium (MH, Sigma Aldrich, St. Louis, MO, USA) was used to grow the strains at 37 °C ± 1 °C for 24 h.

Table 5. Antigenic formula of *Salmonella* serovars used in the study.

Name	O-Antigens	H-Antigens
Salmonella Typhimurium	1,4,[5],12	I:1,2
Salmonella Typhimurium monophasic variant	1,4,[5],12	i:-
Salmonella Infantis	6,7,14	r:1,5

Note. Numbers underlined and in square brackets are in accordance with the international nomenclature for *Salmonella* spp.

4.3. GC-MS Analysis

Analyses were performed on a 7890A gas chromatograph coupled with a 5975C network mass spectrometer (GC-MS) (Agilent Technologies, Waldbronn, Germany). Compounds were separated on an Agilent Technologies HP-5 MS cross-linked poly–5% diphenyl–95% dimethyl polysiloxane (30 m × 0.25 mm i.d., 0.25 µm film thickness) capillary column. The column temperature was initially set at 45 °C, then increased at a rate of 2 °C/min up to 100 °C, then raised to 250 °C at a rate of 5 °C/min, and finally held for 5 min. The injection volume was 0.1 µL, with a split ratio 1:20. Helium was used as the carrier gas, at a flow rate of 0.7 mL/min. The injector, transfer line, and ion-source temperature was 250, 280, and 230 °C, respectively. MS detection was performed with electron ionization (EI) at 70 eV, operating in the full-scan acquisition mode in the m/z range 40–400. The EOs were diluted 1:20 (*v/v*) with n-hexane before GC-MS analysis.

4.4. GC-FID Analysis

Analyses were carried out on a gas chromatograph coupled with a flame ionization detector (FID) Agilent Technologies 7820A. Compounds were separated on an Agilent Technologies HP-5 cross-linked poly–5% diphenyl–95% dimethyl polysiloxane (30 m × 0.32 mm i.d., 0.25 mm film thickness) capillary column. The temperature programme was the same as described in Section 4.3. The injection volume was 0.1 µL in the split mode 1:20. Helium was used as the carrier gas at a flow rate of 1.0 mL/min. The injector and detector temperature were set at 250 and 300 °C, respectively. The EOs and the reference standards were diluted 1:20 (*v/v*) with n-hexane before GC-FID analysis. The analyses were performed in triplicate.

4.5. Qualitative and Semi-Quantitative Analysis

Compounds were identified by comparing the retention times of the chromatographic peaks with those of authentic reference standards run under the same conditions, and by comparing the linear retention indices (LRIs) relative to C8-C40 n-alkanes obtained on the HP-5 column under the above-mentioned conditions with the literature [32]. Peak enrichment by co-injection with authentic reference compounds was also carried out. Comparison of the MS-fragmentation pattern of the target analytes with those of pure components was performed, by using the National Institute of Standards and Technology (NIST version 2.0d, 2005) mass-spectral database. Semi-quantification was calculated as the relative percentage amount of each analyte; in particular, the values were expressed as the percentage peak area relative to the total composition of each EO obtained by GC-FID analysis.

4.6. Antimicrobial Susceptibility Testing against Antibiotics

To investigate the antimicrobial susceptibility to amicacin/clavulonic acid (AMC), piperacillin/tazobactam (TZP), cefotaxime (CTX), ceftazidime (CAZ), cefepime (FEP), ertapenem (ETP), imiprenem (IPM), meropenem (MEM), amikacin (AMK), gentamicin (GEN), ciprofloxacin (CIP), and trimethoprim/sulfamethoxazole (SXT), we performed antimicrobial susceptibility testing (AST) with the VITEK® 2 system according to the manufacturer's instructions, using the software version 7.01 and the AST-N379 cards for Gram-negative bacteria. To test the antimicrobial susceptibility against ciprofloxacin, we performed AST by the Broth Micro Dilution method according to the 2006 ISO 20776-1 procedure. MIC results were categorized as susceptible (S), susceptible by increased exposure (I), and resistant (R) according to the European Committee on Antimicrobial Susceptibility Testing (EUCAST) breakpoints (version 10.0) [33].

4.7. Broth Microdilution Susceptibility Testing against Natural Products

The broth microdilution (BMD) susceptibility test according to the European Committee on Antimicrobial Susceptibility Testing (EUCAST) international guidelines was performed. Muller Hilton broth (Oxoid, Basingstoke, Hampshire, UK) was used to test the antimicrobial activity of GR-OLI and *O. vulgare* EO against the *Salmonella* spp. strains. The BMD test was performed on a 96-well plate by adding 100 µL of a cell suspension equal to 5×10^5 CFU/mL to a final volume of 200 µL. Scalar dilutions, between 16% *v/v* (equal to 40 µL of EOs content/ mL) and 0.125% *v/v* (equal to 0.3 µL of EOs content/mL) of GR-OLI and between 4% (40 µL /mL) and 0.03% (0.3 µL /mL) of *O. vulgare* EO, were tested. A concentration surfactant (Tween 80, Sigma Aldrich, Saint Louis, MO, USA) corresponding to that contained in the GR-OLI was tested together with *O. vulgare* EO to facilitate its solubilization in the hydrophilic medium. Plates were incubated overnight at 37 °C. After this period, MIC values were determined by spectrophotometric reading at 450 nm (EL808, Biotek, Winooski, VT, USA). To evaluate the MBC, 5 µL of the content of each well were seeded on standard medium agar plates, which were incubated for 24 h at 37 °C. Surfactants were tested separately. The MIC is defined as the lowest concentration that completely inhibits the growth of a given organism compared with the growth in the substance-free control; whereas the MBC is defined as the lowest concentration determining the death of 99.9% or more of the initial inoculum. Each test was performed in triplicate, and in each experiment suitable positive controls and blank were added. Surfactants were tested separately.

4.8. Biofilm Assay

All isolates were grown overnight in MH broth (Sigma Aldrich, Saint Louis, MO). To allow the formation of biofilm, cells were diluted in Luria Bertani broth (LB, Sigma Aldrich, Saint Louis, MO, USA) to a turbidity of 0.5 McFarland, corresponding to 5×10^8. To study the activity of both GR-OLI and *O. vulgare* EO on the biofilm formation, both natural compounds were added in triplicate at the maximum concentration of 0.5% and 0.125% and at minimum concentration of 0.125% and 0.03%, respectively. The suspension was inoculated in polystyrene 96-well plates (Thermo Fisher Scientific, Waltham, MA, USA) and incubated at 37 °C for 48 h. No treated cells were added as a positive control in triplicate. Wells were then washed three times with PBS and the resultant biofilms were stained with crystal violet (CV) staining (Sigma-Aldrich, Saint Louis, MO, USA) for 30 min. The stained biofilms were washed in PBS and 100 µL of ethanol were added to each well for one minute to completely dissolve the CV. Then, the ethanol was transferred into a new 96-well plate to determine the absorbance at 560 nm. To test for disaggregation of biofilm, bacterial cells were prepared as described before without adding substances. After 48 h of incubation, biofilm was washed three times with PBS and cells fixated in acetone for 10 min. GR-OLI and *O. vulgare* were diluted in PBS at the maximum and minimum concentration aforementioned and added to the biofilm for another 24 h at 37 °C. Biofilm was then quantified as already described. Both tests were conducted in triplicate and repeated twice.

4.9. Cell Adhesion Assay

Two *Salmonella* spp. strains sensitive to almost all antibiotics (S33 and S42 resistant to only ciprofloxacin) and six multi-resistant strains (S24 and S32 *S.* Typhimurium, S19 and S29 monophasic *S.* Typhimurium and S40 and S41 S. Infantis) were randomly selected and used to study their adhesive capacity on human Caucasian colon adenocarcinoma cells (Caco-2) in the presence or absence of two concentrations of GR-OLI or *O. vulgare*.

The CACO-2 cell line was cultured in Dulbecco's modified Eagle medium (DMEM) (Gibco, Grand Island, NY, USA) supplemented with 10% heat-inactivated fetal calf serum (Integro B.V., Zaandam, The Netherlands), 1% nonessential amino acids (Gibco, Grand Island, NY, USA), and 1 mM glutamine (Gibco, Grand Island, NY, USA) and incubated at 37 °C with 5% CO_2. Differentiated CACO-2 cells were prepared by seeding cells 5 to 10 time in 250-mL flasks (Costar, Oneonta, NY, USA) at 1.6×10^7 cells/mL in DMEM, with all supplements and then transferred to 24-well tissue culture plates at 1.6×10^5 cells/mL. The culture medium was replaced every three days. Overnight grown cultures of *Salmonella* isolates were diluted (1:100) in the presence of 0.125% *v/v* and 0.5% *v/v* of GR-OIL (APA-CT, Forlì, Italy) and grown at 37 °C to an OD660 of 0.8. For each strain, an inoculum was added without adding the formulate as the control. Bacteria were harvested by centrifugation and suspended in Dulbecco's Modified Eagle's Medium to a final concentration of 1×10^7 CFU/mL. Then, a 1-mL bacterial suspension of each strain was added to the wells (1:100 MOI). Plates were incubated for 4 h at 37 °C. After incubation, monolayers were rinsed three times with PBS (phosphate buffer solution) and cells were gently scraped with a cell scraper (Falcon, Reynosa, Tamaulipas, Mexico) and harvested with PBS and washed by centrifugation twice. The adherent bacteria were quantified by plating serial dilutions on LB agar plates and counting CFU. The inoculum was plated to determine viable counts. The assay was performed in triplicate and repeated twice.

4.10. Growth Curves

As described in the broth microdilution susceptibility testing method, a suspension of 5×10^5 cfu/mL of the same strains used for the cell adhesion assay was seeded in a 96-well plate together with *O. vulgare* EO or GR-OLI at the MIC and sub-MIC concentrations or only with culture medium (growth control). Strains were incubated at 37 °C and monitored overnight by detecting OD450 every 30 min for 20 h. A statistical comparison between the OD450 detected at 10, 15, and 20 h of treated and untreated samples was made to quantify the extent of growth inhibition for each treatment.

4.11. Checkerboard Titration Method

Four strains of S. Infantis (S26, S35, S36, S37), all multi-resistant to amoxicillin/clavulanic acid, cefotaxime, cefepime, and ciprofloxacin, were tested using the checkerboard titration method. Then, 96-well microplates were used, each one containing MH broth with concentrations ranging from 12.5% *v/v* (equal to 31 µL of EOs content /mL) to 0.19% *v/v* (equal to 0.5 µL of EOs content /mL) for GR-OLI and from 128 to 0.125 µg/mL for amoxicillin/clavulanic acid or from 16 to 0.03 µg/mL for cefotaxime or from 4 to 0.005 µg/mL for ciprofloxacin and a combination of GR-OLI and one of the aforementioned antibiotics in a checkerboard style. The final inoculum was 5×10^5 cfu/well. The microplates were incubated for 24 h at 37 °C. After the incubation period, the MBCs were evaluated by sowing 5 µL of the contents of each well on nutrient agar and incubating it at 37 °C for 24 h. The FIC value could not be evaluated as due to the turbidity of the contents of the wells, it was not possible to define the MIC values, while the FBC index were calculated in compliance with international guidelines (EUCAST, 2000). Synergism was defined as FBC index <0.5; additivity FBC index between

0.5 and 1; indifference FBC index between 1 and 2; and antagonism FBC index > 2 [33] (EUCAST, 2000). Each experiment was performed in triplicate, independently.

4.12. Statistical Analysis

Relative data of biofilm inhibition and disaggregation (OD values) and of cell adhesion assays (CFU values) were plotted as means ± standard errors (SE). Means whose SE bars did not overlap were considered significantly different. Relative data of growth curves at 10, 15, and 20 h after treatment with *O. vulgare* EO and GR-OLI, which were shown to satisfy the conditions for ANOVA, and were subjected to one-way ANOVA within each *Salmonella* strain and time after treatment. The lowest significant difference (LSD) test at $p < 0.05$ was used to separate levels in strain/time combinations significant at the ANOVA.

5. Conclusions

In conclusion, although the current European legislation admits the use of EOs as flavouring for animal foods, no previous study shows their potential ability to break down the intestinal biofilm of *Salmonella* spp. in livestock, when used at the concentrations admitted for flavouring. Furthermore, our results strongly confirm our hypothesis, since sub-MIC concentrations of both natural compounds interfere with microbial adhesion to intestinal target cells. Finally, sub-MIC concentrations of GR-OLI were able to reactivate the sensitivity of multi-resistant *Salmonella* spp. strains to ciprofloxacin, one of the most used antibiotics in veterinary practices. Our in vitro data, although needing confirmation with in vivo studies, lay the foundations for a new potential use of essential oil-based flavours in the fight against zoonosis. Indeed, although there are many studies on the antimicrobial activity of EOs, in the last 5 years, there is a paucity of articles aimed at evaluating the antimicrobial activity of single EOs on Salmonella spp. strains isolated from chicken and pig farms, and no clinical trial about the in vivo activity of EOs, or a mixture of these, on Salmonella spp. colonization has been published. In this article, for the first time, the potential preventive use of an EOs mixture used in feed as "flavourings" (GR-OLI) against *Salmonella* spp. strains directly isolated from chicken and pig farms has been shown.

Author Contributions: Conceptualization, M.D.V., P.M., F.B. and Y.Y.; Methodology, M.D.V., M.C., L.B.; Formal Analysis, M.D.V., M.C., C.M., S.B.; Investigation, M.D.V., L.B.; Resources, G.T., L.F.; Data Curation, M.D.C., M.C., L.B.; Writing—Original Draft Preparation, M.D.V., M.C., L.B., M.S. (Maurizio Scozzoli); Writing—Review & Editing, P.M., F.B., M.S. (Maurizio Sanguinetti). All authors have read and agreed to the published version of the manuscript.

Funding: This research received no external funding.

Acknowledgments: Lori Morrison is gratefully acknowledged for manuscript proofreading.

Conflicts of Interest: The authors declare no conflict of interest.

References

1. Foley, S.L.; Lynne, A.M.; Nayak, R. Salmonella challenges: Prevalence in swine and poultry and potential pathogenicity of such isolates. *J. Anim. Sci.* **2008**, *86*, E149–E162. [CrossRef] [PubMed]
2. Boyen, F.; Haesebrouck, F.; Maes, D.; Van Immerseel, F.; Ducatelle, R.; Pasmans, F. Non-typhoidal Salmonella infections in pigs: A closer look at epidemiology, pathogenesis and control. *Vet. Microbiol.* **2008**, *130*, 1–19. [CrossRef] [PubMed]
3. European Food Safety Authority European Food Safety Authority and European Centre for Disease Prevention and Control (EFSA and ECDC). The European Union One Health 2018 Zoonoses Report. *EFSA J.* **2019**, *17*, 5926.
4. EFSA. Salmonella control in poultry flocks and its public health impact. *EFSA J.* **2019**, *17*. [CrossRef]
5. Pornsukarom, S.; Thakur, S. Horizontal Dissemination of Antimicrobial Resistance Determinants in Multiple Salmonella Serotypes following Isolation from the Commercial Swine Operation Environment after Manure Application. *Appl. Environ. Microbiol.* **2017**, *83*. [CrossRef]

6. EFSA. Opinion of the scientific panel on biological hazards on "risk assessment and mitigation options of Salmonella in pig production". *EFSA J.* **2006**, *4*, 341.
7. European Union Ban on Antibiotics as Growth Promoters in Animal Feed Enters into Effect. 2006. Available online: https://ec.europa.eu/commission/presscorner/detail/en/IP_05_1687 (accessed on 31 October 2020).
8. Thacker, P.A. Alternatives to antibiotics as growth promoters for use in swine production: A review. *J. Anim. Sci. Biotechnol.* **2013**, *4*, 35. [CrossRef]
9. Omonijo, F.A.; Ni, L.; Gong, J.; Wang, Q.; Lahaye, L.; Yang, C. Essential oils as alternatives to antibiotics in swine production. *Anim. Nutr. (Zhongguo Xu Mu Shou Yi Xue Hui)* **2018**, *4*, 126–136. [CrossRef]
10. International Organization for Standardization. *ISO 9235:2013(en): Aromatic Natural Raw Materials—Vocabulary*; International Organization for Standardization (ISO): Geneva, Switzerland, 2013.
11. Amerah, A.M.; Mathis, G.; Hofacre, C.L. Effect of xylanase and a blend of essential oils on performance and Salmonella colonization of broiler chickens challenged with Salmonella Heidelberg. *Poult. Sci.* **2012**, *91*, 943–947. [CrossRef]
12. Walia, K.; Argüello, H.; Lynch, H.; Leonard, F.C.; Grant, J.; Yearsley, D.; Kelly, S.; Duffy, G.; Gardiner, G.E.; Lawlor, P.G. Effect of strategic administration of an encapsulated blend of formic acid, citric acid, and essential oils on Salmonella carriage, seroprevalence, and growth of finishing pigs. *Prev. Vet. Med.* **2017**, *137*, 28–35. [CrossRef]
13. Zhang, S.; Shen, Y.R.; Wu, S.; Xiao, Y.Q.; He, Q.; Shi, S.R. The dietary combination of essential oils and organic acids reduces Salmonella enteritidis in challenged chicks. *Poult. Sci.* **2019**, *98*, 6349–6355. [CrossRef]
14. Si, W.; Gong, J.; Tsao, R.; Zhou, T.; Yu, H.; Poppe, C.; Johnson, R.; Du, Z. Antimicrobial activity of essential oils and structurally related synthetic food additives towards selected pathogenic and beneficial gut bacteria. *J. Appl. Microbiol.* **2006**, *100*, 296–305. [CrossRef] [PubMed]
15. Oussalah, M.; Caillet, S.; Salmiéri, S.; Saucier, L.; Lacroix, M. Antimicrobial effects of alginate-based films containing essential oils on Listeria monocytogenes and Salmonella typhimurium present in bologna and ham. *J. Food Prot.* **2007**, *70*, 901–908. [CrossRef]
16. Fratianni, F.; De Martino, L.; Melone, A.; De Feo, V.; Coppola, R.; Nazzaro, F. Preservation of chicken breast meat treated with thyme and balm essential oils. *J. Food Sci.* **2010**, *75*, M528–M535. [CrossRef] [PubMed]
17. Casco, G.; Taylor, T.M.; Alvarado, C. Evaluation of novel micronized encapsulated essential oil-containing phosphate and lactate blends for growth inhibition of Listeria monocytogenes and Salmonella on poultry bologna, pork ham, and roast beef ready-to-eat deli loaves. *J. Food Prot.* **2015**, *78*, 698–706. [CrossRef]
18. Boskovic, M.; Djordjevic, J.; Ivanovic, J.; Janjic, J.; Zdravkovic, N.; Glisic, M.; Glamoclija, N.; Baltic, B.; Djordjevic, V.; Baltic, M. Inhibition of Salmonella by thyme essential oil and its effect on microbiological and sensory properties of minced pork meat packaged under vacuum and modified atmosphere. *Int. J. Food Microbiol.* **2017**, *258*, 58–67. [CrossRef] [PubMed]
19. EFSA. Panel on Additives and Products or Substances used in Animal Feed (FEEDAP). Safety and efficacy of an essential oil from Origanum vulgare ssp. hirtum (Link) Ietsw for all animal species. *EFSA* **2019**, *17*, e05909.
20. Rodriguez-Garcia, I.; Silva-Espinoza, B.A.; Ortega-Ramirez, L.A.; Leyva, J.M.; Siddiqui, M.W.; Cruz-Valenzuela, M.R.; Gonzalez-Aguilar, G.A.; Ayala-Zavala, J.F. Oregano Essential Oil as an Antimicrobial and Antioxidant Additive in Food Products. *Crit. Rev. Food Sci. Nutr.* **2016**, *56*, 1717–1727. [CrossRef]
21. Numpaque, M.A.; Oviedo, L.A.; Gil, J.H.; García, C.M.; Durango, D.L. Thymol and carvacrol: Biotransformation and antifungal activity against the plant pathogenic fungi Colletotrichum acutatum and Botryodiplodia theobromae. *Trop. Plant Pathol.* **2011**, *36*, 3–13. [CrossRef]
22. Khan, A.; Ahmad, A.; Ahmad Khan, L.; Padoa, C.J.; van Vuuren, S.; Manzoor, N. Effect of two monoterpene phenols on antioxidant defense system in Candida albicans. *Microb. Pathog.* **2015**, *80*, 50–56. [CrossRef]
23. Gumus, T.; Demirci, A.S.; Sagdic, O.; Arici, M. Inhibition of heat resistant molds: Aspergillus fumigatus and Paecilomyces variotii by some plant essential oils. *Food Sci. Biotechnol.* **2010**, *19*, 1241–1244. [CrossRef]
24. Guarda, A.; Rubilar, J.F.; Miltz, J.; Galotto, M.J. The antimicrobial activity of microencapsulated thymol and carvacrol. *Int. J. Food Microbiol.* **2011**, *146*, 144–150. [CrossRef] [PubMed]
25. Van de Vel, E.; Sampers, I.; Raes, K. A review on influencing factors on the minimum inhibitory concentration of essential oils. *Crit. Rev. Food Sci. Nutr.* **2019**, *59*, 357–378. [CrossRef] [PubMed]
26. WHO Website. Available online: https://www.who.int/ (accessed on 31 October 2020).
27. CDC Website. Available online: https://www.cdc.gov/ (accessed on 31 October 2020).

28. Scallan, E.; Hoekstra, R.M.; Angulo, F.J.; Tauxe, R.V.; Widdowson, M.-A.; Roy, S.L.; Jones, J.L.; Griffin, P.M. Foodborne illness acquired in the United States—major pathogens. *Emerg. Infect. Dis.* **2011**, *17*, 7–15. [CrossRef]
29. International Organization for Standardization (ISO). *ISO 6579:2002: Microbiology of Food and Animal Feeding Stuffs–Horizontal Method for the Detection of Salmonella spp.–Amendment 1: Annex D: Detection of Salmonella spp. in Animal Faeces and in Environmental Samples from the Primary Production Stage*; International Organization for Standardization (ISO): Geneva, Switzerland, 2007.
30. International Organization for Standardization (ISO). *ISO/TR 6579-3:2014: Microbiology of the Food Chain–Horizontal Method for the Detection, Enumeration and Serotyping of Salmonella. Part 3: Guidelines For serotyping of Salmonella spp.*; International Organization for Standardization (ISO): Geneva, Switzerland, 2014.
31. Grimont, P.A.D.; Weill, F.X. Antigenic Formulae of the Salmonella Serovars. Available online: https://www.pasteur.fr/ip/portal/action/WebdriveActionEvent/oid/01s-000036-089 (accessed on 31 October 2020).
32. Adams, R. Identification of Essential Oil Components by Gas Chromatography/Quadrupole Mass Spectroscopy. *Carol Stream* **2007**, *16*, 65–120.
33. Rodriguez-Tudela, J.L.; Arendrup, M.C.; Barchiesi, F.; Bille, J.; Chryssanthou, E.; Cuenca Estrella, M.; Dannaoui, E.; Denning, D.W.; Donnelly, J.P.; Dromer, F.; et al. EUCAST definitive document EDef 7.1: Method for the determination of broth dilution MICs of antifungal agents for fermentative yeasts. *Clin. Microbiol. Infect.* **2008**, *14*, 398–405. [CrossRef] [PubMed]

antibiotics

MDPI

Article

Pimenta Oil as a Potential Treatment for *Acinetobacter baumannii* Wound Infection: In Vitro and In Vivo Bioassays in Relation to Its Chemical Composition

Maha M. Ismail [1,*], Reham Samir [1], Fatema R. Saber [2,*], Shaimaa R. Ahmed [2,3] and Mohamed A. Farag [2,4]

1 Microbiology and Immunology Department, Faculty of Pharmacy, Cairo University, Kasr El-Aini Street, Cairo 11562, Egypt; Reham.samer@pharma.cu.edu.eg

2 Department of Pharmacognosy, Faculty of Pharmacy, Cairo University, Kasr el-Aini Street, Cairo 11562, Egypt; shaimaa.ahmed@pharma.cu.edu.eg (S.R.A.); mohamed.farag@pharma.cu.edu.eg (M.A.F.)

3 Department of Pharmacognosy, College of Pharmacy, Jouf University, Sakaka 2014, Saudi Arabia

4 Department of Chemistry, School of Sciences & Engineering, The American University in Cairo, New Cairo 11835, Egypt

* Correspondence: maha.ismail@pharma.cu.edu.eg (M.M.I.); Fatema.Saber@pharma.cu.edu.eg (F.R.S.); Tel./Fax: +20-3628426 (ext. 00202) (F.R.S.)

Received: 11 September 2020; Accepted: 4 October 2020; Published: 7 October 2020

Abstract: Bacterial biofilm contributes to antibiotic resistance. Developing antibiofilm agents, more favored from natural origin, is a potential method for treatment of highly virulent multidrug resistant (MDR) bacterial strains; The potential of *Pimenta dioica* and *Pimenta racemosa* essential oils (E.Os) antibacterial and antibiofilm activities in relation to their chemical composition, in addition to their ability to treat *Acinetobacter baumannii* wound infection in mice model were investigated; *P. dioica* leaf E.O at 0.05 $\mu g \cdot mL^{-1}$ efficiently inhibited and eradicated biofilm formed by *A. baumannii* by 85% and 34%, respectively. Both *P. diocia* and *P. racemosa* leaf E.Os showed a bactericidal action against *A. baumanii* within 6h at 2.08 $\mu g \cdot mL^{-1}$. In addition, a significant reduction of *A. baumannii* microbial load in mice wound infection model was found. Furthermore, gas chromatography mass spectrometry analysis revealed qualitative and quantitative differences among *P. racemosa* and *P. dioica* leaf and berry E.Os. Monoterpene hydrocarbons, oxygenated monoterpenes, and phenolics were the major detected classes. β-Myrcene, limonene, 1,8-cineole, and eugenol were the most abundant volatiles. While, sesquiterpenes were found as minor components in *Pimenta* berries E.O; Our finding suggests the potential antimicrobial activity of *Pimenta* leaf E.O against MDR *A. baumannii* wound infections and their underlying mechanism and to be further tested clinically as treatment for MDR *A. baumannii* infections.

Keywords: *Acinetobacter baumannii*; MDR; biofilm; antimicrobial; *Pimenta*; *Myrtaceae*; wound infection; eugenol; 1,8-cineole; GC/MS

1. Introduction

Wound healing is a complex biological process in the human body, achieved through programmed phases. Potential factors, including infection, can interfere with these phases and impair healing. There is increasing evidence that bacteria can delay healing processes by living within biofilm communities in which the bacteria are protected from the host immune system and to further develop antibiotic drug resistance. Referring to the previous findings, wound infection is detrimental to wound healing and there is a great demand for the development of more effective treatment for infected and poorly healing wounds [1–3].

Acinetobacter baumannii is one of the most important opportunistic nosocomial pathogens. This pathogen was recorded, with high incidence, among immune-compromised patients as the causative agent of wound infections [2]. Based on its pathogenicity, including biofilm formation and antimicrobial resistance, *A. baumannii* is regarded as a multidrug-resistant pathogen (MDR) that usually complicates wound infection prognosis [4], the phenomenon that warrants further investigation of both nosocomial and community-acquired infections [5]. The World Health Organization recognizes antimicrobial resistance as one of the three most important issues facing human health [6]. The need to develop new spectrum antibiotics to tackle multidrug resistance is still ongoing. Plant-based antimicrobials present a potential source of medication that is successful in treating infectious diseases while minimizing many of the side effects that are frequently associated with synthetic antimicrobials [7,8].

Pimenta genus is an important myrtaceous taxa that encompasses 15 species, mostly found in the Americas Caribbean area, and commonly used for several medicinal purposes [9,10]. *Pimenta dioica* and *Pimenta racemosa* are the most recognized in that genus to exhibit potential pharmacological effects owing to its rich essential oil composition [11]. Traditionally, different *Pimenta* plant parts have been utilized for the alleviation of common cold, viral infections, bronchitis, dental and muscle aches, rheumatic pains, and arthritis. Furthermore, its berry oil is reported to exhibit antimicrobial, antiseptic, anesthetic, and analgesic properties [12]. Essential oils of *Pimenta* are ethno-pharmacologically relevant in traditional management of microbial infections [5,13]. *Pimenta* spp. antimicrobial and antiviral activities have been previously reported against several microbes including *Pseudomonas aeruginosa, Escherichia coli, Staphylococcus aureus*, and antifungal activity against *Candida albicans, Aspergillus niger*, and *Abisidia corymbifera*. *P. racemosa* is used in folk medicine for the treatment of influenza, pneumonia, fever rheumatism, toothache, and abdominal pains [14–16].

Studies on *P. dioica* leaf essential oil showed that eugenol, methyl eugenol, β-caryophyllene, and myrcene were the most abundant constituents, while limonene, 1,8-cineole, terpinolene, β-caryophyllene, β-selinene, and methyl eugenol were the major constituents in other studies [13,17,18].

In the preset study, antimicrobial and antibiofilm activities of *Pimenta* essential oils (E.Os) grown in Egypt were evaluated against *A. baumannii* using in vitro and in vivo models to assess their role as antimicrobial agents in wound infections. To the best of our knowledge, this is the first study reporting on the in vitro and in vivo antimicrobial efficacy of *Pimenta* oils against *A. baumannii* pathogen.

2. Results and Discussion

A major goal of this study was to assess the potential antimicrobial effect for *Pimenta* essential oils (E.Os) represented by its different organ or species type. The minimum inhibitory concentration (MIC) values of the E.Os obtained from *Pimenta racemosa* and *Pimenta dioica* leaves and berries against 15 bacterial isolates were in the range of 0.51–5.2 $\mu g \cdot mL^{-1}$ (Table 1). Among the tested plant parts, leaves' E.Os demonstrated the strongest antimicrobial activities in terms of smaller MIC values and were thus chosen for further in vivo wound infection. Eugenol exhibited the strongest antimicrobial activity with the lowest MIC values (ranging from 0.1 to 5.2 $\mu g \cdot mL^{-1}$ as compared to all tested E.Os). While *P. dioica* berry oil displayed the weakest antimicrobial activity (average MIC value 3.38 $\mu g \cdot mL^{-1}$) compared to other E.Os. Besides, MIC values of β-myrcene, a monoterpene hydrocarbon was higher than 4 $\mu g.mL^{-1}$ against all tested strains.

Although, the order of activity according to the average MIC values was as follows:

Eugenol > *P. dioica* leaf/*P. racemosa* leaf > *P. racemosa* berry > cefepime > *P. dioica* berry.

No statistically significant difference was observed among the activities of the four tested E.Os, eugenol, while a statistically significant difference existed between cefepime and each of the tested E.Os and eugenol (Kruskal–Wallis test, Dunn's multiple comparisons test, p-value < 0.0001).

Table 1. The minimum inhibitory concentration values (MIC), in μg·mL^{-1}, of tested *Pimenta* essential oils, eugenol, and cefepime antibiotic tested against the clinical isolates of *Acinetobacter baumanii* (AB), the results are represented as the mean of three replicates ± standard deviation.

Isolate	*P. dioica* Leaf [a]	*P. racemosa* Leaf [a]	*P. dioica* Berry [a]	*P. racemosa* Berry [a]	Eugenol [a]	Cefepime [b]
AB-1	0.69 ± 0.2	0.52 ± 0.0	0.69 ± 0.2	0.69 ± 0.2	0.10 ± 0.0	10 ± 0.0
AB-2	0.86 ± 0.2	4.84 ± 0.5	5.18 ± 0.0	0.69 ± 0.2	0.20 ± 0.0	5.0 ± 0.0
AB-3	5.18 ± 0.0	1.03 ± 0.0	1.03 ± 0.0	1.03 ± 0.0	0.53 ± 0.0	0.1 ± 0.0
AB-5	0.86 ± 0.2	0.86 ± 0.2	0.86 ± 0.2	0.86 ± 0.2	0.21 ± 0.0	5.0± 0.0
AB-6	0.69 ± 0.2	0.52 ± 0.0	0.69 ± 0.2	0.69 ± 0.2	0.53 ± 0.0	0.6 ± 0.0
AB-8	0.52 ± 0.0	0.52 ± 0.0	5.18 ± 0.0	0.69 ± 0.2	0.21 ± 0.0	6.7 ± 2.8
AB-9	0.52 ± 0.0	0.52 ± 0.0	4.49 ± 1.1	1.03 ± 0.0	0.53±0.0	5.0 ± 0.0
AB-10	5.18 ± 0.0	0.69 ± 0.2	5.18 ± 0.0	0.69 ± 0.2	0.53 ± 0.0	0.1 ± 0.0
AB-13	1.03 ± 0.0	0.52 ± 0.0	0.52 ± 0.0	1.03 ± 0.0	0.45 ± 0.0	1.3 ± 0.0
AB-15	1.03 ± 0.0	1.03 ± 0.0	5.18 ± 0.0	1.03 ± 0.0	5.3 ± 0.0	5.0 ± 0.0
AB-11Q	0.86 ± 0.2	1.03 ± 0.0	5.18 ± 0.0	1.03 ± 0.0	0.53 ± 0.0	0.1 ± 0.0
AB-14T	1.03 ± 0.0	4.66 ± 0.8	5.18 ± 0.0	5.18 ± 0.0	0.78 ± 0.2	0.1 ± 0.0
AB-16Q	1.03 ± 0.0	1.03 ± 0.0	5.18 ± 0.0	5.87 ± 1.1	0.88 ± 0.2	0.3 ± 0.0
AB-7T	4.49 ± 1.1	5.18 ± 0.0	5.18 ± 0.0	5.18 ± 0.0	0.53 ± 0.0	1.0 ± 0.30
ATCC 19606	0.69 ± 0.2	1.03 ± 0.0	0.86 ± 0.2	1.03 ± 0.0	0.47 ± 0.0	0.1 ± 0.0
Average MIC values (μg·mL^{-1})	1.65	1.60	3.38	1.78	0.79	2.70

([a,b]) A statistically significant difference exists between the antimicrobial activity of cefepime and each of the tested E.Os and eugenol (Kruskal-Wallis test-Dunn's multiple comparisons test, p-value < 0.0001).

Semiquantitative determination of biofilm formation by the examined five *A. baumannii* strains revealed that AB-R strain is a strong biofilm forming (BF), and both AB-8 and AB-16 are moderate BF while AB-11 and AB-13 strains are negative BF. Therefore, the effect of E.Os on biofilm formation and eradication was tested on three strains; AB-R, AB-8, and AB-16.

Unexpectedly, no inhibition activity was detected against the strong biofilm formed by the standard *A. baumanii* strain (AB-R). However, all *Pimenta* oils, as well as eugenol, could inhibit biofilm formation by AB-8 and AB-16, at 0.1× MIC (0.05 and 0.10 μg·mL^{-1}, respectively), except for *P. racemosa* leaf E.O, which could inhibit biofilm formed by AB-8 only at 0.05 μg·mL^{-1} (0.1× MIC). In detail, *P. dioica* leaf oil was found superior as antibiofilm relative to eugenol and other tested oils (Figure 1a). Comparing leaf to berry oil, *Pimenta* berry oils showed stronger effect as antibiofilm than that of *P. racemosa* leaf oil. The observed antibiofilm activity of *Pimenta* oils was higher than that of standard eugenol, though with no significant difference (one-way ANOVA, followed by Tukey's multiple comparisons p-value < 0.0001).

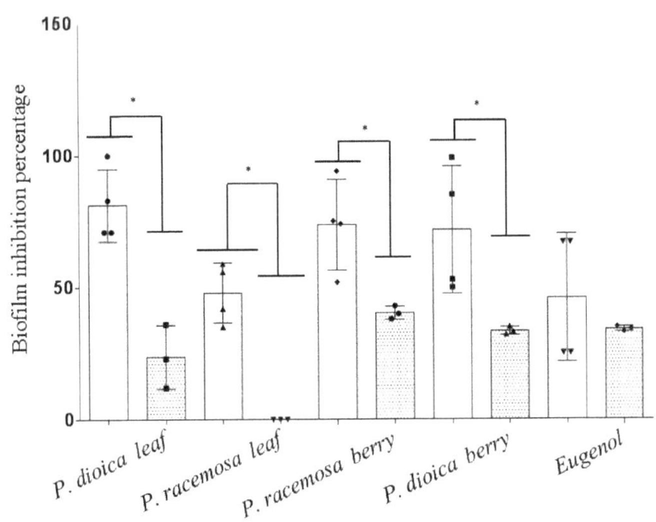

(a)

Figure 1. *Cont.*

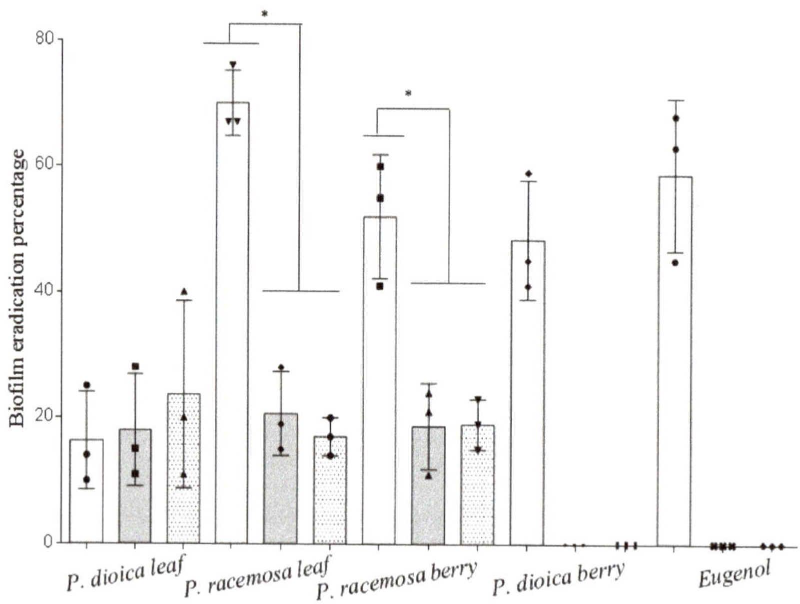

(b)

Figure 1. (a) Biofilm inhibition percentages of *Pimenta* essential oils/eugenol at a dose of 0.1× MIC against MDR AB-8 (white) and AB-16 (light shade) isolates. (b) Biofilm eradication percentages of the tested E.Os/eugenol at a dose of 0.1× MIC against MDR AB-8 (white), AB-R (gray), and AB-16 (light shade) isolates. The symbols in each column (•, ■, ▲, ◆, ▼, ×, +) represent each data point of the triplicate experiments performed for each E.O and eugenol against the tested bacterial isolates. Data represent the means of biofilm inhibition/eradication percentages ± SD, $n = 3$ (one-way ANOVA, followed by Tukey's multiple comparisons, p-value < 0.0001). (*) Means statistically significant difference exists between columns.

The same concentrations of all tested E.Os and eugenol, which inhibited biofilm formation, could also eradicate biofilm formed by AB-8, while only *P. dioica* leaf only, *P. racemosa* leaf and berry E.Os could eradicate biofilm formed by AB-R and AB-16 strains. Eradication of AB-8 strain biofilm by the E.Os and eugenol was more statistically significant than the eradication effect on biofilm formed by other strains (AB-R and AB-16), except for *P. dioica* leaf E.O (one-way ANOVA followed by Tukey's multiple comparisons p-value < 0.0001) (Figure 1b). The reported biofilm eradication activity of *Pimenta* E.Os herein is in agreement with Vázquez-Sánchez et al. [19] who demonstrated the eradication of 24h-old biofilm of *L. monocytogenes* by *Pimenta pseudochariophyllus* E.O at doses of 2.75% and 3% within 30 min of application. In another study, the same E.O could also eradicate 24h-old *Staphylococcus aureus* biofilm at doses of 1.75% and 3%. Eugenol and 1,8-cineol were among the main reported components of this E.O [20].

In this study, eugenol could eradicate the biofilm formed by AB-8 strain only and did not affect AB-R or AB-16 preformed biofilms. In a previous study performed by Yadav et al. [21], it was reported that eugenol (at 0.5× MIC) could eradicate MRSA and MSSA preformed biofilms, these authors reported that within 6 h, eugenol could reduce biomass and cell viability of the preformed biofilms by two mechanisms, the first is by lysis of bacterial cells within biofilms, and the second mechanism is via disruption of the cell-to-cell connections leading to biofilm organization disassembly.

The antimicrobial and antibiofilm activity of *Pimenta* spp. E.Os presented herein, is in agreement with recent studies reporting the potential of myrtaceous plants as efficient inhibitors of bacterial quorum sensing and to possess pronounced antimicrobial and antibiofilm activities [18,22]. A study performed by Vasavi et al. [23] revealed a strong antiquorum sensing activity for the ethyl acetate fraction of ethanolic extract of *P. dioica*, this fraction could inhibit AHL-mediated violacein pigment production in *Chromobacterium violaceum*, ATCC12472, at dose-dependent manner, the authors related this effect to polyphenolic compounds like the E.O components of the tested extract.

The minimum bactericidal concentration (MBC) values of *Pimenta* oils were estimated at different dose levels, where all of the tested oils exhibited bactericidal effect after 24 h incubation. *P. dioica* leaf oil exhibited the most pronounced bactericidal action at 1 μg·mL^{-1} (2× MIC) against the tested AB-8 isolate (Table 2). Both *P. racemosa* leaf and berry E.Os exhibited the same bactericidal activity at 2.08 and 2.76 μg·mL^{-1}, respectively. The bactericidal activity of eugenol was inferior to that of *P. dioica* leaf oil. In addition, *P. dioica* berries E.O at 41.4 μg·mL^{-1} (8× MIC) results in two-log cycle colony count reduction as compared to the control untreated bacteria. While, as a consequence, higher concentration, than 41.4 μg·mL^{-1}, of this E.O was required to achieve a complete bactericidal activity which was not tested.

Table 2. The determined MBC values of *Pimenta* spp. leaf and berry essential oils tested against *A. baumanii* isolate-8 (AB-8) and their MIC values equivalents.

Essential Oils/Standard Compound	MBC Value	MIC Values Equivalents
P. dioica leaf	1.0 μg·mL^{-1}	2× MIC
P. racemosa leaf	2.0 μg·mL^{-1}	4× MIC
P. racemosa berry	2.76 μg·mL^{-1}	4× MIC
P. dioica berry *	More than 41.4 μg·mL^{-1}	More than 8× MIC
Eugenol	1.2 μg·mL^{-1}	6× MIC

* Concentration equivalent to 8× MIC caused only 2-log cycle reduction of the bacterial load.

GC/MS analysis was employed to account for volatiles' compositions in tested *Pimenta* E.Os and mediating for the difference observed in the antimicrobial effects, results are depicted in Table 3. It was revealed that monoterpene hydrocarbons prevail in both *P. racemosa* and *P. dioica* leaf oils, constituting up to 64.4% predominance, with *β*-myrcene as the major identified component accounting for 39.6% and 44.1%, respectively (Figure 2a,b, Table 3).

Next, limonene was identified as a second major monoterpene hydrocarbon in *P. racemosa* leaf and berry at 14–15%, and to less extent in *P. dioica* berry oil (4.6%). In contrast, 1,8-cineole recorded the highest levels in *P. dioica* leaf oil (18.8%) compared to its berries, this was previously reported in the Cuban oil of *P. dioica* leaf (14.69%) [24]. 1,8-Cineole, among reported volatiles, is likely to mediate for the observed antimicrobial effects, in agreement with previously reported its potent antimicrobial effect against *Streptococcus mutans* [25]. In a recent study, Merghni et al. [26] recorded high MIC values displayed by 1,8-cineole against MRSA strains (≥29 mm) and suggestive for its positive effect against MDR bacteria. The antibiofilm activity of *P.dioica* leaf oil could be correlated to its enrichment in 1,8-cineole (18.80%). A study performed by Santos and Rao (2001) [27] reported the ability of 1,8-cineole to alleviate gastric mucosa injury and damage induced by absolute ethanol in rats. Stojkovic et al. [28] investigated the inhibitory effect of 1,8-cineole on apple rot disease caused by *Aspergillus niger* at concentrations of 3% and 6% of 1,8-cineole applied for 3 days, 100% inhibition of apple rot disease was observed. Merghni et al. [26] also reported 1,8-cineole biofilm inhibition effect against *S. aureus* ATCC 6538. Our results suggest possible synergistic antibiofilm activities of volatile oil constituents in *Pimenta* spp.

Eugenol dominated *P. dioica* berries essential oil composition (65.6%). Eugenol was found as a second predominant constituent in *P. racemosa* leaf and berry E.Os at comparable level of 27–31%. In the present study, it was observed that eugenol could inhibit *A. baumannii* biofilm formation at 0.02 μg·mL^{-1} which is in agreement with a study conducted by Kim et al. [29]. Where, more than 75% inhibition of *E. coli* O157:H7 biofilm was observed by eugenol, *P. racemose*, and *P. dioica* berries oils at slightly higher levels

(0.052 μg·mL^{-1}), though a different organism was tested. Noteworthy, eugenol is well-characterized as an antimicrobial agent against oral pathogens, infective diseases, and foodborne infections [30,31].

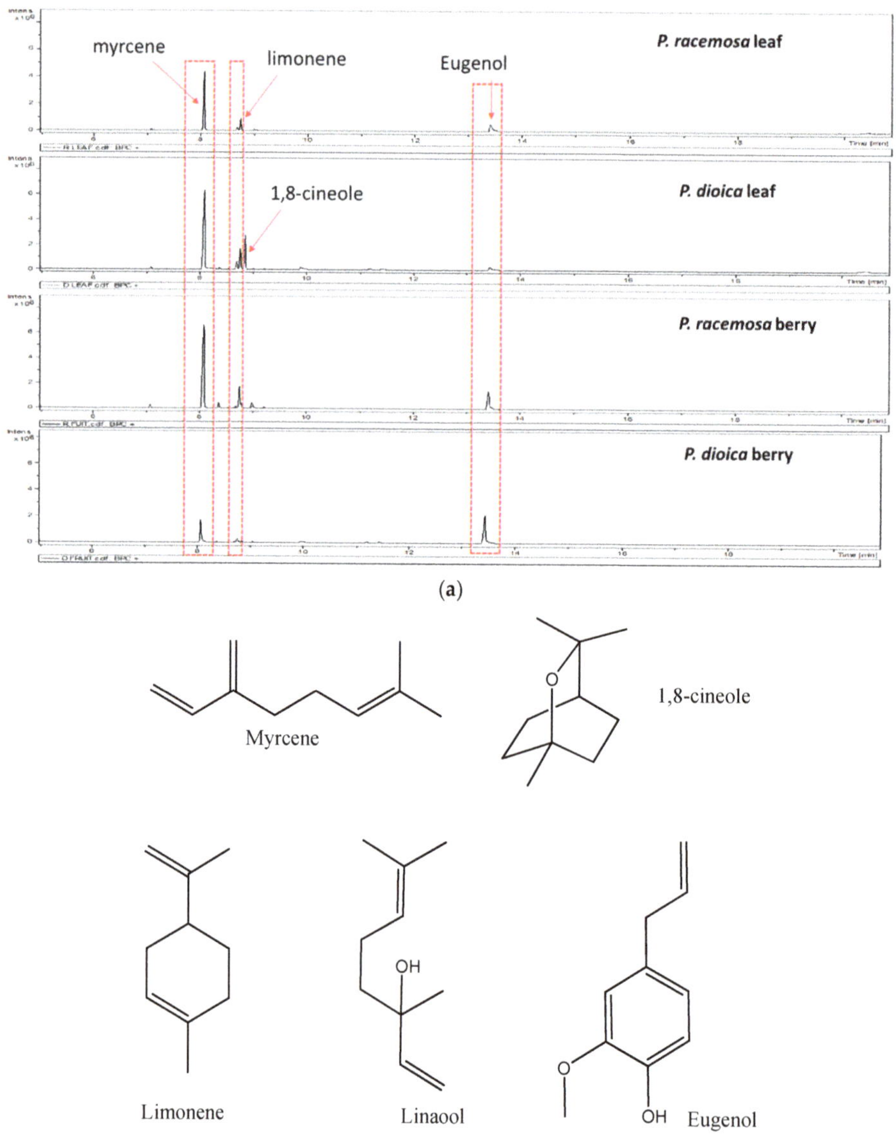

(b)

Figure 2. (**a**) Representative GC/MS chromatograms of *P. racemosa* and *P. dioica* leaf and berry essential oils. (**b**) Structures of the major volatile constituents detected in *P. racemosa* and *P. dioica* essential oils.

Further, sesquiterpene hydrocarbons were only detected in *Pimenta* berries oils at low levels, reaching 0.3% and 1.4% in *P. racemosa* and *P. dioica* berries, respectively. Major sesquiterpenes included δ-Cadinene in *P. dioica* berry oil followed by β-caryophyllene, α-humulene, germacrene D, and α-cubebene.

Although cefepime is considered to be the drug of choice for treatment of *A. baumanii* infections [32], its MIC values varied between 0.1 and 10 μg·mL^{-1} which can be attributed to the variable susceptibilities of the MDR isolates used herein.

Table 3. Relative percentile levels of volatiles identified in *Pimenta* berry and leaf essential oils using GC/MS.

RI * Calculated.	RI Reported	Name	Class	Essential Oil			
				P. racemosa Leaf	*P. dioica* Leaf	*P. racemosa* Berry	*P. dioica* Berry
1186	**1201**	**Decanal**	**aldehyde/ketone**	0.0	0.03	0.0	0.0
			Total aldehyde/ketone	0.0	0.03	0.0	0.0
906	932	α-Pinene		1.3	1.0	1.5	0.2
965.2	988	*β*-Myrcene		39.6	44.1	42.3	13.9
978.8	1003	*p*-Mentha-1(7),8-diene		0.9	0.2	2.9	0.0
980.9	1005	α-Phellandrene		1.0	0.6	0.6	1.0
991.2	1014	α-Terpinene		0.6	0.4	0.9	0.0
993	1014	4-carene		0.0	2.1	0.0	0.4
1001	1020	*p*-Cymene		2.3	2.1	0.9	0.4
1004.2	1024	Limonene		15.5	11.7	14.3	4.6
1019	1032	*β*-*cis*-Ocimene		2.8	0.6	4.6	1.1
1033	1054	*γ*-Terpinene		0.2	0.4	0.7	0.4
1063	1085	*p*-Mentha-2,4(8)-diene		0.1	0.0	0.4	0.3
			Total Monoterpene hydrocarbons	64.4	63.3	69.1	22.2
1009.5	1026	1,8-cineol	**Oxygenated monoterpene**	0.0	18.8	0.0	1.5
1080	1095	*β*-Linalool		2.1	5.3	1.6	3.6
1163	1174	Terpinen-4-ol		0.8	1.5	1.0	1.4
1181.5	1186	*α*-Terpineol		0.3	2.5	0.0	2.2
1250	1247	Chavicol		1.5	0.0	0.2	1.5
1250.1	1264	Geranial		0.0	0.0	0.0	0.6
			Total oxygenated Monoterpene	4.7	28.1	2.8	10.8
1334.5	1356	Eugenol	**Phenols**	31.0	8.6	27.7	65.6
1350	1345	α-Cubebene	**Sesquiterpene hydrocarbon**	0.0	0.0	0.0	0.1
1349	1374	α-Copaene		0.0	0.0	0.3	0.0
1396	1417	*β*-Caryophyllene		0.0	0.0	0.0	0.3
1430	1452	α-Humulene		0.0	0.0	0.0	0.2
1444.8	1484	Germacrene D		0.0	0.0	0.0	0.2
1478	1522	δ-Cadinene		0.0	0.0	0.0	0.6
			Total sesquiterpenes	0.0	0.0	0.3	1.4
			Total	100	100	100	100

RI *, retention index on DB-5-MS column relative to n-alkanes C8-C20.

Accordingly, killing kinetics of AB-8 strain by *P. dioica* and *P. racemosa* leaf oils were further investigated. Within 6h of incubation, both oils could kill AB-8 strain at 2 μg·mL^{-1} (4× MIC) (Figure 3), while longer time (16h) was needed for *P. dioica* leaf oil to kill the bacterium at 1 μg·mL^{-1} (2× MIC). In contrast, the E.O of *P. racemosa* leaf, at 1 μg·mL^{-1}, could not kill the bacteria nor achieve any level of bacterial colony count reduction.

The observed bactericidal action of eugenol is attributed to several mechanisms, involving alteration of bacterial membrane permeability causing leakage of ions leading to cell death [33], and in correlation to its phenolic nature which facilitates its binding to target proteins on the bacterial cell [34]. On the other hand, the monoterpene hydrocarbons *viz.*, *β*-myrcene, *α*-pinene, and limonene are generally known to exert less antibacterial effects in comparison to oxygenated monoterpenes and phenolics [35] and might act to synergize the main action of cineole and eugenol in *Pimenta* oil, thus accounting for the superior antibiofilm effect of the oil compared to eugenol standard tested alone (Figure 1). The lipophilic nature of terpene hydrocarbons greatly affects bacterial membrane permeability and might improve uptake of E.O components into bacterial cells enhancing its penetration into the bacterial biofilms and thus exerting its antimicrobial and antibiofilm action and further potentiating the effect of other concomitant antibiotics [36,37].

According to the obtained MIC, MBC values, and antibiofilm activities, both *P. dioica* and *P. racemosa* leaf oils were selected to test their in vivo effectiveness as antimicrobial agents in a mouse model of wound infection as they showed the lowest MIC and MBC values together with strong antibiofilm activity (Table 1, Figures 1 and 3).

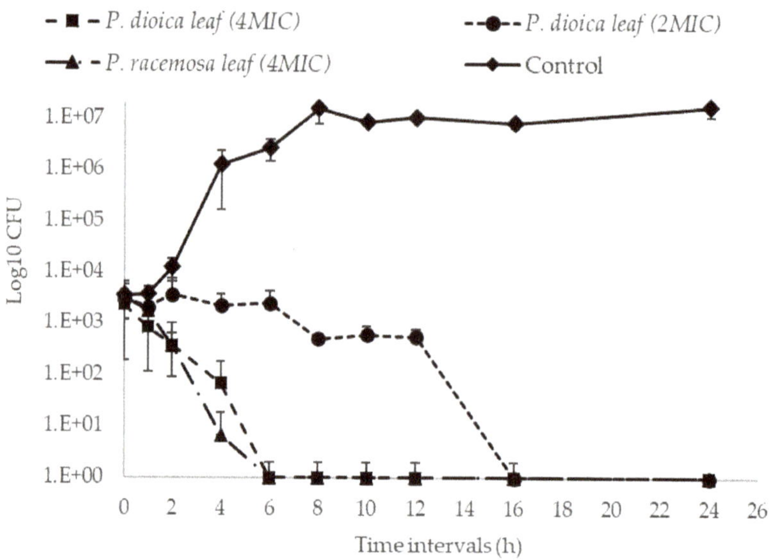

Figure 3. The killing kinetics of *Pimenta dioica* and *Pimenta racemosa* leaf essential oils tested against AB-8 isolate. (■) *P. dioica* leaf E.O. tested at concentration of 2.0 μg·mL^{-1} (4× MIC), (•) *P. dioica* leaf E.O. tested at concentration of 1.0 μg·mL^{-1} (2× MIC), (▲) *P. racemosa* leaf E.O. tested at concentration of 2.0 μg·mL^{-1} (4× MIC). Data is represented by means of the number of recovered colonies counted at each time point ± SD, $n = 3$.

Acinetobacter is regarded as one of the pathogens involved in nosocomial infections and outbreaks. In a previous epidemiological study, skin colonization with *Acinetobacter* spp. was found at a higher rate in hospitalized patients than in healthy people [38], thus a wound infection animal model was tested.

Interestingly, the microbial load of infected wounds in the mice groups treated with each of the E.Os and cefepime demonstrated a statistically significant reduction compared to other groups including the group treated by eugenol (one-way ANOVA followed by Tukey's multiple comparisons test, p-value = 0.0047). This is suggestive that the E.O is more active than standard eugenol due to possible synergistic action of volatile oil constituents. Three log cycle reduction was observed in the total bacterial count of groups treated with E.Os compared to infected PBS-treated control group (Figure 4). *P. dioica* leaf oil did not show a significantly higher activity than that of *P. racemosa* on reducing bacterial count recovered from both treatments. No significant differences were recorded among the infected untreated group and groups treated with either almond oil vehicle or eugenol indicating that eugenol was less active in this in vivo assay. Nevertheless, Kim et al. [29] showed that eugenol could prolong the lifespan of a nematode worm infected with EHEC. Yadav et al. [21]

reported eugenol, at sub-MIC, to be effective in reducing *S. aureus* colonization by 88% in a rat middle ear infection model.

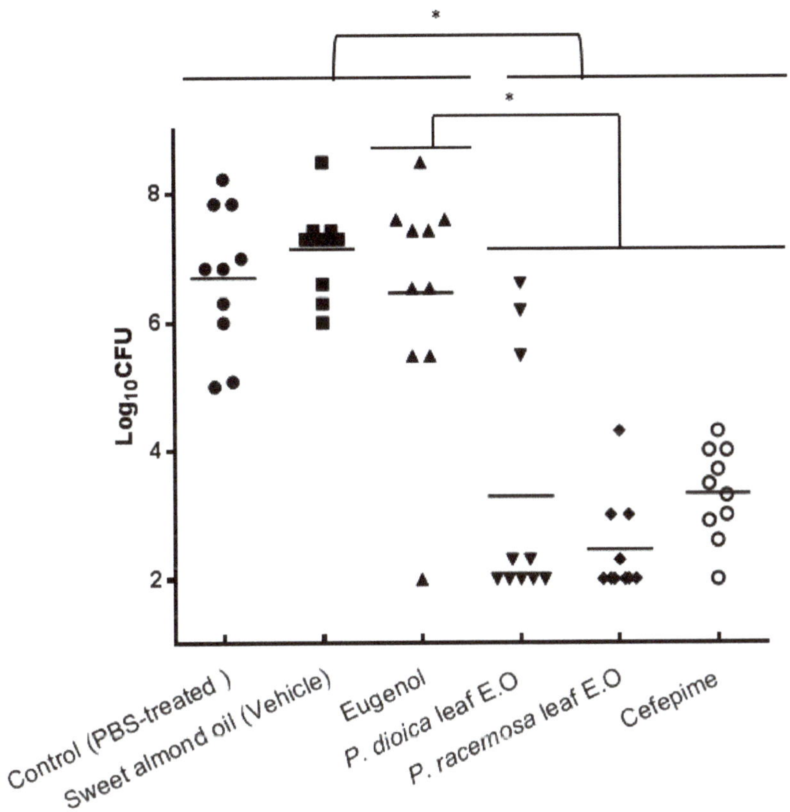

Figure 4. The antimicrobial effect of *Pimenta dioica* and *Pimenta racemosa* leaf essential oils and eugenol at 5.2 $\mu g \cdot mL^{-1}$ (10× MIC) concentration and cefepime at 25 $\mu g \cdot g^{-1}$ of mouse weight, tested against AB-8 isolate in a skin wound infection mouse model. Data is represented by the mean colony counts recovered from each of the six infected groups ± SD, six days post-infection. (•) data points of control infected PBS-treated group, (■) data points of sweet almond-treated group (negative (vehicle) control group), (▲) data points of eugenol-treated group, (▼) data points of *P. dioica* leaf E.O-treated group, (◆) data points of *P. racemosa* leaf E.O-treated group, (○) data points of cefepime-treated group (positive drug control group). $n = 10$ mice per group. (One-way ANOVA followed by Tukey's multiple comparisons test, p-value = 0.0047). (*) Means statistically significant difference exists between groups.

The antimicrobial activity of *P. dioica* leaf oil could be attributed to its 1,8-cineole enriched content in addition to eugenol [39]. Cefepime showed a better reduction of microbial load than that observed due to *P. racemosa* leaf E.O and a comparable effect to *P. dioica* leaf E.O but these differences were statistically nonsignificant (one-way ANOVA followed by Tukey's multiple comparisons test, p-value = 0.0047).

Another study performed by Karumathil et al. [40] revealed the in vitro ability of eugenol to reduce *A. baumannii* adhesion and invasion to human keratinocytes and also to inhibit biofilm formation in an in vitro collagen matrix wound model.

Nevertheless, few studies have investigated the in vivo antimicrobial effects of plant-derived E.Os against *A. baumannii* infections and likely to exert a stronger effect being composed of complex volatiles compared to a single chemical component. This study is the first to report the in vivo antimicrobial effect of *P. dioica* and *P. racemosa* E.Os in an *A. baumannii* wound infection model. In the same context, Tsai et al. [41] screened 30 Chinese herbs for antimicrobial effect against extensively drug-resistant *A. baumannii* isolates. Where, only *Scutellaria barbata* aqueous extract showed the highest activity in vitro and is likely to be mediated by polar type compounds in contrast to lipophilic nature of E.O reported herein and suggesting that different chemically based metabolites could evoke antimicrobial action against *A. baumannii*.

3. Materials and Methods

3.1. Plant Material

Leaves and berries of *Pimenta dioica* (L.) Merr. and *Pimenta racemosa* (Mill.) J.W. Moore were collected from Zohria Botanical Garden in May 2017. They were identified by Mrs. Therese Labib, Consultant of Plant Taxonomy at the Ministry of Agriculture and Orman Botanical Garden, Giza, Egypt. A voucher specimen of the plant material was deposited at Pharmacognosy Department, Faculty of Pharmacy, Cairo University, Egypt.

3.2. Preparation of Essential Oil

Leaf and berry of *Pimenta* species under investigation were hydro-distilled separately in a Clevenger-type apparatus for 4h, according to the procedure described in the Egyptian pharmacopeia (2005) [42]. The obtained oils were dehydrated by filtration through anhydrous sodium sulfate and kept in a refrigerator for subsequent GC/MS analysis and antimicrobial assays.

3.3. GC/MS Analysis of the Volatile Oil Composition

E.Os. prepared from *P. dioica* and *P. racemosa* leaf and berry were subjected to GC/MS analysis. The injection volume was 1 µL. The instrument was controlled by the Shimadzu Class-5000 Version 2.2 software containing a NIST62 (National Institute of Standards and Technology) MS library. Volatile components were separated on a DB5-MS column (30 m length, 0.25 mm inner diameter, and 0.25 µm film thickness (J&W Scientific, Santa Clara, CA, USA). Injections were made in the split mode for 30 s, and the gas chromatograph was operated under the following conditions: injector 220 °C and column oven 40 °C for 3 min, then programmed at a rate of 12 °C/min to 180 °C, kept at 180 °C for 5 min, and finally ramped at a rate of 40 °C/min to 220 °C and kept for 2 min, He carrier gas at 1 mL/min. The transfer line and ion-source temperatures were adjusted at 230 and 180 °C, respectively. The HP quadrupole mass spectrometer was operated in the electron ionization mode at 70 eV. The scan range was set at 40–500 *m/z*.

The percentages of different components in each oil sample were determined by computerized peak area measurements relative to each other. Volatile components were identified using the procedure described in Farag and Wessjohann [43]. The peaks were first deconvoluted using AMDIS software (www.amdis.net) and identified by its retention indices (RI) relative to n-alkanes (C6–C20), mass spectrum matching to NIST, WILEY library database. Results are depicted in Table 2.

3.4. Bacterial Isolates and Culture Conditions

Fourteen MDR clinical isolates of *Acinetobacter baumanii*, in addition to one reference strain (ATCC 19606), were used in this study. They were isolated from patients in Kasr El-Ainy hospital, Cairo, Egypt [44]. The strains were grown aerobically on Luria–Bertani (L.B) agar/broth (Oxoid, UK) with shaking at 180 rpm at 37 °C for 24 h.

3.5. Determination of the Minimum Inhibitory Concentration (MIC) by Agar Microdilution Technique

MIC was determined according to the method described by Golus et al. [45]. Briefly, freshly prepared two-fold serial dilutions of the test essential oil (E.O.)/standard volatiles (eugenol and myrcene) were prepared in dimethyl sulfoxide (DMSO). In total, 1 µL of each E.O dilution was dispensed into the

U-shaped bottom 96-well microplates. Aliquots of 100 µL of the molten L.B agar medium (at ≈60 °C) were mixed with E.O. in each well before solidification. Then, 2 µL of a freshly prepared bacterial suspension (in sterile normal saline (10^7 CFU·mL^{-1})) was inoculated onto the surface of the solidified mixture (L.B agar medium + E.O/standard compound).

The lowest concentration showing no visible bacterial growth after incubation at 30 °C for 24 h were recorded. Negative and positive controls (L.B agar medium + DMSO and L.B agar medium + DMSO + bacteria, respectively) and antibiotic control (cefepime) were included. The test was performed in triplicate.

3.6. *Effect of Essential Oils on Biofilm Formation and Eradication*

Five *A. baumannii* strains, namely AB-8, AB-R, AB-11, AB-13, and AB-16, were semiquantitatively examined for their biofilm forming ability in a 96-well flat bottom microtiter plate according to a formula reported by Naves et al. [46]:

$$\mathbf{BF} = \mathbf{AB} - \mathbf{CW}$$

where:

BF, biofilm formation

AB, stained bacteria cells attached to the wells

CW, stained control wells

Bacterial isolate is considered a strong, moderate, weak, or negative biofilm forming if BF is ≥0.300, 0.200–0.299, 0.100–0.199, and <0.100, respectively [46].

Then, strong or moderate biofilm forming strains were selected to perform biofilm inhibition/eradication assays.

3.6.1. Biofilm Inhibition Assay

The assay was performed according to Hussein et al. [47]. To estimate the ability of E.O/standard to inhibit biofilm formation by bacteria, an overnight culture of the corresponding isolate in L.B broth, adjusted at OD_{600} of 1, was diluted 1:100 in fresh LB broth. Then, 200 µL of each culture were dispensed in the wells of 96-well flat-shaped bottom microplates. A total of 1 µL of sub-MIC concentrations of each E.O./standard compound (0.5× MIC, 0.25× MIC, and 0.1× MIC) was mixed with LB broth in the wells and then incubated overnight at 37 °C. Growth was monitored by recording the optical density at 600 nm using microplate reader (BioTek Synergy 2, Winooski, VTUSA). Plates were then washed gently twice with phosphate-buffered saline to remove planktonic cells without disturbing the biofilm. Plates were left to completely dry in a laminar air flow cabinet. For biofilm visualization, 200 µL of 0.5% crystal violet solution were added to each well and left still for 30 min. Plates were then washed twice with distilled water to remove excess stain and left to dry completely. To extract the color of crystal violet, 150 µL of 99% ethanol were added per well and plates were left for 15 min with gentle shaking. The extracted color was measured colorimetrically at 570 nm in order to estimate the extent of biofilm formation. Negative and positive controls (broth only and broth inoculated with bacteria, respectively) were included. The test was performed in triplicate.

3.6.2. Biofilm Eradication Assay

Test strains were allowed to form mature biofilm for 24 h in the wells of 96-well flat-shaped bottom microplates under the same conditioned mentioned in biofilm inhibition assay without any E.O or inhibitory compound. After that, the spent media containing the planktonic cells were discarded and fresh L.B broth with different concentrations of the test E.O./standard compound (0.5× MIC, 0.25× MIC, and 0.1× MIC) were added to the mature biofilm in the wells. The plates were further incubated for 24 h at 37 °C, then they were washed and stained for biofilm visualization according to the steps mentioned in biofilm inhibition assay. Negative and positive controls (broth only and broth inoculated with bacteria, respectively) were included. The test was performed in triplicate.

3.7. Determination of the Minimum Bactericidal Concentration (MBC) by Broth Microdilution Technique

Amounts of 100 µL of L.B broth were dispensed into the U-shaped bottom 96-well microplates, concentrations equivalent to 1× MIC, 2× MIC, 4× MIC, 6× MIC, and 8× MIC of each E.O/standard compound, against *A. baumannii* strain-8 (AB-8), were tested. In total, 2 µL of a freshly prepared bacterial suspension (10^7 CFU·mL^{-1}) was inoculated into the mixture, plates were then incubated at 37 °C for 24 h. Viable colony count on L.B agar medium was performed to determine the MBC at which no colonies appear. Negative and positive controls were included (L.B broth + DMSO and L.B broth + DMSO + bacteria, respectively). The test was performed in triplicate.

3.8. Kill Kinetics Assay

Time–kill kinetics of *P. dioica* and *P. racemosa* leaf E.O/standard compound against AB-8 strain was performed according to the method of [48]. The killing kinetics of the E.O were assayed at the bactericidal concentrations. First, 2 µL of 10^7 CFU·mL^{-1} suspension of AB-8 strain was incubated with MBCs of E.Os in L.B broth for up to 24 h. Then, samples of each test were withdrawn at time intervals of 0, 1, 2, 4, 6, 8, 10, 12, 16, and 24 h, diluted and subjected to viable colony count on L.B agar medium. Plates were then incubated at 37 °C for 24 h and then the visible colonies were counted. Positive and negative controls of DMSO + AB-8 in L.B broth and DMSO in L.B broth, respectively, were included. The assay was performed in triplicate.

3.9. In Vivo Wound Infection Animal Model

3.9.1. Ethical Statement

All experiments involving animals were conducted according to the ethical policies and procedures approved by the ethics committee of the Faculty of Pharmacy, Cairo University, Cairo, Egypt (Approval no. MI-2364).

3.9.2. Experimental Design and Induction of Infection

The animal model was performed according to Wang et al. [49] as follows: 60 adult 6–8 weeks old male mice weighing 25–35 g were obtained from the Modern Veterinary Office for Laboratory Animals, Giza, Egypt. Animals were kept in cages under well-defined and standardized conditions (humidity and temperature-controlled room; 12-h light and 12-h dark cycle). The mice were initially examined to exclude any sign of skin inflammation and were fed with a standard dry food and water on demand.

Back hair was clipped from the cervical to mid-lumbar dorsum, and the skin was rinsed with ethanol. The skin on the shaved back of the mice was lifted with forceps and a 1.0 × 1.0 cm full thickness excisional wound was made by removing full thickness skin with a scissors. Infection was induced by adding aliquots of 10 µL freshly prepared suspension of isolate AB-8 in LB broth containing 10^7–10^8 CFU·mL^{-1} into the wound and allowed to be absorbed (the experiment was performed twice, once using 10^7 and then using 10^8 CFU·mL^{-1}. Each time, the used inoculum was determined by viable colony count at time zero). *P. dioica* and *P. racemosa* leaf E.O/eugenol were prepared as 10× MIC which is equivalent to 5× MBC (5.2 µg·mL^{-1}) in sweet almond oil as the vehicle. Cefepime was used as a positive drug control at a dose of 25 µg·g^{-1} of mice weight [31].

Twenty-four hours post-infection, animals were divided into six groups, 10 mice each, and treatment proceeded for 6 days.

Each group received 25 μL of respective treatment:

- Group 1: *P. dioica* leaf E.O. dissolved in sweet almond oil (5.2 $\mu g{\cdot}mL^{-1}$).
- Group 2: *P. racemosa* leaf E.O. dissolved in sweet almond oil (5.2 $\mu g{\cdot}mL^{-1}$).
- Group 3: Eugenol dissolved in sweet almond oil (2 $\mu g{\cdot}mL^{-1}$)
- Group 4: Cefepime solution (25 μg per gram of mouse weight).
- Group 5: Vehicle (sweet almond oil).
- Group 6: Phosphate-buffered saline.

Animals were then sacrificed and the wounded skin was removed for viable count technique (counts recovered from groups were compared), in brief, the wounded skins were cut into small pieces using sterile scalpels and then homogenized with PBS. Then, 20 μL aliquots of each suspension were 10-fold serially diluted using PBS, then, 10 μL of each dilution was spotted onto the surface of L.B agar medium, incubated at 37 °C for 24 h and then, the recovered colonies were counted and groups compared.

3.10. Statistical Analysis

One-way ANOVA, followed by Tukey's multiple comparisons and Kruskal–Wallis test, Dunn's multiple comparisons test (p-value < 0.05) were performed using GraphPad Prism 6.01 (GraphPad Software, Inc., San Diego, CA, USA).

4. Conclusions

Chemical characterization of *Pimenta racemosa* and *Pimenta dioica* leaf and berry E.Os revealed the abundance of oxygenated monoterpenes and phenolics' which are likely to contribute to the oil antimicrobial effect. The strong antimicrobial and antibiofilm activities of both *P. dioica* and *P. racemosa* leaf oils pose them for future incorporation in treatment of MDR *A. baumanii* infections. Our future perspective will focus on designing and characterization of a suitable pharmaceutical nano-formulation containing the *Pimenta* E.Os, to be further up-scaled to be incorporated in antimicrobial and antibiofilm adjuvant therapy after sufficient clinical trials.

Author Contributions: Conceptualization, M.A.F., F.R.S. and M.M.I.; methodology, M.A.F., M.M.I., F.R.S., R.S. and S.R.A.; software, M.A.F., F.R.S. and M.M.I.; validation, F.R.S., M.M.I. and R.S.; formal analysis, M.M.I. and F.R.S.; investigation, F.R.S., M.M.I. and R.S.; resources, M.A.F.; data curation, M.A.F., F.R.S, M.M.I. and R.S., writing—original draft preparation, F.R.S., M.M.I., R.S. and S.R.A.; writing—review and editing M.A.F., F.R.S., M.M.I., R.S. and S.R.A.; visualization, M.A.F., F.R.S., M.M.I. and R.S.; supervision, M.A.F.; project administration, M.A.F. All authors have read and agreed to the published version of the manuscript.

Funding: This research received no external funding, this work was supported by Faculty of Pharmacy, Cairo University.

Acknowledgments: The authors would like to acknowledge Mona Tawfik Al-Kashef and Omneya Helmy, Microbiology and Immunology Faculty of Pharmacy, Cairo University, Egypt, who generously donated the clinical bacterial isolates used in this study.

Conflicts of Interest: The authors declare no conflict of interest.

References

1. Guo, S.A.; DiPietro, L.A. Factors affecting wound healing. *J. Dent. Res.* **2010**, *89*, 219–229. [CrossRef] [PubMed]
2. Mihu, M.R.; Sandkovsky, U.; Han, G.; Friedman, J.M.; Nosanchuk, J.D.; Martinez, L.R. The use of nitric oxide releasing nanoparticles as a treatment against Acinetobacter baumannii in wound infections. *Virulence* **2010**, *1*, 62–67. [CrossRef] [PubMed]
3. Edwards, R.; Harding, K.G. Bacteria and wound healing. *Curr. Opin. Infect. Dis.* **2004**, *17*, 91–96. [CrossRef] [PubMed]
4. Howard, A.; O'Donoghue, M.; Feeney, A.; Sleator, R.D. *Acinetobacter baumannii*: An emerging opportunistic pathogen. *Virulence* **2012**, *3*, 243–250. [CrossRef]
5. Peleg, A.Y.; Seifert, H.; Paterson, D.L. *Acinetobacter baumannii*: Emergence of a successful pathogen. *Clin. Microbiol. Rev.* **2008**, *21*, 538–582. [CrossRef]

6. Bassetti, M.; Ginocchio, F.; Mikulska, M. New treatment options against gram-negative organisms. *Crit. Care* **2011**, *15*, 501–515. [CrossRef]
7. Anand, U.; Jacobo-Herrera, N.; Altemimi, A.; Lakhssassi, N. A comprehensive review on medicinal plants as antimicrobial therapeutics: Potential avenues of biocompatible drug discovery. *Metabolites* **2019**, *9*, 258. [CrossRef]
8. Parekh, J.; Jadeja, D.; Chanda, S. Efficacy of aqueous and methanol extracts of some medicinal plants for potential antibacterial activity. *Turk. J. Biol.* **2006**, *29*, 203–210.
9. Morton, J.F. *Atlas of Medicinal Plants of Middle America: Bahamas to Yucatan*; Charles C. Thomas: Springfield, IL, USA, 1981.
10. Chaverri, C.; Cicció, J.F. Leaf and fruit essential oil compositions of *Pimenta guatemalensis* (Myrtaceae) from Costa Rica. *Rev. De Biol. Trop.* **2015**, *63*, 303–311. [CrossRef]
11. Contreras-Moreno, B.Z. Chemical Composition of Essential Oil of Genus Pimenta (Myrtaceae). In *Potential of Essential Oils*; El-Shemy, H., Ed.; IntechOpen Limited: London, UK, 2018; p. 21. [CrossRef]
12. Seidemann, J. *World Spice Plants: Economic Usage, Botany, Taxonomy*; Springer Science & Business Media: Berlin, Germany, 2005.
13. Oussalah, M.; Caillet, S.; Saucier, L.; Lacroix, M. Antimicrobial effects of selected plant essential oils on the growth of a *Pseudomonas putida* strain isolated from meat. *Meat Sci.* **2006**, *73*, 236–244. [CrossRef]
14. Tucker, A.O.; Maciarello, M.J.; Adams, R.P.; Landrum, L.R.; Zanoni, T.A. Volatile leaf oils of Caribbean Myrtaceae. I. Three varieties of *Pimenta racemosa* (Miller) J. Moore of the Dominican Republic and the commercial bay oil. *J. Essent. Oil Res.* **1991**, *3*, 323–329. [CrossRef]
15. Ayedoun, A.M.; Adeoti, B.S.; Setondji, J.; Menut, C.; Lamaty, G.; Bessiére, J.-M. Aromatic plants from Tropical West Africa. IV. Chemical composition of leaf oil of *Pimenta racemosa* (Miller) JW Moore var. racemosa from Benin. *J. Essent. Oil Res.* **1996**, *8*, 207–209. [CrossRef]
16. Alitonou, G.A.; Noudogbessi, J.-P.; Sessou, P.; Tonouhewa, A.; Avlessi, F.; Menut, C.; Sohounhloue, D. Chemical composition and biological activities of essential oils of *Pimenta racemosa* (Mill.) JW Moore. from Benin. *Int. J. Biosci.* **2012**, *2*, 1–12.
17. Tucker, A.O.; Maciarello, M.J.; Landrum, L.R. Volatile leaf oils of Caribbean Myrtaceae. II. *Pimenta dioica* (L.) Merr. of Jamaica. *J. Essent. Oil Res.* **1991**, *3*, 195–196. [CrossRef]
18. Prasad, M.A.; Zolnik, C.P.; Molina, J.J. Leveraging phytochemicals: The plant phylogeny predicts sources of novel antibacterial compounds. *Future Sci. OA* **2019**, *5*, FSO407. [CrossRef] [PubMed]
19. Vázquez-Sánchez, D.; Galvão, J.A.; Ambrosio, C.M.; Gloria, E.M.; Oetterer, M. Single and binary applications of essential oils effectively control *Listeria monocytogenes* biofilms. *Ind. Crop. Prod.* **2018**, *121*, 452–460. [CrossRef]
20. Vázquez-Sánchez, D.; Galvão, J.A.; Mazine, M.R.; Gloria, E.M.; Oetterer, M. Control of *Staphylococcus aureus* biofilms by the application of single and combined treatments based in plant essential oils. *Int. J. Food Microbiol.* **2018**, *286*, 128–138. [CrossRef]
21. Yadav, M.K.; Chae, S.-W.; Im, G.J.; Chung, J.-W.; Song, J.-J. Eugenol: A phyto-compound effective against methicillin-resistant and methicillin-sensitive *Staphylococcus aureus* clinical strain biofilms. *PLoS ONE* **2015**, *10*, e0119564. [CrossRef]
22. Shehabeldine, A.M.; Ashour, R.M.; Okba, M.M.; Saber, F.R. *Callistemon citrinus* bioactive metabolites as new inhibitors of methicillin-resistant *Staphylococcus aureus* biofilm formation. *J. Ethnopharmacol.* **2020**, *254*, 112669. [CrossRef]
23. Vasavi, H.S.; Arun, A.B.; Rekha, P.D. Inhibition of quorum sensing in *Chromobacterium violaceum* by *Syzygium cumini* L. and *Pimenta dioica* L. *Asian Pac. J. Trop. Biomed.* **2013**, *3*, 954–959. [CrossRef]
24. Hernández Díaz, L.; Rodríguez Jorge, M.; García, D.; Pino Alea, J. Actividad antidermatofítica in vitro de aceites esenciales. *Rev. Cuba. Plantas Med.* **2003**, *8*.
25. Goldbeck, J.C.; do Nascimento, J.E.; Jacob, R.G.; Fiorentini, Â.M.; da Silva, W.P. Bioactivity of essential oils from *Eucalyptus globulus* and *Eucalyptus urograndis* against planktonic cells and biofilms of *Streptococcus mutans*. *Ind. Crop. Prod.* **2014**, *60*, 304–309. [CrossRef]
26. Merghni, A.; Noumi, E.; Hadded, O.; Dridi, N.; Panwar, H.; Ceylan, O.; Mastouri, M.; Snoussi, M. Assessment of the antibiofilm and antiquorum sensing activities of *Eucalyptus globulus* essential oil and its main component 1, 8-cineole against methicillin-resistant *Staphylococcus aureus* strains. *Microb. Pathog.* **2018**, *118*, 74–80. [CrossRef] [PubMed]

27. Santos, F.; Rao, V. 1, 8-cineol, a food flavoring agent, prevents ethanol-induced gastric injury in rats. *Digest. Dis. Sci.* **2001**, *46*, 331–337. [CrossRef] [PubMed]
28. Stojković, D.; Soković, M.; Glamočlija, J.; Džamić, A.; Ćirić, A.; Ristić, M.; Grubišić, D. Chemical composition and antimicrobial activity of *Vitex agnus-castus* L. fruits and leaves essential oils. *Food Chem.* **2011**, *128*, 1017–1022.
29. Kim, Y.-G.; Lee, J.-H.; Gwon, G.; Kim, S.-I.; Park, J.G.; Lee, J. Essential oils and eugenols inhibit biofilm formation and the virulence of *Escherichia coli* O157: H7. *Sci. Rep.* **2016**, *6*, 36377. [CrossRef]
30. Dhara, L.; Tripathi, A. Antimicrobial activity of eugenol and cinnamaldehyde against extended spectrum beta lactamase producing enterobacteriaceae by in vitro and molecular docking analysis. *Eur. J. Integr. Med.* **2013**, *5*, 527–536. [CrossRef]
31. Marchese, A.; Barbieri, R.; Coppo, E.; Orhan, I.E.; Daglia, M.; Nabavi, S.F.; Izadi, M.; Abdollahi, M.; Nabavi, S.M.; Ajami, M. Antimicrobial activity of eugenol and essential oils containing eugenol: A mechanistic viewpoint. *Crit. Rev. Microbiol.* **2017**, *43*, 668–689. [CrossRef]
32. Cunha, B.A. Optimal therapy for multidrug-resistant *Acinctobacter baumannii*. *Emerg. Infect. Dis.* **2010**, *16*, 170–171. [CrossRef]
33. Devi, K.P.; Sakthivel, R.; Nisha, S.A.; Suganthy, N.; Pandian, S.K. Eugenol alters the integrity of cell membrane and acts against the nosocomial pathogen *Proteus mirabilis*. *Arch. Pharmacal Res.* **2013**, *36*, 282–292. [CrossRef]
34. Burt, S. Essential oils: Their antibacterial properties and potential applications in foods—A review. *Int. J. Food Microbiolology* **2004**, *94*, 223–253. [CrossRef] [PubMed]
35. Trombetta, D.; Castelli, F.; Sarpietro, M.G.; Venuti, V.; Cristani, M.; Daniele, C.; Saija, A.; Mazzanti, G.; Bisignano, G. Mechanisms of antibacterial action of three monoterpenes. *Antimicrob. Agents Chemother.* **2005**, *49*, 2474–2478. [CrossRef] [PubMed]
36. Liu, Q.; Niu, H.; Zhang, W.; Mu, H.; Sun, C.; Duan, J. Synergy among thymol, eugenol, berberine, cinnamaldehyde and streptomycin against planktonic and biofilm-associated food-borne pathogens. *Lett. Appl. Microbiolol.* **2015**, *60*, 421–430. [CrossRef] [PubMed]
37. Freitas, P.R.; de Araújo, A.C.J.; dos Santos Barbosa, C.R.; Muniz, D.F.; da Silva, A.C.A.; Rocha, J.E.; de Morais Oliveira-Tintino, C.D.; Ribeiro-Filho, J.; da Silva, L.E.; Confortin, C. GC-MS-FID and potentiation of the antibiotic activity of the essential oil of *Baccharis reticulata* (ruiz & pav.) pers. and α-pinene. *Ind. Crop. Prod.* **2020**, *145*, 112106. [CrossRef]
38. Seifert, H.; Dijkshoorn, L.; Gerner-Smidt, P.; Pelzer, N.; Tjernberg, I.; Vaneechoutte, M. Distribution of *Acinetobacter* species on human skin: Comparison of phenotypic and genotypic identification methods. *J. Clin. Microbiol.* **1997**, *35*, 2819–2825. [CrossRef] [PubMed]
39. Aldoghaim, F.S.; Flematti, G.R.; Hammer, K.A. Antimicrobial activity of several cineole-rich Western Australian Eucalyptus essential oils. *Microorganisms* **2018**, *6*, 122. [CrossRef]
40. Karumathil, D.P.; Surendran-Nair, M.; Venkitanarayanan, K. Efficacy of *Trans*-cinnamaldehyde and Eugenol in Reducing *Acinetobacter baumannii* Adhesion to and Invasion of Human Keratinocytes and Controlling Wound Infection In Vitro. *Phytother. Res.* **2016**, *30*, 2053–2059. [CrossRef]
41. Tsai, C.-C.; Lin, C.-S.; Hsu, C.-R.; Chang, C.-M.; Chang, I.-W.; Lin, L.-W.; Hung, C.-H.; Wang, J.-L. Using the Chinese herb *Scutellaria barbata* against extensively drug-resistant *Acinetobacter baumannii* infections: In vitro and in vivo studies. *Bmc Complementary Altern. Med.* **2018**, *18*, 96. [CrossRef]
42. Egyptian Pharmacopeia. *Central Administration of Pharmaceutical Affairs (CAPA)*, 4th ed.; Ministry of Health and Population: Cairo, Egypt, 2005; pp. 31–33.
43. Farag, M.A.; Wessjohann, L.A. Volatiles profiling in medicinal licorice roots using steam distillation and solid-phase microextraction (SPME) coupled to chemometrics. *J. Food Sci.* **2012**, *77*, C1179–C1184. [CrossRef]
44. Kashif, M.T.; Yassin, A.S.; Hosny, A.E.-D.M. Detection of AmpC *beta*-lactamases using sodium salicylate. *J. Microbiol. Methods* **2012**, *91*, 354–357. [CrossRef]
45. Golus, J.; Sawicki, R.; Widelski, J.; Ginalska, G. The agar microdilution method–a new method for antimicrobial susceptibility testing for essential oils and plant extracts. *J. Appl. Microbiol.* **2016**, *121*, 1291–1299. [CrossRef] [PubMed]
46. Naves, P.; Del Prado, G.; Huelves, L.; Gracia, M.; Ruiz, V.; Blanco, J.; Rodríguez-Cerrato, V.; Ponte, M.; Soriano, F. Measurement of biofilm formation by clinical isolates of *Escherichia coli* is method-dependent. *J. Appl. Microbiol.* **2008**, *105*, 585–590. [CrossRef] [PubMed]

47. Hussein, S.H.; Samir, R.; Aziz, R.K.; Toama, M.A. Two putative MmpL homologs contribute to antimicrobial resistance and nephropathy of enterohemorrhagic *E. coli* O157: H7. *Gut Pathog.* **2019**, *11*, 15. [CrossRef] [PubMed]
48. Stratton, C.; Weeks, L.; Aldridge, K. Comparison of kill-kinetic studies with agar and broth microdilution methods for determination of antimicrobial activity of selected agents against members of the Bacteroides fragilis group. *J. Clin. Microbiol.* **1987**, *25*, 645–649. [CrossRef] [PubMed]
49. Wang, X.; Ghasri, P.; Amir, M.; Hwang, B.; Hou, Y.; Khilili, M.; Lin, A.; Keene, D.; Uitto, J.; Woodley, D.T. Topical application of recombinant type VII collagen incorporates into the dermal–epidermal junction and promotes wound closure. *Mol. Ther.* **2013**, *21*, 1335–1344. [CrossRef] [PubMed]

antibiotics

MDPI

Article

Chemical Profiling and Discrimination of Essential Oils from Six *Ferula* Species Using GC Analyses Coupled with Chemometrics and Evaluation of Their Antioxidant and Enzyme Inhibitory Potential

Fadia S. Youssef [1], Munira A. Mamatkhanova [2], Nilufar Z. Mamadalieva [2,3], Gokhan Zengin [4], Salima F. Aripova [2], Elham Alshammari [5] and Mohamed L. Ashour [1,*]

1 Department of Pharmacognosy, Faculty of Pharmacy, Ain Shams University, Cairo 11566, Egypt; fadiayoussef@pharma.asu.edu.eg
2 Institute of the Chemistry of Plant Substances, Academy of Sciences of RUz, Mirzo Ulugbek str. 77, Tashkent 100170, Uzbekistan; munir_05@mail.ru (M.A.M.); nmamadalieva@yahoo.com (N.Z.M.); salima_aripova@mail.ru (S.F.A.)
3 Department of Bioorganic Chemistry, Leibniz Institute of Plant Biochemistry, Weinberg 3, D-06120 Halle (Saale), Germany
4 Department of Biology, Science Faculty, Selcuk University, 42130 Konya, Turkey; gokhanzengin@selcuk.edu.tr
5 Department of Pharmacy Practice, College of Pharmacy, Princess Nourah bint Abdulrahman University, Riyadh 11671, Saudi Arabia; ejalshammari@pnu.edu.sa
* Correspondence: ashour@pharma.asu.edu.eg; Tel.: +20-10-68-222-354; Fax: +20-22-405-1107

Received: 22 July 2020; Accepted: 12 August 2020; Published: 14 August 2020

Abstract: The differences in the composition of essential oils obtained from the aerial parts of six *Ferula* species viz., *F. caratavica* (*Fc*), *F. kuchistanica* (*Fk*), *F. pseudoreoselinum* (*Fp*), *F. samarcandica* (*Fs*), *F. tenuisecta* (*Ft*) and *F. varia* (*Fv*) were detected both qualitatively and semi-quantitatively using GC-MS and GC-FID analyses. One hundred and six metabolites were identified that account for 92.1, 96.43, 87.43, 95.95, 92.90 and 89.48% of *Fc*, *Fk*, *Fp*, *Fs*, *Ft* and *Fv* whole essential oils, respectively. The data from the GC-MS analyses were subjected to unsupervised pattern recognition chemometric analysis utilizing principal component analysis (PCA) to improve the visualization of such differences among the six species. *Fk* and *Ft* are very closely related to each other and were gathered together in one cluster. The antioxidant potential was assessed in vitro using different assays including 2,2′-azino-bis (3-ethylbenzothiazoline-6-sulphonic acid) (ABTS), cupric reducing antioxidant capacity (CUPRAC), ferric reducing power (FRAP) and phosphomolybdenum (PM) assays. *Ft* and *Fp* exhibited the most notable antioxidant properties as evidenced by their pronounced activities in most of the antioxidant assays performed, followed by *Fc*. *Fk* showed the most effective tyrosinase inhibitory potential, which was estimated as 119.67 mgKAE/g oil, while *Fp* exhibited the most potent α-amylase inhibitory potential, which was equivalent to 2.61 mmol ACAE/g oil. Thus, it was concluded that *Ferula* species could serve as a promising natural antioxidant drug that could be included in different products and spices to alleviate hyperglycemia and used as a natural ingredient in pharmaceutical cosmetics to counteract hyperpigmentation.

Keywords: *Ferula*; GC; essential oils; chemometrics; antioxidant activity; enzyme inhibition

1. Introduction

Essential oils comprise a mixture of secondary metabolites, which are biosynthesized by aromatic plants as natural protectants [1]. The role of essential oils is not restricted to protection as they also offer many therapeutic benefits to humans that can exceed the benefits provided by the dried herbs on their own [2]. Recently, they have become well known as a part of traditional medicine for the treatment of a plethora of human ailments, in aromatherapy, as well as in spices with high nutritive value [3]. In addition, many essential oils as well as plant extracts have shown significant antioxidant potential [4–6]. New sources of medicinal agents that are effective and safe as well as selective has recently become the main target in drug discovery. Medicinal plants in general, and their volatile constituents in particular, act as a very important sources for the production of a huge number of biologically active agents, which are attractive chemical leads that are promising therapeutic agents for the alleviation of many ailments [7,8]. Many biological activities have been ascribed to the volatile constituents obtained from a variety of plants such as antinociceptive, anticancer, antiphlogistic, antiviral, antioxidant, antimicrobial, antimycotic, antiparasitic and insecticidal activities [9]. Moreover, the volatile constituents of plants are highly popular in the food, cosmetic and pharmaceutical industries because of their broad acceptance by consumers, relative safety, and their potential multipurpose effect [10,11].

The Apiaceae family is well-known for its rich aromatic plants, which are categorized under approximately 112 genera and nearly 316 species. Anise, chervil, celery, coriander, cumin, caraway, dill, fennel, ferula and galabanum are significant members of this family and they are characterized by their notable odor owing to the presence of considerable amounts of essential oils or the oleoresin predominant in their different organs [3]. These plants are widely used for culinary purposes either for their aroma or as nutrients [12].

Ferula constitutes the third largest genus in the Apiaceae family with nearly 180 species. The members of this genus are very popular for their essential oils, which are recognized as having many biological activities including antibacterial, antifungal, antiviral, antispasmodic, anticonvulsant, and antioxidant activity as well as having high nutritive value [13,14].

This study aimed to investigate the contents of the essential oil from six *Ferula* species growing in Uzbekistan, namely, *F. caratavica* (*Fc*), *F. kuchistanica* (*Fk*), *F. pseudoreoselinum* (*Fp*), *F. samarcandica* (*Fs*), *F. tenuisecta* (*Ft*) and *F. varia* (*Fv*) using GC analyses. Discrimination of these species was carried by coupling the data obtained from GC-analyses with chemometrics employing unsupervised pattern recognition techniques represented by principal component analysis (PCA). Furthermore, the antioxidant potential of the different essential oil samples using different assays, namely, 2,2′-azino-bis (3-ethylbenzothiazoline-6-sulphonic acid) (ABTS), cupric reducing antioxidant capacity (CUPRAC), ferric reducing power (FRAP), and the phosphomolybdenum (PM) assay were evaluated *in vitro*. In addition, an evaluation of the possible enzymatic inhibitory activities of essential oils against tyrosinase and α-amylase was done using standard in vitro bioassays.

2. Results and Discussion

2.1. Qualitative and Semi-quantitative Determinations by GC-MS and GC-FID

The differences in the composition of the essential oils obtained from the aerial parts of *Fc*, *Fk*, *Fp*, *Fs*, *Ft* and *Fv* were detected both qualitatively and semi-quantitatively using GC-MS and GC-FID analyses, respectively. All of the essential oils are yellow in color and possess a characteristic odor. Characterization of the essential oils using GC analyses revealed the presence of 106 metabolites (Table 1, Figures 1 and 2) that account for 92.10, 96.43, 87.43, 95.95, 92.90 and 89.48% of *Fc*, *Fk*, *Fp*, *Fs*, *Ft* and *Fv* whole essential oils, respectively. Twenty-nine compounds were detected in *Fc* with α-pinene (21.17%), 10,13 docosadienoic acid methyl ester (15.20%), β-caryophyllene oxide (13.23%) and caryophyllene (10.88%) representing the predominant compounds. Meanwhile, thirty-nine compounds were identified in *Fk* essential oil with α-pinene (36.79%) and verbenol (8.49%) being the major compounds. In *Fp*, forty-five compounds were characterized with 4-terpineol (16.28%), α-pinene (10.99%), β-myrcene (6.04%), β-caryophyllene oxide (5.69%), p-cymen-8-ol (5.36%) and spathulenol (5.34%) as the main metabolites in the oil. Furthermore, 15 compounds were determined in *Fs* oil with the main compounds, palmitic acid, β-myrecene, heptacosane, octacosane, hexacosane and pentcosane accounting for 39.09, 10.75, 10.27, 9.60, 8.99 and 6.29%, respectively. For *Ft*, 62 compounds were detected of which α-pinene (42.0%), camphene (8.34%) and α-cadinol (8.14%) exist in high percentages in the oil. Finally, 25 compounds were identified in the *Fv* oil with 10,13 docosadienoic acid methyl ester (69.61%) constituting the major component (Figure 3). From the data shown in Table 1, it was concluded that monoterpenes are the predominate class of essential oil metabolites in *Fc*, *Fk* and *Ft*, where they represents 24.90, 42.91 and 61.95%, respectively, while oxygenated monoterpenes are the dominant class of metabolites in *Fp* (35.60%), and they also exist in a high percentage in *Fk* (34.82%). On the contrary, fatty acids are highly predominate in *Fs* and *Fv* and account for 82.55 and 79.84%, respectively.

Table 1. Composition of volatile oil in the aerial parts of *F. caratavica* (*Fc*), *F. kuchistanica* (*Fk*), *F. pseudoreoselinum* (*Fp*), *F. samarcandica* (*Fs*), *F. tenuisecta* (*Ft*) and *F. varia* (*Fv*).

	Compound	RI		Content, (%)						Identification Methods
		Cal.	Rep.	*Fc*	*Fk*	*Fp*	*Fs*	*Ft*	*Fv*	
1.	n-Nonane	889	900	0.37	-	0.16	-	-	-	MS, RI,
2.	Tricyclene	913	913	-	-	-	-	0.37	-	MS, RI,
3.	3-Thujene	919	919	-	-	0.50	-	0.36	-	MS, RI, AU
4.	α-Pinene	925	925	21.17	36.79	10.99	1.5	42.0	-	MS, RI, AU
5.	Camphene	941	941	2.91	0.47	-	-	8.34	-	MS, RI,
6.	Sabinene	970	970	-	-	0.94	-	0.35	-	MS, RI,
7.	β-Pinene	973	973	0.82	1.88	3.11	-	3.59	-	MS, RI, AU
8.	6-Methyl-5-heptene-2-one	986	986	-	0.35	-	-	-	-	MS, RI, AU
9.	β-Myrcene	989	989	-	0.19	6.04	10.75	1.90	-	MS, RI,
10.	n-Decane	998	1000	1.00	0.16	0.47	2.44	0.06	0.62	MS, RI,
11.	α-Phellandrene	1003	1003	-	1.81	tr	-	-	-	MS, RI, AU
12.	(3E)-3-Hexenyl acetate	1007	1006	-	-	-	-	-	1.40	MS, RI,
13.	3-Carene	1009	1009	-	tr	0.97	-	0.07	-	MS, RI,
14.	2-Carene	1016	1018	-	-	tr	-		-	MS, RI,
15.	β-Cymene	1024	1025	-	tr	2.59	-	0.39	-	MS, RI,
16.	Limonene	1028	1028	-	1.77	0.83	1.15	4.80	-	MS, RI, AU
17.	τ-Terpinene	1059	1059	-	-	0.30	-	-	-	MS, RI, AU
18.	Linalool oxide	1074	1074	-	0.25	-	-	-	-	MS, RI,
19.	Terpinolene	1089	1089	-	-	-	-	tr	-	MS, RI,
20.	Dehydro-p-cymene	1090	1090	-	-	-	-	0.15	-	MS, RI,
21.	n-Undecane	1098	1100	tr	1.60	-	1.56	-	0.43	MS, RI,
22.	β-Linalool	1100	1100	tr	tr	2.03	-	0.54	-	MS, RI, AU
23.	*cis-p*-Menth-2,8-dienol	1108	1104	-	0.80	1.78	-	0.40	-	MS, RI,
24.	Fenchol	1116	1117	-	-	-	-	0.04	-	MS, RI,
25.	6-Camphenol	1128	1131	-	2.75		-	0.05	-	MS, RI,
26.	Limonene oxide	1135	1133	-	-	-	-	tr	-	MS, RI,
27.	4-Isopropenyl-1-methyl-2-cyclohexen-1-ol	1137	1142	-	0.32	0.49	-	0.12	-	MS, RI,
28.	L-pinocarveol	1141	1141	-	3.90	0.74		0.33		MS, RI,
29.	Verbenol	1148	1148	-	8.49	-	-	1.25	-	MS, RI,
30.	*trans*-2-Nonenal	1160	1161	-	-	-	-	0.03	0.73	MS, RI,
31.	3-Pinanone	1163	1160	-	tr	-	-	-	-	MS, RI,
32.	Verbenone	1165	1173	-	1.43	-	-	-	-	MS, RI,
33.	Borneol	1169	1169	-	-	-	-	0.11	-	MS, RI,
34.	4-Terpineol	1179	1179	-	0.39	16.28	-	0.36	-	MS, RI,
35.	p-Cymen-8-ol	1186	1186	tr	2.85	5.36	-	0.80	0.49	MS, RI,
36.	α-Terpineol	1193	1193	-	0.52	5.00	-	0.41	-	MS, RI,
37.	Myrtenol	1199	1199	-	2.30	0.65	-	0.21	-	MS, RI,
38.	*cis*-Geraniol	1210	1210	-	-	0.35	-	tr	-	MS, RI,
39.	Verbenone	1214	1214	-	3.95	1.11	-	0.18	-	MS, RI,
40.	Fenchyl acetate	1222	1223	-	4.51	-	-	-	-	MS, RI,
41.	*cis*-Carveol	1225	1220	-	tr	-	-	0.46	-	MS, RI,
42.	β-Citronellol	1230	1230	-	-	-	-	0.05	-	MS, RI,
43.	*trans*-Carveol	1234	1229	-	-	-	-	0.04	-	MS, RI,
44.	Thymol methyl ether	1237	1237	-	0.19	0.23	-	0.03	-	MS, RI,
45.	D-Carvone	1249	1249	-	1.68	-	-	tr	-	MS, RI,

Table 1. *Cont.*

	Compound	RI		Content, (%)						Identification Methods
		Cal.	Rep.	*Fc*	*Fk*	*Fp*	*Fs*	*Ft*	*Fv*	
46.	Nerol	1252	1251	-	tr	-	-	tr	-	MS, RI,
47.	Bornyl acetate	1290	1290	-	0.49	0.39	-	0.31	-	MS, RI,
48.	(-)-trans-Pinocarvyl acetate	1305	297	-	-	1.19	-	-	-	MS, RI,
49.	Carvacrol	1306	1306	-	tr	-	-	-	-	MS, RI,
50.	α-Cubebene	1353	1353	-	-	-	-	0.16	-	MS, RI,
51.	D-longifolene	1370	1370	-	tr	-	-	-	-	MS, RI,
52.	α-Copaene	1380	1380	-	0.38	0.41	-	0.18	0.39	MS, RI,
53.	β-Gurjunene	1386	1388	-	4.81	-	-	-	-	MS, RI,
54.	β-Bourbonene	1390	1390	-	-	-	-	-	tr	MS, RI,
55.	β-Elemene	1395	1395	-	-	-	-	0.24	tr	MS, RI,
56.	Jasmone	1403	1399	-	tr	-	-	-	-	MS, RI,
57.	β-Caryophyllene	1425	1425	10.88	-	0.91	-	0.08	tr	MS, RI, AU
58.	τ-Elemene	1438	1438	-	-	-	-	0.12	-	MS, RI,
59.	Patchoulene	1440	1440	tr	-	0.38	-		-	MS, RI,
60.	Alloaromadendrene	1447	1442	-	-	-	-	-	0.63	MS, RI,
61.	Geranyl acetone	1456	1455	-	4.48	0.58	-	-	-	MS, RI,
62.	α-Humulene	1461	1461	2.98	tr	-	-	0.04	-	MS, RI,
63.	τ-Muurolene	1467	1467	-	0.27	-	-	0.21	0.87	MS, RI,
64.	α-Curcumene	1487	1486	-	tr	-	-		-	MS, RI,
65.	Germacrene D	1489	1489	-	-	-	-	0.13	-	MS, RI,
66.	β-Eudesmene	1495	1495	-	-	-	-	0.20	-	MS, RI,
67.	β-Guaiene	1503	1500	0.65	-	tr	-	0.38	tr	MS, RI,
68.	α-Muurolene	1508	1508	-	-	0.35	-	0.74	-	MS, RI,
69.	Cuparene	1514	1513	3.09	0.23	-	-	-	-	MS, RI,
70.	α-Selinene	1514	1517	-	tr	-	-	0.07	0.34	MS, RI,
71.	τ-Cadinene	1523	1521	-	-	-	-	0.75	0.50	MS, RI,
72.	β-Cadinene	1524	1529	tr	-	-	-	-	-	MS, RI,
73.	δ-Cadinene	1531	1531	1.37	-	-	tr	3.07	-	MS, RI,
74.	Elemol	1557	1577	-	-	-	-	tr	-	MS, RI,
75.	Nerolidol	1566	1564	1.70	-	-	-	1.74	-	MS, RI,
76.	Germacrene B	1572	1569	-	-	-	-	-	1.07	MS, RI,
77.	Germacrene D-4-ol	1585	1583	-	-	-	-	0.75	-	MS, RI,
78.	Spathulenol	1587	1587	-	-	5.34	-	0.38	0.65	MS, RI,
79.	Globulol	1590	1590	-	1.06	3.25	-	0.18	-	MS, RI,
80.	Caryophyllene oxide	1594	1594	13.23	tr	5.69	-		2.14	MS, RI, AU
81.	Guaiol	1602	1602	-	-	-	-	-	0.53	MS, RI,
82.	Cubenol	1606	1605	-	tr	-	-	1.13	-	MS, RI,
83.	β-Eudesmol	1612	1613	-	-	-	-	0.20	-	MS, RI,
84.	τ-Eudesmol	1631	1631	-	-	0.66	-		-	MS, RI,
85.	τ-Muurolol	1652	1652	2.03	-	-	tr	3.49	1.02	MS, RI,
86.	δ-Cadinol	1656	1656	tr	2.82	-	tr	0.79	-	MS, RI,
87.	τ-Muurolol	1665	1661	4.64	-	-	-		-	MS, RI,
88.	α-Eudesmol	1666	1662	-	2.54	-	-		1.01	MS, RI,
89.	α-Cadinol	1669	1669	-	-	-	-	8.14	-	MS, RI,
90.	Cedr-8-en-13-ol	1682	1688	2.17	-	-	-		-	MS, RI,
91.	α-Bisabolol	1692	1692	-	-	-	-	0.56	-	MS, RI,
92.	Farnesol	1726	1725	0.82	-	-	-	1.13	-	MS, RI,
93.	Hexadecanal	1817	1819	-	-	-	-	-	1.16	MS, RI,
94.	Hexahydrofarnesyl acetone	1845	1845	-	tr	-	-	-	-	MS, RI,
95.	Palmitic acid	1977	1975	-	-	-	39.03	-	-	MS, RI, AU
96.	trans-9-Octadecen-1-ol	2068	2068	-	-	1.31	-	-	-	MS, RI,
97.	Heptadecanoic acid ethyl ester	2082	2082	1.06	-	-	-	-	-	MS, RI,
98.	*trans*-Phytol	2120	2122	2.69	-	0.42	-	-	-	MS, RI,
99.	Docosane	2200	2200	-	-	0.60	1.47	-	-	MS, RI,
100.	Tricosane	2301	2300	-	-	0.34	tr	-	2.66	MS, RI,
101.	Tetracosane	2395	2400	-	-	0.51	4.49	-	-	MS, RI,
102.	10,13 Docosadienoic acid methyl ester	2449	2449	15.2	-	-	-	-	69.61	MS, RI,
103.	Pentacosane	2498	2500	0.66	-	0.95	6.26	-	-	MS, RI,
104.	Hexacosane	2598	2600	0.63	-	1.10	8.99	-	3.23	MS, RI,
105.	Heptacosane	2697	2700	1.26	-	1.35	10.27	-	-	MS, RI,
106.	Octacosane	2790	2800	0.77	-	1.13	9.60	-	tr	MS, RI,
	Monoterpene hydrocarbons			24.9	42.91	26.27	13.40	61.95	-	
	Oxygenated monoterpene			tr	34.82	35.60	-	5.69	0.49	
	Sesquiterpene hydrocarbons			18.97	5.69	2.05	tr	6.37	3.80	
	Oxygenated sesquiterpene			24.59	10.9	15.52	tr	18.49	5.35	
	Others			23.64	2.11	7.87	82.55	0.40	79.84	
	Total			92.10	96.43	87.31	95.95	92.90	89.48	

Compounds were identified based on a comparison of the compounds' mass spectral data and retention indices with those of the NIST Mass Spectral Library (December 2011), the Wiley Registry of Mass Spectral Data, 8th edition and by comparison with the authentic standard (AU). The content (%) was calculated using the normalization method based on the GC-FID data generated from the average of three independent chromatographic runs.

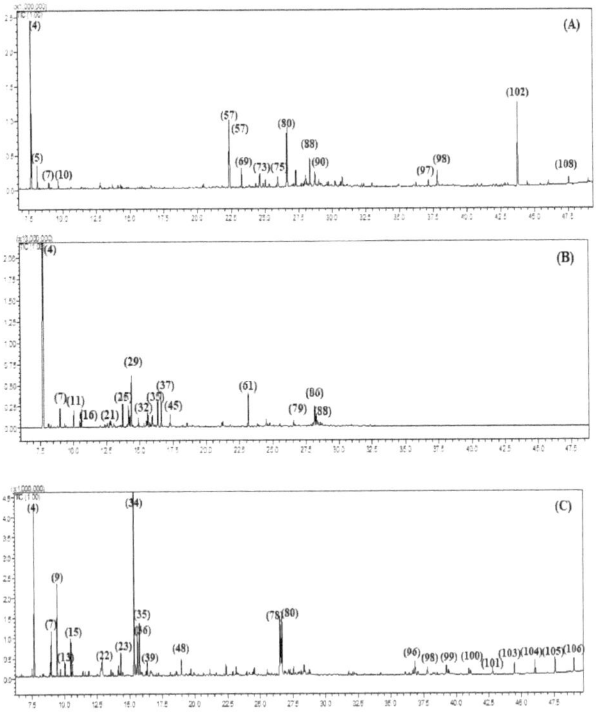

Figure 1. GC-MS chromatograms of *F. caratavica* (**A**), *F. kuchistanica* (**B**) and *F. pseudoreoselinum* (**C**).

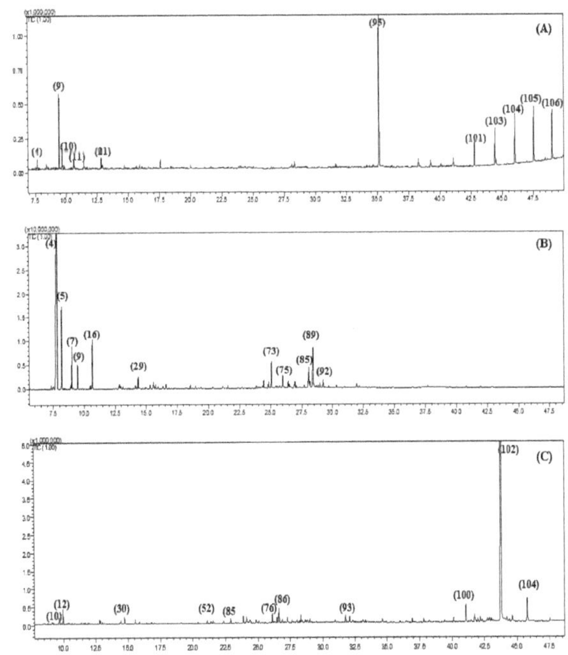

Figure 2. GC-MS chromatograms of *F. samarcandica* (**A**), *F. tenuisecta* (**B**) and *F. varia* (**C**).

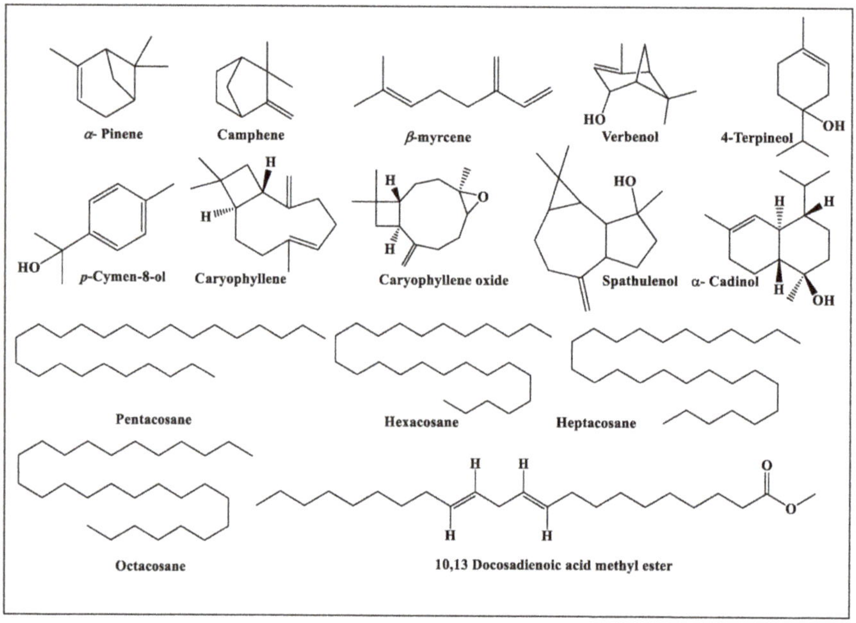

Figure 3. Main secondary metabolites in the *Ferula* species.

2.2. Chemometric Analysis

It is extremely difficult to identify the qualitative and quantitative differences between the *Ferula* species under evaluation with the naked eye. So, the data obtained from GC analyses were subjected to unsupervised pattern recognition chemometric analysis utilizing PCA to improve the visualization of these differences. The results of the PCA, as represented by the obtained score plot shown in Figure 4A effectively discriminated the six *Ferula* species into five clusters along the first component (PC1) and the second component (PC2) that account for 57% and 30%, respectively, or 87% of the total variance. From the obtained results, it is obvious that both *Fk* and *Ft* are very closely related to each other as they are gathered together in one cluster in the lower left quadrant. However, PC1 successfully discriminated between *Fk* and *Ft* with negative values of PC1 as they are located in the lower left quadrant and *Fc* and *Fv*, which show positive values of PC1 are located in the lower right quadrant. Meanwhile, PC2 significantly discriminated between *Fk* and *Ft*, which show negative values of PC2 as they are located in the lower left quadrant and between *Fs* and *Fp*, which show positive values of PC2 and are located in the upper left quadrant. Furthermore, both PC1 and PC2 significantly discriminated between *Fc* and *Fv*, which show positive values for PC1 and negative values for PC2 as they are located in the lower right quadrant and between *Fs* and *Fp*, displaying negative values for PC1 and positive values for PC2 as they are located in the upper left quadrant. The major discriminatory signals are α-pinene, 10,13-docosadienoic acid methyl ester and palmitic acid as revealed in the loading plot shown in Figure 4B.

The Pearson correlation coefficient (r) between the essential oil contents of different studied samples indicated that *Fc* had a highly significant positive correlation with *Ft* ($r = 0.71$), *Fk* ($r = 0.58$), *Fv* ($r = 0.47$) and *Fp* ($r = 0.35$), while a non-significant negative correlation was observed between *Fc* and *Fs* (the highest correlations were observed between *Ft* and *Fk* ($r = 0.89$, $p < 0.001$), between *Fc* and *Ft* ($r = 0.71$, $p < 0.001$), and between *Fc* and *Fk* ($r = 0.58$, $p < 0.001$) as seen in Table 2. These data indicate that three samples, *Ft*, *Fk*, and Fc have highly similar essential oil content.

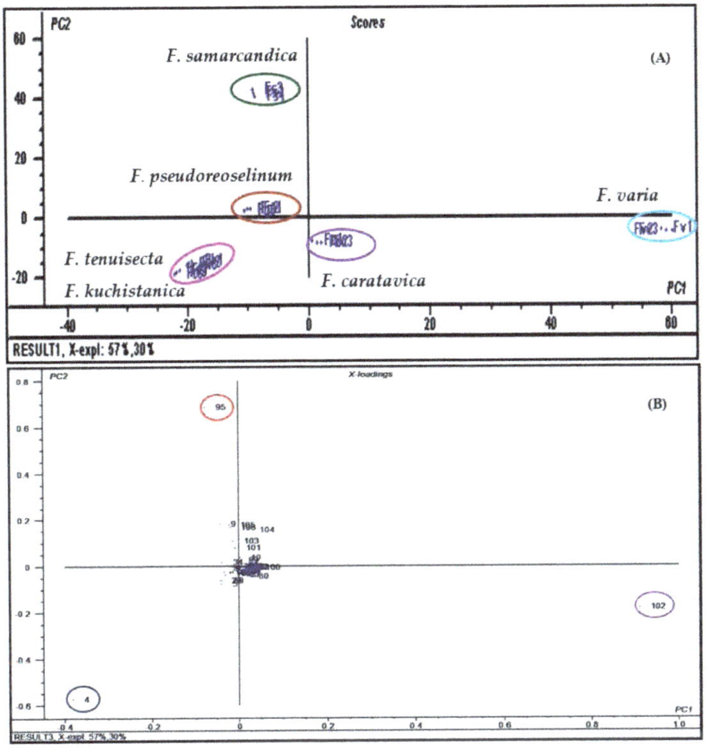

Figure 4. Score plot (**A**) and loading plot (**B**) of GC data obtained from *F. caratavica*, *F. kuchistanica*, *F. pseudoreoselinum*, *F. samarcandica*, *F. tenuisecta* and *F. varia* essential oil analyses using principal component analysis (PCA). In the loading plot, compounds are given numbers as in Table 1 where the major discriminatory signals are α-pinene **(4)**, palmitic acid **(95)** and 10,13-docosadienoic acid methyl ester **(102)**.

Table 2. The Pearson correlation matrix of the essential oils content of different samples.

	Fc	*Fk*	*Fp*	*Fs*	*Ft*	*Fv*
Fc	-	0.58 ***	0.35 ***	−0.02	0.71 ***	0.47 ***
Fk	0.58 ***	-	0.43 ***	−0.02	0.89 ***	−0.03
Fp	0.35 ***	0.43 ***	-	0.05	0.45 ***	−0.03
Fs	−0.02	−0.02	0.05	-	−0.002	−0.02
Ft	0.71 ***	0.89 ***	0.45 ***	−0.002	-	−0.03
Fv	0.47 ***	−0.03	−0.03	−0.02	−0.03	-

The data is represented as the r value of the correlation coefficient and *** is the level of significance, $p < 0.001$.

2.3. Biological Evaluation

2.3.1. Antioxidant Potential of Different *Ferula* Species

The antioxidant potential of the different essential oil samples was performed in vitro using the 2,2′-azino-bis(3-ethylbenzothiazoline-6-sulfonic acid) (ABTS), the cupric ion reducing antioxidant capacity (CUPRAC), The ferric reducing antioxidant power (FRAP) and the phosphomolybdenum method (PM) assays. The results displayed in Table 3 reveal that most of the samples showed considerable antioxidant potential in the performed assays. *Fc* (41.36 mgTE/g oil) exhibited the most antioxidant activity in ABTS assays, followed by *Fk* (29.12 mgTE/g oil) and *Ft* (28.03 mgTE/g oil). However, in CUPRAC assay, *Fp* (289.45 mgTE/g oil) showed the most superior antioxidant potential followed by *Ft* (278.87 mgTE/g oil) and *Fk* (120.43 mgTE/g oil). Furthermore, *Ft* exhibited the most significant antioxidant power in both FRAP and PM assays with antioxidant activity equivalent to 136.81 mgTE/g oil and 78.66 mmolTE/g oil, respectively, followed by *Fp*, which showed antioxidant potential of 121.64 mgTE/g oil and 50.86 mmolTE/g oil in FRAP and PM assays, respectively. Thus, it can be concluded that the essential oil from both *Ft* and *Fp* exhibited the most notable antioxidant properties as evidenced by their pronounced activities in most of the performed antioxidant assays, followed by *Fc*. α-Pinene, the predominant compound in *Ft* and *Fp* has previously been shown to possess notable antioxidant activity [15]. Additionally, the significant antioxidant activity found in this study, which can be interpreted as a result of the synergistic action between the different components that exist in the oils, was in accordance with that previously reported for many other *Ferula* species such as *F. microcolea*, *F. orantalis* and *F. communis*. Various mechanisms can be used to interpret antioxidant potential including the prohibition of chain initiation, peroxide decomposition, obstruction of continual hydrogen removal as well as the scavenging of free radical and uniting transition metal ion catalysts [3,16,17]. Additionally, α-pinene, the main constituent in both *Ft* and *Fp*, has previously been shown to be a potent antioxidant in both DPPH and FRAP assays, displaying EC_{50} values equal to 310 and 238 μg/mL, respectively [18].

Table 3. Antioxidant activities of the essential oil samples of *Ferula* species using the 2,2′-azino-bis(3-ethylbenzothiazoline-6-sulfonic acid) (ABTS), the cupric ion reducing antioxidant capacity (CUPRAC), The ferric reducing antioxidant power (FRAP) and the phosphomolybdenum method (PM) assays.

Samples	ABTS (mgTE/g Oil)	CUPRAC (mgTE/g Oil)	FRAP (mgTE/g Oil)	PM (mmolTE/g Oil)
F. caratavica (Fc)	41.36 ± 1.27 ^{a}	83.54 ± 3.13 ^{c}	47.34 ± 0.65 ^{e}	5.59 ± 0.01 ^{f}
F. kuchistanica (Fk)	29.12 ± 0.85 ^{b}	120.43 ± 9.36 ^{b}	80.74 ± 0.25 ^{c}	36.42 ± 0.07 ^{c}
F. pseudoreoselinum (Fp)	22.68 ± 1.03 ^{c}	289.45 ± 7.30 ^{a}	121.64 ± 0.01 ^{b}	50.86 ± 0.07 ^{b}
F. samarcandica (Fs)	11.84 ± 1.37 ^{d}	74.39 ± 4.73 ^{c,d}	43.21 ± 0.48 ^{f}	14.37 ± 0.04 ^{e}
F. tenuisecta (Ft)	28.03 ± 3.89 ^{b}	278.87 ± 8.51 ^{a}	136.81 ± 1.98 ^{a}	78.66 ± 0.15 ^{a}
F. varia (Fv)	7.04 ± 0.47 ^{e}	65.90 ± 1.66 ^{d}	55.00 ± 0.18 ^{d}	15.33 ± 0.07 ^{d}

Values are reported as mean ± S.D of three parallel measurements. TE: Trolox equivalents. Different superscripts (a–f) indicate significant differences in the tested *Ferula* species ($p < 0.05$).

2.3.2. Tyrosinase and α-Amylase Inhibitory Potential

Tyrosinase enzyme is an oxidase enzyme containing copper that assists in the completion of the first two steps of mammalian melanogenesis, which leads to undesirable hyperpigmentation. Thus, the search for effective tyrosinase inhibitors has recently become vital so that they can be incorporated in cosmetics for effective skin whitening and to counteract hyperpigmentation [19]. *Fk* showed the most effective tyrosinase inhibitory potential, which was estimated as 119.67 mgKAE/g oil followed by *Fv*, which showed an inhibitory potential equivalent to 118.42 mgKAE/g oil, where KAE is a Kojic acid equivalent, a potent tyrosinase inhibitory drug. *Fv* oil is rich in 10,13 docosadienoic acid methyl ester, a polyunsaturated fatty acid, which greatly accounts for its promise as a tyrosinase inhibitor [20]. The underlying tyrosinase inhibitory mechanism mainly relies on the essential oils being rich in components that possess a hydrophobic portion that competitively inhibits the active sites of tyrosinase enzyme with subsequent interference of melanin synthesis. This inhibition may be achieved via interaction with Cu^{+2} that exists in the active sites of tyrosinase in addition to the prohibition of tautomerization to dopachrome triggered by the oil, which behaves as a reducing agent and blocks of the oxidation reaction during the formation of melanin intermediates during the conversion of tyrosinase/DOPA into melanin, thus reducing skin pigmentation [21].

The α-amylase enzyme is critical in assisting in the catalysis of the first steps in the conversion of starch into maltose, and subsequently to glucose [22,23]. Nowadays, α-amylase inhibitors are used in therapeutic approaches to counteract hyperglycemia. *Fp* and *Fv* exhibited the most potent α-amylase inhibitory potential as evidenced by their pronounced inhibitory activity, which was equivalent to 2.61 and 1.40 mmol ACAE/g oil, respectively, in which ACAE is the acarbose equivalent, a potent α-amylase inhibitor (Figure 5). 4-Terpineol as well as α-pinene, which predominate the essential oil of *Fp*, were previously reported to possess considerable α-amylase inhibitory activity [24]. Similarly, the potent α-amylase inhibitory potential is mainly due to the synergistic action between the different components, which is in accordance to different previously reported studies that confirmed the α-amylase inhibitory effect of different terpenes and different *Ferula* species such as *F. gummosa* essential oil [24,25].

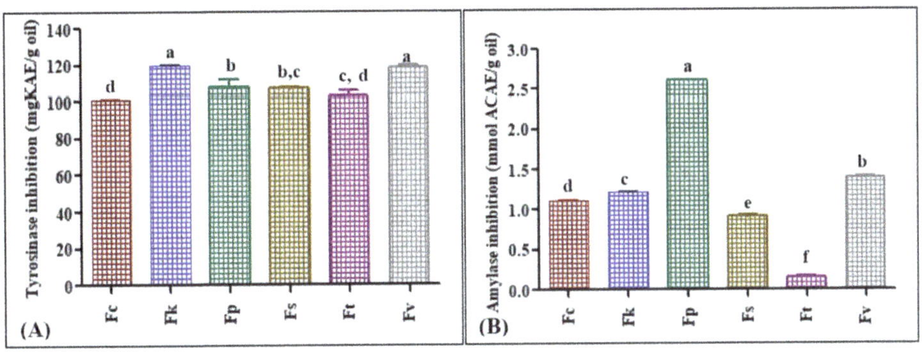

Figure 5. In vitro tyrosinase inhibition (**A**) and α-amylase inhibition (**B**) of the essential oil of different *Ferula* species, *F. caratavica* (*Fc*), *F. kuchistanica* (*Fk*), *F. pseudoreoselinum* (*Fp*), *F. samarcandica* (*Fs*), *F. tenuisecta* (*Ft*) and *F. varia* (*Fv*). Different letters (a–f) indicate significant differences in the tested *Ferula* species ($p < 0.05$).

3. Materials and Methods

3.1. Plant Material

Aerial parts (flowers, leaves and stems) of *F. caratavica* Regel & Schmalh (N2004), *F. pseudoreoselinum* (Regel & Schmalh.) Koso-Pol., p.p. (N1489), *F. tenuisecta* Korovin (N1488) were collected from the Tashkent region of Uzbekistan. *F. varia* (Schrenk ex Fisch., C.A.Mey. & Avé-Lall.) Trautv. (N1407) was collected from the Bukhara region (Uzbekistan), while *F. kuchistanica* Korovin (N1425) and *F. samarcandica* Korovin (N1919) were collected from the Samarkand region of Uzbekistan. The plants were collected during the flowering stage in June–July 2018. Their taxonomic authentication was accomplished by Dr. A. Nigmatullaev at the Institute of the Chemistry of Plant Substances (Tashkent, Uzbekistan).

3.2. Preparation of Essential Oil Samples

All the plant materials were air-dried in the shade for 7 days at room temperature and powdered using a mortar and pestle to get particles of a uniform, reduced size. Preparation of the essential oil samples was achieved by hydrodistillation of the air-dried aerial parts of the different *Ferula* species, *F. caratavica (Fc)*, *F. kuchistanica (Fk)*, *F. pseudoreoselinum* (*Fp*), *F. samarcandica* (*Fs*), *F.* tenuisecta *(Ft)* and *F. varia* (*Fv*) for 2 h by Clevenger-type apparatus. Anhydrous Na_2SO_4 was used to dehydrate the prepared essential oils, yielding 0.4, 0.7, 0.3, 0.3, 0.8 and 0.5 % v/w of dry weight for *Fc*, *Fk*, *Fp*, *Fs*, *Ft* and *Fv*, respectively. Then the various oil samples were maintained at −30 °C in dark-colored stoppered glasses until their analyses [26,27].

3.3. GC-FID and GC-MS Analyses

A Shimadzu GC-17A gas chromatograph (Shimadzu Corporation, Kyoto, Japan) with an FID detector and DB-5 fused-bonded cap column (Phenomenex; 29 m × 0.25 mm i.d., film thickness 0.25 µm; Torrance, California, USA) was utilized for the semi-quantitative determination of the different components of the essential oils using the normalization method to get the relative percentage of each component and applying GC-FID data that is highly sensitive using GC solution® software ver. 2.4 (Shimadzu Corporation, Kyoto, Japan). The areas under the peaks (AUP) were determined using three independent runs where the total area is considered as 100%. Meanwhile, the Shimadzu GC-2010 plus gas chromatograph (Shimadzu Corporation, Kyoto, Japan) supplied with Rtx-5MS (Restek, Bellefonte, PA, USA) in addition to a quadrupole mass spectrometer was used for the identification of the essential oil different metabolites. Instrument settings were adjusted according to what was previously reported [28,29]. The Wiley Registry of Mass Spectral Data 8th edition, NIST MassSpectral Library (December 2011), and previously reported data were employed to confirm the identity of the compounds and the retention indexes were calculated to corroborate the identification of the volatile compounds [30,31].

3.4. Chemometric and ANOVA Analysis

To examine the differences between the essential oils' components prepared from different *Ferula* species, the data collected from the different GC-MS spectra were subjected to chemometric analysis of unsupervised pattern recognition represented by PCA, which was processed by employing Unscrambler 9.7 (CAMO SA, Oslo, Norway) [28,32]. Meanwhile, other statistical analyses used for biological assessment were performed using ANOVA assay (with Tukey's test, significant value: $p < 0.05$) and Xlstat 2017 software.

3.5. Biological Evaluation

3.5.1. Determination of the Antioxidant Potential

The antioxidant activity of the different essential oil samples from different *Ferula* species was evaluated using ABTS, CUPRAC, FRAP and PM assays. These assays were performed following the methods described by Mamadalieva et al. [33]. The antioxidant activities were reported as Trolox equivalents and the samples were analyzed in triplicate.

3.5.2. Determination of Enzyme Inhibitory Effects

The possible inhibitory potential of the essential oil samples was investigated against tyrosinase and α-amylase enzymes using standard in vitro bioassays as previously reported by Mamadalieva et al. [33] in which all the samples were analyzed in triplicate. Results are expressed in mgKAE/g oil for tyrosinase inhibitory activity and in mmol ACAE/g oil for α-amylase inhibition.

4. Conclusions

The essential oils obtained from different *Ferula* species, *F. caratavica*, *F. kuchistanica*, *F. pseudoreoselinum*, *F. samarcandica*, *F. tenuisecta* and *F. varia* showed significant variation as revealed by GC analyses. Furthermore, this variation became more clearly observable when coupled with a chemometric approach as represented by PCA used as an unsupervised pattern recognition technique. Additionally, the obtained essential oils showed notable antioxidant as well as tyrosinase and α-amylase inhibitory activities with variable degrees, which is mainly related to the differences in the secondary metabolites that predominate in the oils. Thus, it was concluded that the different *Ferula* species could serve as a promising natural antioxidant drug that could be included in different products and used as spices to alleviate hyperglycemia and as a natural ingredient in pharmaceutical cosmetics to counteract hyperpigmentation. Chemometric study based on gathering the different biological activities of many additional *Ferula* species will be considered. It is recommended that further in vivo studies such as animal and bioavailability studies be carried out to confirm the obtained results.

Author Contributions: F.S.Y., identification of the essential oil compounds, chemometric analysis, writing the whole manuscript; M.A.M., N.Z.M., S.F.A., collection of the plants, isolation of the essential oil samples and revising the manuscript; G.Z., performing the biological studies; E.A. and M.L.A., supervising the study and revising the manuscript. All authors have read and agreed to the published version of the manuscript.

Funding: This research was funded by the Deanship of Scientific Research at Princess Nourah bint Abdulrahman University through the Fast-Track Research Funding Program.

Conflicts of Interest: The authors declare no conflict of interest.

References

1. Rehman, R.; Hanif, M.A.; Mushtaq, Z.; Al-Sadi, A.M. Biosynthesis of essential oils in aromatic plants: A review. *Food Rev. Int.* **2016**, *32*, 117–160. [CrossRef]
2. Sharifi-Rad, J.; Sureda, A.; Tenore, G.C.; Daglia, M.; Sharifi-Rad, M.; Valussi, M.; Tundis, R.; Sharifi-Rad, M.; Loizzo, M.R.; Ademiluyi, A.O. Biological activities of essential oils: From plant chemoecology to traditional healing systems. *Molecules* **2017**, *22*, 70. [CrossRef] [PubMed]
3. Sahebkar, A. Biological activities of essential oils from the genus *Ferula* (Apiaceae). *Asian Biomed. Res. Rev. News.* **2010**, *4*, 835–847. [CrossRef]
4. Burits, M.; Bucar, F. Antioxidant activity of *Nigella sativa* essential oil. *Phytother. Res.* **2000**, *14*, 323–328. [CrossRef]
5. Koleva, I.I.; Van Beek, T.A.; Linssen, J.P.; Groot, A.d.; Evstatieva, L.N. Screening of plant extracts for antioxidant activity: A comparative study on three testing methods. *Phytochem. Anal.* **2002**, *13*, 8–17. [CrossRef]
6. Lee, S.E.; Hwang, H.J.; Ha, J.-S.; Jeong, H.-S.; Kim, J.H. Screening of medicinal plant extracts for antioxidant activity. *Life Sci.* **2003**, *73*, 167–179. [CrossRef]
7. Moniruzzaman; Shahinuzzaman; Haque, A.; Khatun, R.; Yaakob, Z. Gas chromatography mass spectrometry analysis and in vitro antibacterial activity of essential oil from Trigonella foenum-graecum. *Asian Pac. J. Trop. Biomed.* **2015**, *5*, 1033–1036. [CrossRef]
8. Rashid, S.; Rather, M.A.; Shah, W.A.; Bhat, B.A. Chemical composition, antimicrobial, cytotoxic and antioxidant activities of the essential oil of *Artemisia indica* Willd. *Food Chem.* **2013**, *138*, 693–700. [CrossRef]
9. Adorjan, B.; Buchbauer, G. Biological properties of essential oils: An updated review. *Flav. Frag. J.* **2010**, *25*, 407–426. [CrossRef]
10. Sun, J.; Wang, X.; Wang, P.; Li, L.; Qu, W.; Liang, J. Antimicrobial, antioxidant and cytotoxic properties of essential oil from *Dictamnus angustifolius*. *J. Ethnopharmacol.* **2015**, *159*, 296–300. [CrossRef]
11. Martins, C.D.M.; Nascimento, E.A.D.; de Morais, S.A.; de Oliveira, A.; Chang, R.; Cunha, L.; Martins, M.M.; Martins, C.H.G.; Moraes, T.D.S.; Rodrigues, P.V. Chemical constituents and evaluation of antimicrobial and cytotoxic activities of *Kielmeyera coriacea* Mart. & Zucc. essential oils. *Evid.-Based Complement. Altern. Med.* **2015**, *2015*, 842047.
12. Safari, O.; Sarkheil, M.; Paolucci, M. Dietary administration of *Ferula* (*Ferula asafoetida*) powder as a feed additive in diet of koi carp, *Cyprinus carpio* koi: Effects on hemato-immunological parameters, mucosal antibacterial activity, digestive enzymes, and growth performance. *Fish. Physiol. Biochem.* **2019**, *45*, 1277–1288. [CrossRef] [PubMed]
13. Mohammadhosseini, M.; Venditti, A.; Sarker, S.D.; Nahar, L.; Akbarzadeh, A. The genus *Ferula*: Ethnobotany, phytochemistry and bioactivities—A review. *Ind. Crops Prod.* **2019**, *129*, 350–394. [CrossRef]
14. Iranshahy, M.; Iranshahi, M. Traditional uses, phytochemistry and pharmacology of *Asafoetida* (*Ferula assafoetida* oleo-gum-resin)—A review. *J. Ethnopharmacol.* **2011**, *134*, 1–10. [CrossRef] [PubMed]
15. Him, A.; Ozbek, H.; Turel, I.; Oner, A.C. Antinociceptive activity of alpha-pinene and fenchone. *Pharmacologyonline* **2008**, *3*, 363–369.
16. Amiri, H. Chemical composition and antioxidant activity of essential oil and methanolic extracts of *Ferula microcolea* (Boiss.) Boiss (Apiaceae). *Int. J. Food Prop.* **2014**, *17*, 722–730. [CrossRef]

17. Nguir, A.; Mabrouk, H.; Douki, W.; Ismail, M.B.; Jannet, H.B.; Flamini, G. Chemical composition and bioactivities of the essential oil from different organs of *Ferula communis* L. growing in Tunisia. *Med. Chem. Res.* **2016**, *25*, 515–525. [CrossRef]
18. Bouzenna, H.; Hfaiedh, N.; Giroux-Metges, M.-A.; Elfeki, A.; Talarmin, H. Potential protective effects of alpha-pinene against cytotoxicity caused by aspirin in the IEC-6 cells. *Biomed. Pharmacother.* **2017**, *93*, 961–968. [CrossRef]
19. Chang, T.-S. An updated review of tyrosinase inhibitors. *Int. J. Mol. Sci.* **2009**, *10*, 2440–2475. [CrossRef]
20. Guo, Y.-J.; Pan, Z.-Z.; Chen, C.-Q.; Hu, Y.-H.; Liu, F.-J.; Shi, Y.; Yan, J.-H.; Chen, Q.-X. Inhibitory effects of fatty acids on the activity of mushroom tyrosinase. *Appl. Biochem Biotechnol.* **2010**, *162*, 1564–1573. [CrossRef]
21. Taherkhani, M. Chemical constituents, total phenolic content, antimicrobial, antioxidant and radical scavenging properties, chelating ability, tyrosinase inhibition and in vitro cytotoxic effects of *Artemisia aucheri* herbs. *Pharm. Chem. J.* **2017**, *50*, 736–745. [CrossRef]
22. Youssef, F.S.; Ashour, M.L.; Ebada, S.S.; Sobeh, M.; El-Beshbishy, H.A.; Singab, A.N.; Wink, M. Antihyperglycaemic activity of the methanol extract from leaves of *Eremophila maculata* (Scrophulariaceae) in streptozotocin-induced diabetic rats. *J. Pharm. Pharmacol.* **2017**, *69*, 733–742. [CrossRef] [PubMed]
23. Thabet, A.A.; Youssef, F.S.; El-Shazly, M.; El-Beshbishy, H.A.; Singab, A.N.B. Validation of the antihyperglycaemic and hepatoprotective activity of the flavonoid rich fraction of *Brachychiton rupestris* using in vivo experimental models and molecular modelling. *Food Chem. Toxicol.* **2018**, *114*, 302–310. [CrossRef] [PubMed]
24. Jelenkovic, L.; Jovanovic, V.S.; Palic, I.; Mitic, V.; Radulovic, M. In vitro screening of α-amylase inhibition by selected terpenes from essential oils. *Trop. J. Pharm. Res.* **2014**, *13*, 1421–1428. [CrossRef]
25. Heydari-Majd, M.; Rezaeinia, H.; Shadan, M.R.; Ghorani, B.; Tucker, N. Enrichment of zein nanofibre assemblies for therapeutic delivery of Barije (*Ferula gummosa* Boiss) essential oil. *J. Drug Deliv. Sci. Technol.* **2019**, *54*, 101290. [CrossRef]
26. Mamadalieva, N.Z.; Youssef, F.S.; Ashour, M.L.; Akramov, D.K.; Sasmakov, S.A.; Ramazonov, N.S.; Azimova, S.S. A comparative study on chemical composition and antimicrobial activity of essential oils from three *Phlomis* species from Uzbekistan. *Nat. Prod. Res.* **2019**, 1–6. [CrossRef]
27. Youssef, F.S.; Hamoud, R.; Ashour, M.L.; Singab, A.N.; Wink, M. Volatile oils from the aerial parts of *Eremophila maculata* and their antimicrobial activity. *Chem. Biodivers.* **2014**, *11*, 831–841. [CrossRef]
28. Thabet, A.A.; Youssef, F.S.; El-Shazly, M.; Singab, A.N.B. GC-MS and GC-FID analyses of the volatile constituents of *Brachychiton rupestris* and *Brachychiton discolor*, their biological activities and their differentiation using multivariate data analysis. *Nat. Prod. Res.* **2020**, *34*, 590–594. [CrossRef]
29. Labib, R.; Youssef, F.; Ashour, M.; Abdel-Daim, M.; Ross, S. Chemical composition of *Pinus roxburghii* bark volatile oil and validation of its anti-inflammatory activity using molecular modelling and bleomycin-induced inflammation in *albino* mice. *Molecules* **2017**, *22*, 1384. [CrossRef]
30. Mamadalieva, N.Z.; Youssef, F.S.; Ashour, M.L.; Sasmakov, S.A.; Tiezzi, A.; Azimova, S.S. Chemical composition, antimicrobial and antioxidant activities of the essential oils of three Uzbek Lamiaceae species. *Nat. Prod. Res.* **2019**, *33*, 2394–2397. [CrossRef]
31. Ayoub, I.M.; Youssef, F.S.; El-Shazly, M.; Ashour, M.L.; Singab, A.N.B.; Wink, M. Volatile constituents of *Dietes bicolor* (Iridaceae) and their antimicrobial activity. *Z. Naturforsch. C* **2015**, *70*, 217–225. [CrossRef] [PubMed]
32. Aboulwafa, M.M.; Youssef, F.S.; Gad, H.A.; Sarker, S.D.; Nahar, L.; Al-Azizi, M.M.; Ashour, M.L. Authentication and discrimination of green tea samples using UV–vis, FTIR and HPLC techniques coupled with chemometrics analysis. *J. Pharm. Biomed. Anal.* **2019**, *164*, 653–658. [CrossRef] [PubMed]
33. Mamadalieva, N.Z.; Böhmdorfer, S.; Zengin, G.; Bacher, M.; Potthast, A.; Akramov, D.K.; Janibekov, A.; Rosenau, T. Phytochemical and biological activities of *Silene viridiflora* extractives. Development and validation of a HPTLC method for quantification of 20-hydroxyecdysone. *Ind. Crop. Prod.* **2019**, *129*, 542–548. [CrossRef]

antibiotics

MDPI

Article

Studying the Gene Expression of *Penicillium rubens* Under the Effect of Eight Essential Oils

Zuzana Kisová [1], Andrea Šoltýsová [2,3], Mária Bučková [1], Gábor Beke [1], Andrea Puškárová [1] and Domenico Pangallo [1,*]

1 Institute of Molecular Biology, Slovak Academy of Sciences, Dúbravská cesta 21, 84551 Bratislava, Slovakia; zuzana.kisova@savba.sk (Z.K.); maria.buckova@savba.sk (M.B.); gabor.beke@savba.sk (G.B.); andrea.puskarova@savba.sk (A.P.)

2 Department of Molecular Biology, Faculty of Natural Sciences of Comenius University, Ilkovičova 6, 84215 Bratislava, Slovakia; andrea.soltysova@uniba.sk

3 Institute for Clinical and Translational Research, Biomedical Research Center, Slovak Academy of Sciences, Dúbravská cesta 9, 84505 Bratislava, Slovakia

* Correspondence: domenico.pangallo@savba.sk; Tel.: +421-2-5930-7443

Received: 22 May 2020; Accepted: 16 June 2020; Published: 19 June 2020

Abstract: Essential oils (EOs) are well-known for their beneficial properties against a broad range of microorganisms. For the better understanding of their mechanism of action in fungi, a microarray approach was used in order to evaluate the gene expression of *Penicillium chrysogenum* (recently renamed *P. rubens*) exposed to the indirect contact (vapors) of eight EOs. The selection of assayed EOs was based on their antifungal activity. The extraction of RNA and the microarray hybridization procedure were optimized for the analysis of *P. rubens*. Gene ontology annotation was performed to investigate the functional analysis of the genes. To uncover the metabolic pathway of these differentially expressed genes, they were mapped into the KEGG BRITE pathway database. The transcriptomic analysis showed that, from a total of 12,675 genes, only 551 genes are annotated, and the other 12,124 genes encoded hypothetical proteins. Further bioinformatic analysis demonstrated that 1350 genes were upregulated and 765 downregulated at least with half (four) of the utilizing EOs. A microarray investigation has confirmed the main impact of EOs to metabolic processes in *P. rubens* involved in vital functions. Presumably, this is the first time that a microarray hybridization analysis was performed in order to evaluate the gene expression of *P. rubens* exposed to various EOs.

Keywords: *Penicillium rubens*; essential oils; growth inhibition; RNA microarray; gene expression; metabolic pathway analysis

---

1. Introduction

Nowadays, when antimicrobial resistance becomes a worldwide problem [1], the alternative ways of treatment, such as the use of natural products, are getting the foreground [2]. EOs are synthetized via the secondary metabolism from different parts of plants, such as buds, leaves, flowers, twigs, seeds, fruits, roots, bark or wood, located in secretory tissues [3] (p. 95). From a chemical point of view, EOs are a complex mixture of volatile elements, mainly terpenoids, phenol-derived aromatic compounds and aliphatic compounds, with high antibacterial, antifungal, antiviral, antiparasitical and insecticidal activities [4].

The use of spices and herbs in food preservation or in medicine for their health benefits has a long history [5]. The botanical family *Lamiaceae* has significant representatives for the production of EOs, such as *Origanum vulgare* L., *Mentha piperita* L. and *Thymus vulgaris* L. with a large-scale of antioxidant and antimicrobial characteristics [6]. Tea tree EOs from *Melaleuca alternifolia* Maiden and Betche (*Myrtaceae*) have shown antiseptic and anti-inflammatory effects [7] and medical plant *Eugenia caryophyllata* Thunb., which belongs also to the family *Myrtaceae*, is known for its antifungal, antiviral, antioxidant, anti-inflammatory and anticancer attributes [8]. Previous studies have shown diverse effects of cinnamon plants (*Cinnamomum cassia* L.) of the *Lauraceae* family: antidiabetic, antioxidant, anti-inflammatory, antimicrobial, anticancer and other important activities, such as lipid or cardiovascular disease-lowering compounds [9]. *Cymbopogon flexuosus* Nees ex Steud., worldwide known as lemongrass (from East Indian), belongs to the *Poaceae* family, which is important in EO productions [10,11], possessing antifungal and antimicrobial properties [12,13]. EOs obtained from *Thuja plicata* Donn. (*Cupressaceae*) tree leaves were assigned a broad spectrum of antimicrobial abilities, as well as potential uses in the reduction of sick building syndrome (SBS) [14].

Airborne microorganisms, such as fungi and their spores, are widespread in outdoor and indoor habitats, too [15]. From a health point of view, fungi are representatives of serious health hazards, including respiratory problems, fungal infections (mycoses), irritant effects, allergy reactions and other nonspecific medical troubles [16,17], and could also be a potential problem linked with sick building syndrome—SBS [17,18]. Members of the genus *Penicillium* are considered to be among these colonizers, with diverse roles: decomposition of organic materials by its enzymatic properties, producing toxic secondary metabolites (SM; mycotoxins) and even indoor air quality contaminators [19]. Filamentous fungus *Penicillium rubens* was, according to the study of Wilson and Straus [20], the most detected species associated with SBS.

The fungal contamination in an indoor environment can be solved using fungicides able to kill or inhibit the fungal growth [16]. Synthetic antifungal agents such as bleach, alcohol (100%), quaternary ammonium compounds, formaldehyde and multipurpose industrial disinfectants such as Cavicide® and Virkon® are detergents frequently used for contamination removal [16], for their unpleasant effects on human health are considered as hazardous, which can cause respiratory tract irritation and allergic respiratory and skin reactions. The demand of natural fungicides instead of synthetic chemicals is rising [21] due to concerns regarding human health [22,23]. Replacing synthetic antifungal agents with plant extracts could be an alternative that can reduce, remove or control fungi and molds in a specific environment [23].

Many studies are present in the literature about the effects of EOs against microorganisms [24], but there is little information, especially about fungi, regarding the genetic and metabolism mechanisms that cause these antimicrobial effects. Several authors have described the fungicide activity of EOs as that their components cause the disintegration of fungal hyphae, amplifying membrane permeability [25,26]. An antifungal effect is associated with the lipophilicity of EOs that allows penetration through cell walls and the manipulation with enzymes implicated in the synthesis of cell walls, thus making changes in the morphological characteristics of the fungus [26,27]. In order to extend our knowledge, a key research is to understand the mechanisms of action of EOs to molds. A microarray hybridization approach is a useful method that enables the simultaneous detection of the expression levels of thousands of genes [28]. The expression of an organism's genes is affected by environmental growth conditions [29] and serves to provide information about gene regulation [30]. In this study, we have investigated the fungicide effects of the vapor of eight EOs on the gene expression of *P. rubens* using a microarray strategy.

2. Results

2.1. Inhibition of Mycelial Growth Using EOs Vapors

In present study the inhibition concentrations of eight EOs vapors, for which the microscopic filamentous fungus still survived, were investigated. The microarray hybridizations were carried out with the RNA obtained from the mycelia of *P. rubens* isolated 24 h after exposure *to Cymbopogon flexuosus* (CF), *Cinnamomum cassia* (CC), *Origanum vulgare* (OV), *Thymus vulgaris* (TV) and *Thuja plicata* (TP) EO vapors and 48 h after exposure to *Mentha piperita* (MP), *Eugenia caryophyllata* (EC) and *Melaleuca alternifolia* (MA) EO vapors. The incubation time was dependent on the inhibition ability of EOs to the mycelial growth. In the end of the exponential phase, the samples were collected. A concentration of selected EOs was in the range of 0.01–0.05 μg/mL. Under these conditions, the conidia formation and germination of spores was suppressed and coupled with the activation of stress-responding processes. The RNA extracted from the fungus without exposure to EOs, after 24 and 48 h of cultivation, served as the control.

2.2. Impact of Essential Oils on Genome-Wide Gene Expression

The gene expression profile of *P. rubens* was evaluated by Gene Expression 8 × 15K custom Microarray Slide containing 12,675 genes, from which only 551 are annotated, and the other 12,124 genes encode hypothetical proteins. A microarray-based analysis allowed us to reveal differentially expressed genes (fold change >2) by comparing the expression profiles of genes between the control and the fungus exposed to EOs.

We identified several differently expressed genes (Table 1) for targeted EOs, of which 1430 upregulated genes and 833 downregulated genes were detected in at least four assayed EOs ($n \geq 4$). However, from 2263 genes (1430 upregulated and 833 downregulated), 148 (80 upregulated and 68 downregulated) genes were with mixed regulation, i.e., genes were not just up- or downregulated by at least four EOs, but also, in at least one sample, they were regulated in the other way (e.g., one gene upregulated by four EOs and downregulated by one EO). These genes were discarded from further analysis. The final number of genes, which had the same regulation in at least half of the samples exposed to different EOs ($n \geq 4$): 1350 upregulated and 765 downregulated genes (Table S1).

Table 1. The number of up- or downregulated genes presented in at least half of the samples after exposure to selected essential oils (EOs).

<table>
<tr><th>Essential Oil</th><th>No. Upregulated</th><th>No. Downregulated</th><th>No. of Genes Detected in at Least 4 Assayed EOs ($n \geq 4$)</th></tr>
<tr><td>Thuja plicata</td><td>1730</td><td>1477</td><td rowspan="8">1430
upregulated
833
downregulated</td></tr>
<tr><td>Cinnamomum cassia</td><td>1760</td><td>835</td></tr>
<tr><td>Eugenia caryophyllata</td><td>1818</td><td>1728</td></tr>
<tr><td>Cymbopogon flexuosus</td><td>1831</td><td>1043</td></tr>
<tr><td>Melaleuca alternifolia</td><td>1121</td><td>702</td></tr>
<tr><td>Origanum vulgare</td><td>1609</td><td>1221</td></tr>
<tr><td>Mentha piperita</td><td>2419</td><td>1684</td></tr>
<tr><td>Thymus vulgaris</td><td>1520</td><td>1004</td></tr>
</table>

2.3. Gene Ontology Analysis

To examine the functions of significantly expressed genes (1350 upregulated and 765 downregulated), we performed gene ontology by gene set enrichment analysis (GSEA). In order to classify genes into functional categories, gene ontology-based overrepresentation analyses were executed. We identified our dataset overrepresented in several biological processes (Figures 1A and 2A), cellular components (Figures 1B and 2B) and molecular functions (Figure 1C).

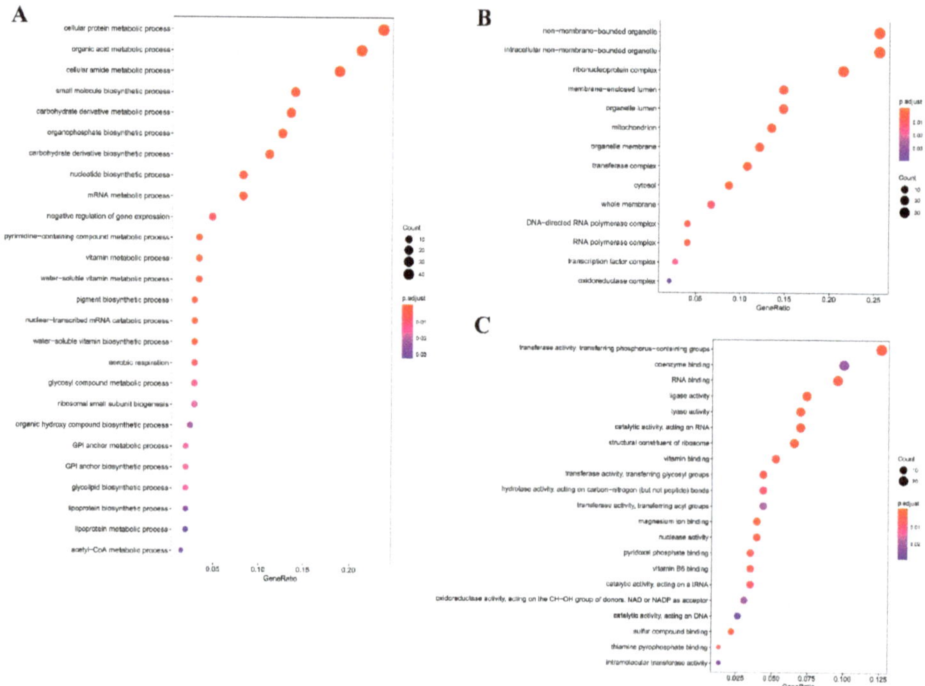

Figure 1. Results of a gene ontology (GO) gene set enrichment analysis (GSEA) of *Penicillium rubens* upregulated genes for different GO domains' (**A**) biological processes, (**B**) cellular components and (**C**) molecular functions. The figures describe the rate of involvement of upregulated genes in the processes (black circles) and, also, about the significance (color changes from blue to red).

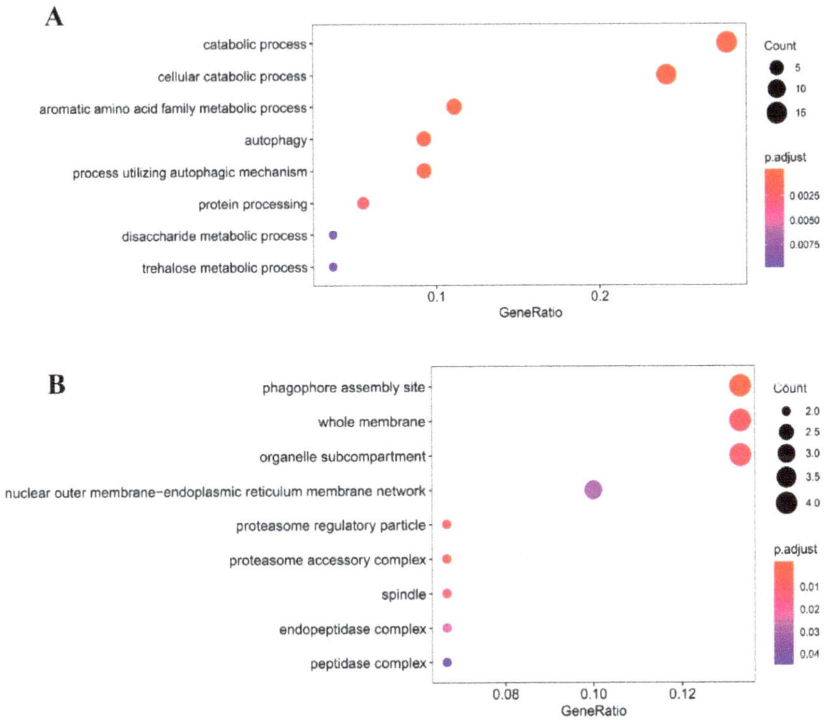

Figure 2. Results of GO GSEA of *P. rubens* downregulated genes for different GO domains' (**A**) biological processes and (**B**) cellular components. No GO molecular functions were detected. The significance, as well as the rate of involvement of downregulated genes, are marked with the same colors and shapes.

2.4. Functional Classes Analysis

To interpret the data closer, we used overlapped the functional classification from the COG, KOG and KEGG BRITE databases [31]; however, many of the *P. rubens* genes were not well-characterized or with unknown functions. Genes with significantly changed expressions were classified. The occurrence of up- or downregulated genes in appropriate functional classes were compared to the overall gene representation in functional classes from the whole genome of the fungus presented in a microarray. We identified significantly dysregulated functional classes (Figure 3), where the increased occurrence of up- or downregulated genes was revealed. Figure 3 depicts the most significantly altered functional classes with significantly increased occurrences of upregulated genes: translation, ribosomal structure and biogenesis ($p = 0.0000018$); RNA processing and modification ($p = 0.0001$) and replication, recombination and repair ($p = 0.01$). The observed upregulation is moreover supported, with significantly less than the expected occurrence of downregulated genes in these classes.

Other significantly altered functional classes with statistically significant increases of upregulated genes were: energy production and conversion ($p = 0.0056$) and coenzyme transport and metabolism ($p = 0.047$) and, with the increase of downregulated genes, inorganic ion transport and metabolism ($p = 0.0007$) and secondary metabolites biosynthesis, transport and catabolism ($p = 0.04$). Significant increases in the occurrences of downregulated genes in the functions of unknown classes ($p = 0.001$) point to a number of genes that may play important roles in the processes of the response of *P. rubens* to EOs exposure. Unfortunately, their functions still remain unknown.

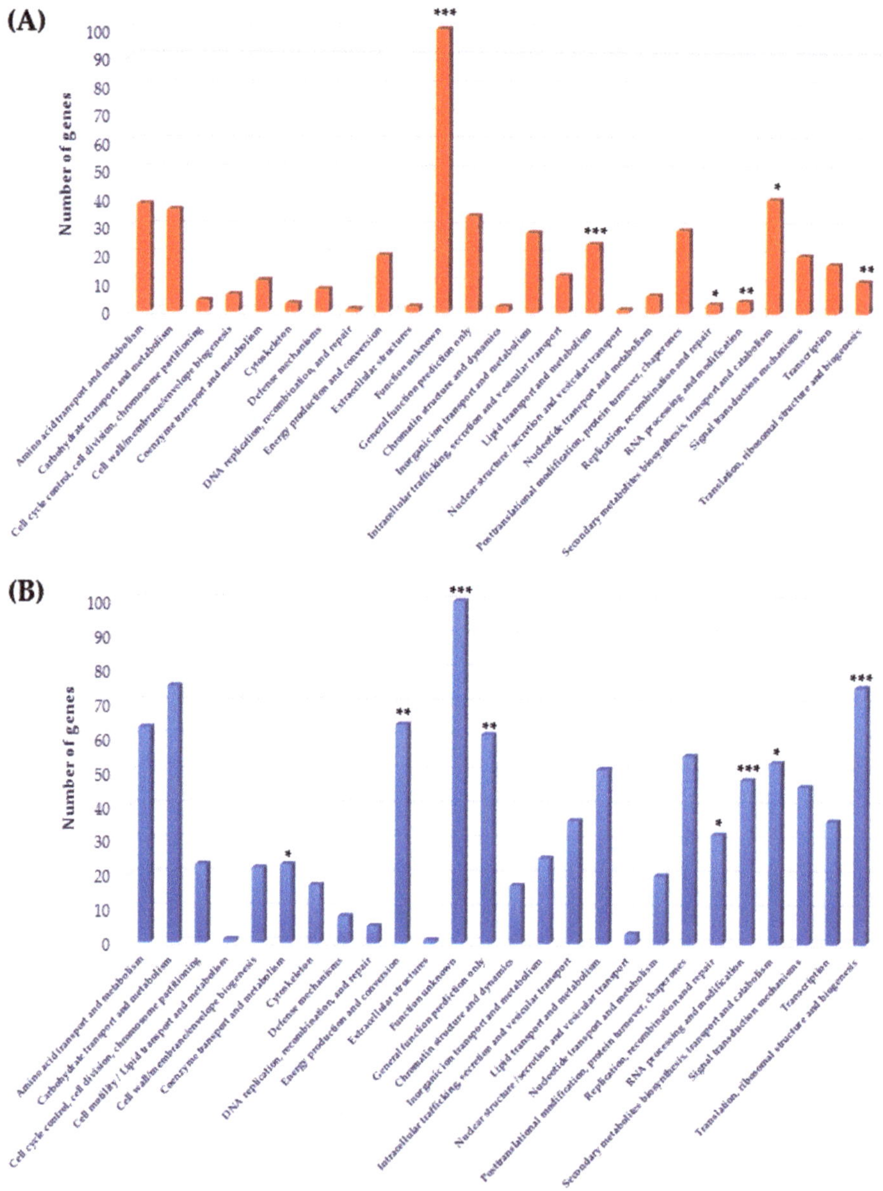

Figure 3. The most significant pathways affected by vapors of eight different EOs. Based on the results of the GO enrichment analysis, data shows changed expression level of genes (**A**) underexpressed and (**B**) overexpressed. * p-value ≤ 0.05, ** p-value ≤ 0.01 and *** p-value ≤ 0.001.

2.5. Metabolic Pathway Analysis

In the present study, considerable alterations in gene expression levels to stress responses on EOs vapors were found in *P. rubens*. We have been identified 99 altered pathways, containing 351 genes from our dataset (Figure S1). The occurrence of genes from our dataset was, together, 534, caused by the occurrence of some genes in more than one metabolic pathway. Commonly, we observed significant alterations of gene expressions in various metabolic pathways: nucleotide transport and metabolism; translation and protein biosynthesis processes; amino acid metabolism; metabolism of cofactor and vitamins; basic cellular processes; basic metabolism pathway and energy production and others (Table S2).

2.5.1. Genes Involved in Protein Synthesis

In order to deeply understand the metabolic pathways of these differentially expressed genes, they have been grouped and systematically studied. Data analysis revealed that around 28% of genes are associated with the biosynthesis of proteins. The representation of genes that encode enzymes involved in amino acid (AA) metabolisms was strongly diverse. Enzymes coding aminoacyl-tRNA synthetases (aaRS) are located in the cytoplasm of the cell. Among these enzymes, tryptophanyl (Pc12g04630), lysyl (Pc12g09250), asparaginyl (Pc12g15910), methionyl (Pc16g02040), prolyl (Pc16g07440) and histidyl (Pc22g02880) were overexpressed compared to valyl (Pc20g09480), which had decreased expression.

In many pathways, like in β-lactam antibiotic production, analysis revealed changes on the transcriptional level in metabolic pathways, where AAs have a key role. The upregulated Pc20g04020 gene encodes a threonine synthase, an enzyme that is involved in threonine biosynthesis. The other gene, which showed a higher expression level, was Pc13g07730. This overexpressed gene is annotated as L-threonine ammonia-lyase. Results showed that 5-aminolevulinate synthase (Pc22g13500) is another upregulated protein involved in these biosynthetic pathways. In contrast, cystathionine gamma-synthase (Pc20g08350) showed a decreased transcriptional level over the aforementioned genes. Taken together, all these data and the other in Table S2, interference with the functioning of the cell is confirmed at the proteomic level.

2.5.2. Transcriptional Modifications of Genes Encoding Carbohydrates

Our results showed that the metabolism of carbohydrates during the exposition of EOs is generally upregulated. From a range of metabolic pathways, encoded by multiple gene clusters, increased transcriptional levels were achieved: at glycolysis (Pc18g01490, Pc12g10630, Pc20g01630, Pc12g13500, Pc20g04410 and Pc12g16040); the pentose phosphate pathway (Pc12g02790, Pc13g14570, Pc12g13500 and Pc20g04410); fructose and mannose metabolisms (Pc16g12970, Pc22g09390, Pc21g05470, Pc20g01550, Pc13g12020, Pc12g09190, Pc12g13500, Pc21g04410, Pc21g04400 and Pc16g10970) and galactose metabolism (Pc12g13500, Pc14g00310, Pc12g07810 and Pc20g04410). The dominant gene in each group is 6-phosphofructokinase (Pc12g13500). It is responsible for transferring phosphorus-containing groups. In contrast, low levels of downregulated genes involved in the carbohydrate metabolism were obtained. As an example, Pc16g08460 the D-arabinitol dehydrogenase (NADP+) and Pc20g15580, the L-glyceraldehyde reductase. The list of all genes with changed transcript levels is displayed in Table S2.

2.5.3. Changed Expression Level of Genes Involved in Fat Metabolism

The functional analysis of the genes revealed numerous altered metabolic pathways involved in lipid metabolism. Most of them remain with unknown functions that cannot be readily assigned. These hypothetical proteins have a certain similarity with conserve proteins that encode genes for fatty acids; steroid biosynthesis and glycerophospholipid, glycerolipid and sphingolipid metabolisms. Among them, Pc22g00420 is annotated as acetyl-CoA-acetyltransferase, also known as acetoacetyl-CoA thiolase, and demonstrated low levels of expression; it has a key role in the regulation of ergosterol synthesis. The other important protein is glycerol-3-phosphate O-acetyltransferase (Pc22g05820). An analysis revealed the overexpression of this gene. In addition to the above information, all alterations are summarized in Table S2.

2.5.4. Gene Related to Secondary Metabolites

A gene-by-gene analysis showed that among the downregulated genes with hypothetical functions are just a few of them that are included in β-lactam resistance, aflatoxin biosynthesis or ABC transporters. The most interesting gene is pc18g01310, which codes for proteins regulating enzymatic reactions at the level of the cell wall and cellular organelles. It belongs to the group of chitin hydrolases -β-N-acetylhexosaminidase, which is released during autolysis into the medium. In total, two genes encoding enzymes involved in the biosynthesis of aflatoxin have shown reduced expression levels. The Pc12g16460 gene is linked by norsolorinic acid ketoreductase and pc22g19340-coding versiconal hemiacetal acetate esterase. In our investigation, EO treatments increased the expression level of the Pc21g18900 gene. It encodes a hypothetical protein included in monobactam biosynthesis and is named as the 4-hydroxy-tetrahydrodipicolinate synthase.

3. Discussion

In a fast-paced world, it is important to expand our knowledge and search for the undiscovered. The results of the present work demonstrate changes in the transcriptomic profile of the mold *P. rubens* in association with stress responses caused by the exposure to EO vapors. According to our literature research, there is no investigation studying the gene expression of *P. rubens* under exposure to the vapors of EOs.

Plant derivates, such as EOs, can modulate, inhibit and even kill the microorganism through specific mechanisms [32]. Unfortunately, the exact mechanism of EO effects is still not entirely clear; their different composition of substances makes them unique. The chemical compositions of assayed EOs are guaranteed by the producer based on the performed a GC/MS analysis and can be checked for each flask of EOs through the web page: https://sourcetoyou.com. EOs possess a wide spectrum of biological activities, including antimicrobial, antiviral, insecticidal and antifungal properties [33].

The regulation of genes responsible for protein synthesis is a limiting factor for cell survival. It can be inhibited in a number of ways, such as by altering the expression level of aminoacyl-tRNA synthetases [34]. In our case, treatments with EOs significantly show their power to decrease (for valyl-aaRS) or increase (tryptophanyl-, lysyl-, asparagynil-, methionyl-, prolyl- and histidyl-aaRS) the expression of several *P. rubens* genes. Modifications of these enzymes lead to changes in the mechanism of the translation of cellular proteins, metabolic and signaling pathways [34].

Amino acid metabolisms are an indispensable process in β-lactam production. Three AAs serve as building blocks, namely α-aminoadipate, L-cysteine and L-valine. A previous analysis demonstrated that the gene Pc20g04020 is considered to be unique for coding the threonine synthase in *P. rubens* [35]. It plays a key role in catalyzing the conversion of O-phospho-L-homoserine into L-threonine and phosphate. In the next step, L-threonine can be converted into another AA, such as L-cysteine. Pc13g07730 is annotated as L-threonine deaminase, which removes an amino group from L-threonine, resulting in the production of 2-oxobutanoate and pyruvate. The corresponding enzyme that catalyzes the conversion of glycine and succinal-CoA into 5-aminolevulinate, CoA and CO_2 is 5-aminolevulinate synthase (Pc22g13500), also named as the pyridoxal 5′-phosphate-dependent (PLP) enzyme [35]. These findings suggested that the influence of applied EOs may play an important role in modulating these signaling pathways and, thus, may regulate processes involved in cysteine production, which is very important for beta lactam production [35].

According to current knowledge, gene Pc20g08350 is coding a cystathionine gamma synthase, which is a transsulfuration enzyme involved in the catalysis of the PLP-dependent γ-replacement of O-succinyl-L-homoserine and L-cysteine, yielding L-cystathione and acetate [35]. Hypothetically, the alteration of the β-lactam production can bring to the modification a synthesis of diverse compounds that can increase the vulnerability of the fungus against different environmental factors.

Carbohydrates have an irreplaceable role in fungi and represent an important part of the cell wall (glucans and chitin), as well in the role of storage polysaccharide (glycogen), disaccharide and sugar alcohols [36]. An interesting gene is 6-phosphofructokinase (Pc12g13500) which has a strong similarity to 6-phosphofructokinase pfkA to *Aspergillus niger* [37]. The Pc16g08460 gene is annotated as D-arabinitol dehydrogenase (NADP+) and has a low expression level. If compared with other fungi, this enzyme is considered as one of the few enzymes capable of utilizing arabitol as a main substrate [38]. The downregulated Pc20g15580 gene, coding L-glyceraldehyde reductase, takes part in a D-galacturonate degradation pathway and plays an important role in the pentose and glucuronate interconversion pathways [39].

The genus *Penicillium* includes species well-known for lipolytic enzyme productions, especially lipases and esterases, which are able to use lipids as carbon sources. They also participate as biocatalysts, which provide the hydrolysis of water-soluble short acyl chain esters as sure as water-insoluble long-chain triacylglycerols [40]. Acetyl-CoA-acetyltransferase (Pc22g00420) has a key role in the regulation of ergosterol synthesis and is the first main catalytic enzyme in the mevalonate pathway. This enzyme regulates the transformation of acetotyl-CoA from two molecules of acetyl-CoA [41]. If the ergosterol production is broken down, it leads to inhibition of the fungus growth. The biosynthesis of triglyceride is necessary to provide the decisive energy molecules, as well as in the biosynthesis of fatty acids and phospholipids [42]. Triglycerol-3-phosphate O-acetyltransferase (Pc22g05820) is an integral component of the membrane that has transferase activity. This upregulated enzyme is the first enzyme catalyzing the acylation of glycerol 3-phosphate in the glycerolipid metabolism [43].

Over the past decades, scientists have further explored the role of enzymes in biological processes [44], performing a wide genome sequencing of the filamentous fungus *P. chrysogenum* Wisconsin 54-1255. A transcriptomic analysis demonstrated the altered transcript levels in many metabolic pathways. Some of our outcomes coincided to their published results. The downregulated gene associated with the secondary metabolite productions was the glycoside hydrolases (Pc18g01310), also named as β-N-acetylhexosaminidase, which has a strong similarity to the hypothetical β-hexosaminidase A precursor in *Bacillus halodurans* [44]. This enzyme is an important part in the chitinolytic system in the cell wall of the growing fungus. It plays a key role in the controlling of the cell wall chitin lysis and, at the same time, in protecting cells from rupture [45].

The next affected upregulated gene is Pc21g18900, which belongs to the DapA family proteins, catalyzing the condensation of (S)-aspartate-beta-semialdehyde ((S)-ASA) and pyruvate to 4-hydroxy-tetrahydrodipicolinate (HTPA) [46]. It is located in the cytoplasm of the cell and has a predicted role in the biosynthesis of the secondary metabolites (monobactams) and lysine [47]. To date, the function of this enzyme is still not well-understood [48].

The study of the gene expression of *P. rubens* exposed to eight different EO vapors was never done before using a microarray system; moreover, we were able to get a big amount of data at once, which is currently almost unknown. This experiment can be considered as pioneering in understanding the effect of several EOs on various *P. rubens* biochemical pathways. The information obtained by this study allowed us to begin to comprehend the antifungal mechanisms of EOs in a more complete way.

4. Materials and Methods

4.1. Essential Oils

The study was performed using the following commercial EOs: arborvitae (TP) from *T. plicata* Donn., cassia (CC) from *C. cassia* L., clove (EC) from *E. caryophyllata* Thunb., lemongrass (CF) from *C. flexuosus* Nees ex Steud., melaleuca (MA) from *M. alternifolia* Maiden, Betche. oregano (OV) from *O. vulgare* L., peppermint (MP) from *M. piperita* L. and thyme (TV) from *T. vulgaris* L. (doTERRA, Pleasant Grove, UT, USA). In order to avoid photo-oxidation, the EOs were preserved in dark glass vials.

4.2. Fungal Strain and Fungicide Activity of the Vapor Phase of EOs

Antifungal activity of eight EOs were investigated against airborne mold *Penicillium rubens Wisconsin* 54-1255 (American Type Culture Collection-ATCC 28089). The fungus was cultivated on Malt

Extract Agar (MEA; Sigma-Aldrich, Saint Louis, MO, USA) at 28 °C. In order to determine the fungicide activity of volatized EOs, a 5-mm square of growing fungal mycelia from MEA was placed into 5 mL of malt extract broth (MEB; Sigma-Aldrich) inside small (60-mm diameter) Petri dishes and incubated at room temperature (24–26 °C). Different quantity of EOs (Table 2) was applied on the inner surface of the Petri dish lid at dose levels of 1 µL/1 mL air space. As controls, served fungus *P. rubens* were placed into MEB medium without the application of EOs. To prevent vapor leakage, the Petri dishes were sealed with parafilm and incubated in the dark for 24 h for TP, CC, CF, OV, and TV; MP; EC and MA for 48 h at 22 °C. When the fungus reached the exponential phase, it was harvested. After cultivation, 50–100 mg of samples were collected by inoculation loop under aseptic conditions into sterile RNase-free plastic tubes, quickly frozen in liquid nitrogen and lysed in 1 mL of commercially available TRIzol™ Reagent (Thermo Fisher Scientific, Waltham, MA, USA). Before storage at −80 °C or directly used to perform RNA isolation, the samples needed to be incubated for 30 min at room temperature.

Table 2. Used essential oils their highest noncytotoxic concentrations.

Essential Oil	Concentration
Thuja plicata (TP)	0.025 µg/mL
Cinnamomum cassia (CC)	0.05 µg/mL
Eugenia caryophyllata (EC)	0.05 µg/mL
Cymbopogon flexuosus (CF)	0.025 µg/mL
Melaleuca alternifolia (MA)	0.05 µg/mL
Origanum vulgare (OV)	0.01 µg/mL
Mentha piperita (MP)	0.025 µg/mL
Thymus vulgaris (TV)	0.05 µg/mL

4.3. RNA Isolation and Quality Control

Total RNA was isolated from fungal mycelia using a Direct-zol™ RNA MiniPrep Plus (Zymo Research, Irvine, CA, USA) kit according to the protocol of the manufacturer. RNA quality was evaluated using the Experion Automated Electrophoresis System for RNA analysis (Bio-Rad Laboratories, Inc., Hercules, CA, USA), and the RNA concentration was measured using Nanodrop ND-2000 (Thermo Fisher Scientific). Total RNA degradation/quality was determined based on the 18S and 28S rRNA ratio. RNA samples with the RQI number (calculated based on the 18S/28S rRNA ratio) above the value 7.5 were selected for subsequent gene expression analysis.

4.4. Microarray Analysis of Gene Expression

Microarray analysis was performed using the total RNA and comparing the gene expression of the control sample with the samples treated with EOs. One-hundred nanograms of total RNA were amplified and labeled with a Low Input Quick Amp Labeling Kit (Agilent, Santa Clara, CA, USA) according to the manufacturer's instructions. After, labeling samples were purified using a GeneJET RNA Purification Kit (Thermo Fisher Scientific) to remove unincorporated nucleotides. Three-hundred nanograms of labeled test samples vs. control samples were mixed and applied onto a *Penicillium rubens* Gene Expression 8x15K custom microarray slide (Agilent Technologies) and hybridized 17 h at 65 °C by rotating the slide at the speed 10 rpm in a hybridization oven (Agilent Technologies). After hybridization, two wash steps were performed (Gene Expression Wash Buffer Kit; Agilent Technologies), and the slide was scanned at a resolution of 2 µm using an Agilent Microarray Scanner.

4.5. Image and Data Analysis

Tagged Image File Format (TIFF) multiscan image was converted and processed using Feature Extraction Software 12.1 (Agilent Technologies) to acquire spot intensities. Acquired data were analyzed in GeneSpring 12.6 GX software to obtain differences in the gene expression. Significant differences in the fold of the gene expression change between the control and treated samples were ≥2.0. Finally, a pathway analysis was performed to reveal the molecular pathways significantly modified in our experiment ($p \leq 0.05$). The R programming language version 3.3.3 [49] and the openxlsx library (github) were used. To perform a KEGG enrichment analysis, the Bioconductor library clusterProfiler [50] was applied to genes with a *p*-value cutoff 0.05 and BH *p*-value adjustment method (Benjamini – Hochberg method) [51]. Heatmaps from the KEGG enrichment analysis were created by Bioconductor libraries complexHeatmap [52] and circlize [53]. Biochemical pathways' graphical representations were obtained using the pathview [54] library from Bioconductor.

5. Conclusions

In summary, the gene expression data obtained in this study reported the transcriptional changes in *P. rubens* genes influenced by the vapors of EOs. The study provides important findings regarding the understanding of the impact of EOs to fungal cells. The effects of EOs create a cascade of events that affect the metabolism and, consequently, the functions and structure of the fungal cell. It is evident that all the metabolic pathways, especially for vital functions such as polysaccharide and carbohydrate metabolisms, fatty acid metabolism, nucleotide and nucleoside metabolisms or the regulation of the production of SM, are deeply influenced by the antifungal properties of these natural compounds.

Each EO has different characteristics, and our study evidenced how some of them have more efficient antifungal activities and some others possess only a growth-inhibiting effect. Therefore, their application depends on the result that we wish to obtain, to kill or to mitigate. As natural extracts, they should be safe and can replace some more dangerous synthetic chemicals in order to disinfect indoor contaminated air. We have achieved precious experiences in the development and optimization of the microarray analysis, so it would be worth utilizing it in order to evaluate the antifungal properties of EO vapors against other indoor airborne fungi and compare the new data with this one.

Supplementary Materials: The following are available online at http://www.mdpi.com/2079-6382/9/6/343/s1: Figure S1. Hierarchical clustering of differentially expressed genes, illustrated by the heatmap. Table S1. List of regulated genes. Table S2: List of genes and their corresponding metabolic pathways.

Author Contributions: Conceptualization, A.Š. and D.P.; methodology, Z.K., M.B., A.Š., A.P. and D.P.; software, A.Š. and G.B.; formal analysis, Z.K., A.Š. and D.P.; investigation, Z.K. and A.Š.; resources, D.P.; data curation, A.Š. and G.B.; writing—original draft preparation, Z.K., A.Š., G.B. and D.P.; writing—review and editing, Z.K., A.Š. and D.P.; visualization, A.Š. and G.B.; supervision, M.B., A.Š. and D.P.; project administration, D.P. and funding acquisition, M.B., A.P. and D.P. All authors have read and agreed to the published version of the manuscript.

Funding: This research was funded by VEGA projects 2/0061/17 (Innovative disinfection strategies: the essential oils effect on microflora and materials of cultural heritage objects) and 2/0059/19 (Combination of nanoparticles and essential oils for mitigating the biodeterioration on various types of building materials).

Conflicts of Interest: The authors declare no conflict of interest. The funders had no role in the design of the study; in the collection, analyses or interpretation of data; in the writing of the manuscript or in the decision to publish the results.

References

1. Prestinaci, F.; Pezzotti, P.; Pantosti, A. Antimicrobial resistance: A global multifaceted phenomenon. *Pathog. Glob. Health* **2015**, *109*, 309–318. [CrossRef] [PubMed]
2. Yap, P.S.X.; Yiap, B.C.; Ping, H.C.; Lim, S.H.E. Essential oils, a new horizon in combating bacterial antibiotic resistance. *Open Microbiol. J.* **2014**, *8*, 6–14. [CrossRef]
3. Schmidt, E. Production of Essential Oils. In *Handbook of Essential Oils. Science, Technology and Applications*; Baser, K.H., Buchbauer, G., Eds.; CRC Press: Boca Raton, FL, USA, 2010; pp. 83–119.
4. Bilia, A.R.; Guccione, C.; Isacchi, B.; Righeschi, C.; Firenzuoli, F.; Bergonzi, M.C. Essential Oils Loaded in Nanosystems: A Developing Strategy for a Successful Therapeutic Approach. *Evid Based Complement Altern. Med.* **2014**, *651593*. [CrossRef] [PubMed]

5. Opara, E.I.; Chohan, M. Culinary Herbs and Spices: Their Bioactive Properties, the Contribution of Polyphenols and the Challenges in Deducing Their True Health Benefits. *Int. J. Mol. Sci.* **2014**, *15*, 19183–19202. [CrossRef] [PubMed]
6. Nieto, G. Biological Activities of Three Essential Oils of the Lamiaceae Family. *Medicines* **2017**, *4*, 63. [CrossRef]
7. Carson, C.F.; Hammer, K.A.; Riley, T.V. *Melaleuca alternifolia* (Tea Tree) oil: A review of antimicrobial and other medicinal properties. *Clin. Microbiol. Rev.* **2006**, *19*, 50–62. [CrossRef]
8. Han, X.; Parker, T.L. Anti-inflammatory activity of clove (*Eugenia caryophyllata*) essential oil on human dermal fibroblasts. *Pharm. Biol.* **2017**, *55*, 1619–1622. [CrossRef]
9. Rao, P.V.; Gan, S.H. Cinnamon: AMultifaceted Medicinal Plant. *Evid Based Complement Altern. Med.* **2014**, 642942. [CrossRef]
10. Guenther, E. *The Essential Oils*; Van Nostrand, Inc.: Princeton, NJ, USA, 1950; p. 703.
11. Akhila, A. *Essential Oil-Bearing Grasses: The Genus Cymbopogon*; CRC Press: Boca Raton, FL, USA, 2010; p. 262.
12. Chao, S.C.; Young, D.G.; Oberg, C.J. Screening for inhibitory activity of essential oils on selected bacteria, fungi and viruses. *J. Essent. Oil Res.* **2000**, *12*, 639–649. [CrossRef]
13. Sharma, P.R.; Mondhe, D.M.; Muthiah, S.; Pal, H.C.; Shahi, A.K.; Saxena, A.K.; Qazi, G.N. Anticancer activity of an essential oil from *cymbopogon flexuosus*. *Chem. Biol. Interact.* **2009**, *179*, 160–168. [CrossRef]
14. Hudson, J.; Kuo, M.; Vimalanathan, S. The Antimicrobial Properties of Cedar Leaf (*Thuja plicata*) Oil; A Safe and Efficient Decontamination Agent for Buildings. *Int. J. Environ. Res. Public Health* **2011**, *8*, 4477–4487. [CrossRef] [PubMed]
15. Shams-Ghahfarokhi, M.; Aghaei-Gharehbolagh, S.; Aslani, N.; Razzaghi-Abyaneh, M. Investigation on distribution of airborne fungi in outdoor environment in Tehran, Iran. *J. Environ. Health Sci. Eng.* **2014**, *12*, 54. [CrossRef] [PubMed]
16. Burge, H. Bioaerosols: Prevalence and health effects in the indoor environment. *J. Allergy Clin. Immun.* **1990**, *86*, 687–701. [CrossRef]
17. Rogawansamy, S.; Gaskin, S.; Taylor, M.; Pisaniello, D. An Evaluation of Antifungal Agents for the Treatment of Fungal Contamination in Indoor Air Environments. *Int. J. Environ. Public Health* **2015**, *12*, 6319–6332. [CrossRef]
18. Cooley, J.D.; Wong, W.C.; Jumper, C.A.; Straus, D.C. Correlation between the prevalence of certain fungi and sick building syndrome. *Occup. Environ. Med.* **1998**, *55*, 579–584. [CrossRef]
19. Visagie, C.M.; Houbraken, J.; Frisvad, J.C.; Hong, S.-B.; Klaassen, C.H.W.; Perrone, G.; Seifert, K.A.; Varga, J.; Yaguchi, T.; Samson, R.A. Identification and nomenclature of the genus *Penicillium*. *Stud. Mycol.* **2014**, *78*, 343–371. [CrossRef]
20. Wilson, S.C.; Straus, D.C. The presence of fungi associated with sick building syndrome in North American zoological institutions. *J. Zoo Wildl. Med.* **2002**, *33*, 322–327. [CrossRef]
21. Mehra, T.; Köberle, M.; Braunsdorf, C.; Mailänder-Sanchez, D.; Borelli, C.; Schaller, M. Alternative approaches to antifungal therapies. *Exp. Dermatol.* **2012**, *21*, 778–782. [CrossRef]
22. Glegg, G.A.; Richards, J.P. Chemicals in household products: Problems with solutions. *Environ. Manag.* **2007**, *40*, 889–901. [CrossRef]
23. Whiley, H.; Gaskin, S.; Schroder, T.; Ross, K. Antifungal properties of essential oils for improvement of indoor air quality: A review. *Rev. Envirion. Health* **2018**, *33*, 63–76. [CrossRef]
24. Reyes-Jurado, F.; Navarro-Cruz, A.R.; Ochoa-Velasco, C.E.; Palou, E.; López-Malo, A.; Ávila-Sosa, R. Essential oils in vapor phase as alternative antimicrobials: A review. *Crit. Rev. Food Sci. Nutr.* **2019**, *18*, 1–10. [CrossRef] [PubMed]
25. Dorman, H.J.D.; Deans, S.G. Antibacterial agents from plants: Antibacterial activity of plant volatile oils. *J. Appl. Microbiol.* **2000**, *88*, 308–316. [CrossRef] [PubMed]
26. Pandey, A.K.; Kumar, P.; Singh, P.; Tripathi, N.N.; Bajpai, V.K. Essential Oils: Sources of Antimicrobials and Food Preservatives. *Front. Microbiol.* **2016**, *7*, 2161. [CrossRef] [PubMed]
27. Cox, S.D.; Mann, C.M.I.; Markham, J.L.; Bell, H.C.; Gustafson, J.E.; Warmington, J.R.; Wyllie, S.G. The mode of antimicrobial action of the essential oil of *Melaleuca alternifolia* (tea tree oil). *J. Appl. Microbiol.* **2000**, *88*, 170–175. [CrossRef] [PubMed]
28. Sealfon, S.C.; Chu, T.T. RNA and DNA Microarrays. *Methods Mol. Biol.* **2011**, *671*, 3–34. [CrossRef] [PubMed]

29. Bajaj, I.; Veiga, T.; van Dissel, D.; Pronk, J.T.; Daran, J.M. Functional characterization of a *Penicillium chrysogenum* mutanase gene induced upon co-cultivation with Bacillus subtilis. *BMC Microbiol.* **2014**, *14*, 114. [CrossRef]
30. Lowe, R.; Shirley, N.; Bleackley, M.; Dolan, S.; Shafee, T. Transcriptomics technologies. *PLoS Comput. Biol.* **2017**, *18*, 13. [CrossRef]
31. Salo, O.V.; Ries, M.; Medema, M.H.; Lankhorst, P.P.; Vreeken, R.J.; Bovenberg, R.A.L.; Driessen, J.M. Genomic mutational analysis of the impact of the classical strain improvement program of β-lactam producing *Penicillium chrysogenum*. *BMC Genom.* **2015**, *16*, 937. [CrossRef]
32. Hyldgaard, M.; Mygind, T.; Meyer, R.L. Essential oils in food preservation: Mode of action, synergies, and interactions with food matrix components. *Front. Microbiol.* **2012**, *3*, 12. [CrossRef]
33. Nazzaro, F.; Frantianni, F.; Coppola, R.; Feo, V. Essential oils and antifugal activity. *Pharmaceuticals* **2017**, *10*, 4. [CrossRef]
34. Rajendran, V.; Kalita, P.; Shukla, H.; Kumar, A.; Tripathi, T. Aminoacyl-tRNA synthetases: Structure, function and drug discovery. *Int. J. Biol. Macromol.* **2018**, *111*, 400–414. [CrossRef] [PubMed]
35. Wu, M.; Crismaru, C.G.; Salo, O.; Bovenberg, R.A.L.; Driessen, A.J.M. Impact of classical strain improvement of *Penicillium rubens* on amino acid metabolism during β-lactam production. *Appl. Environ. Microbiol.* **2020**, *86*, e01561-19. [CrossRef] [PubMed]
36. Rabha, J.; Jha, D.K. Metabolic Diversity of Penicillium. In *Developments in Microbial Biotechnology and Bioengineering-Penicillium System Properties and Applications*; Gupta, V.K., Rodriguez-Couto, S., Eds.; Elsevier: Amsterdam, The Netherlands, 2018; pp. 217–234.
37. Capuder, M.; Šolar, T.; Benčina, M.; Legiša, M. Highly active, citrate inhibition resistant form of *Aspergillus niger* 6-phosphofructo-1-kinase encoded by a modified pfkA gene. *J. Biotechnol.* **2009**, *144*, 51–57. [CrossRef] [PubMed]
38. Link, T.; Lohaus, G.; Heiser, I.; Mendgen, K.; Hahn, M.; Voegele, R.T. Characterization of a novel $NADP^+$-dependent D-arabitol dehydrogenase from the plant pathogen *Uromyces fabae*. *Biochem. J.* **2005**, *389*, 289–295. [CrossRef]
39. Martens-Uzunova, E.S.; Schaap, P.J. An evolutionary conserved D-galacturonic acid metabolic pathway operates across filamentous fungi capable of pectin degradation. *Fungal. Genet. Biol.* **2008**, *45*, 1449–1457. [CrossRef]
40. Chinaglia, S.; Chiarelli, L.R.; Maggi, M.; Rodolfi, M.; Valentini, G.; Picco, A.M. Biochemistry of lipolytic enzymes secreted by *Penicillium solitum* and *Cladosporium cladosporioides*. *Biosci. Biotechnol. Biochem.* **2014**, *78*, 245–254. [CrossRef]
41. Zhong, Z.; Norvienyeku, J.; Yu, J.; Chen, M.; Cai, R.; Hong, Y.; Chen, L.; Zhang, D.; Wang, B.; Zhou, J.; et al. Two different subcellular-localized Acetoacetyl-CoA acetyltransferases differentiate diverse functions in *Magnaporthe oryzae*. *Fungal. Genet. Biol.* **2015**, *83*, 58–67. [CrossRef]
42. Coleman, R.A.; Lee, D.P. Enzymes of triacylglycerol synthesis and their regulation. *Prog. Lipid. Res.* **2004**, *43*, 134–176. [CrossRef]
43. Gao, Q.; Shang, Y.; Huang, W.; Wang, C. Glycerol-3-phosphate acyltransferase contributes to triacylglycerol biosynthesis, lipid droplet formation, and host invasion in *Metarhizium robertsii*. *Appl. Environ. Microbiol.* **2013**, *79*, 7646–7653. [CrossRef]
44. Van der Berg, M.A.; Albang, R.; Albermann, K.; Badger, J.H.; Daran, J.M.; Driessen, A.J.; Garcia-Estrada, C.; Fedorova, N.D.; Harris, D.M.; Heijne, W.H.; et al. Genome sequencing and analysis of the filamentous fungus *Penicillium chrysogenum*. *Nat. Biotechnol.* **2008**, *26*, 1161–1168. [CrossRef]
45. Slámová, K.; Bojarová, P.; Petrásková, L.; Křen, V. β-N-Acetylhexosaminidase: What's in a name? *Biotechnol. Adv.* **2010**, *28*, 682–693. [CrossRef]
46. Desbois, S.; John, U.P.; Perugini, A. Dihydrodipicolinate synthase is absent in fungi. *Biochimie* **2018**, *152*, 73–84. [CrossRef] [PubMed]
47. Domigan, L.J.; Scally, S.W.; Fogg, M.J.; Hutton, C.A.; Perugini, M.A.; Dobson, R.C.J.; Muscroft-Taylor, A.C.; Gerrard, J.A.; Devenish, S.R.A. Characterisation of dihydrodipicolinate synthase (DHDPS) from *Bacillus anthracis*. *Biochim. Biophysic. Acta* **2009**, *1794*, 1510–1516. [CrossRef]
48. Ren, W.; Tao, J.; Shi, D.; Chen, W.; Chen, C. Involvement of a dihydopicolinate synthase gene (FaDHDPS1) in fugal development, pathogenesis and stress responses in *Fusarium asiaticum*. *BMC Microbiol.* **2018**, *18*, 128. [CrossRef]

49. R Core Team. *R: A Language and Environment for Statistical Computing*; R Foundation for Statistical Computing: Vienna, Austria, 2013; Available online: http://www.R-project.org/ (accessed on 14 April 2020).
50. Yu, G.; Wang, L.; Han, Y.; He, Q. clusterProfiler: An R package for comparing biological themes among gene clusters. *OMICS* **2012**, *16*, 284–287. [CrossRef]
51. Benjamini, Y.; Hochberg, Y. Controlling the false discovery rate: A practical and powerful approach to multiple testing. *J. R. Statist. Soc.* **1995**, *57*, 289–300. [CrossRef]
52. Gu, Z.; Eils, R.; Schlesner, M. Complex heatmaps reveal patterns and correlations in multidimensional genomic data. *Bioinformatics* **2016**, *32*, 2847–2849. [CrossRef]
53. Gu, Z.; Gu, L.; Eils, R.; Schlesner, M.; Brors, B. Circlize Implements and enhances circular visualization in R. *Bioinformatics* **2014**, *30*, 2811–2812. [CrossRef]
54. Luo, W.; Brouwer, C. Pathview: An R/Bioconductor package for pathway-based data integration and visualization. *Bioinformatics* **2013**, *29*, 1830–1831. [CrossRef]

antibiotics

MDPI

Article

Morphological, Chemical, and Genetic Characteristics of Korean Native Thyme Bak-Ri-Hyang (*Thymus quinquecostatus* Celak.)

Minju Kim [1], Jun-Cheol Moon [2], Songmun Kim [1,*] and Kandhasamy Sowndhararajan [3,*]

1 School of Natural Resources and Environmental Science, Kangwon National University, Chuncheon 24341, Gangwon-do, Korea; scent@kangwon.ac.kr
2 Agriculture and Life Sciences Research Institute, Kangwon National University, Chuncheon 24341, Gangwon-do, Korea; jhmoon73@gmail.com
3 Department of Botany, Kongunadu Arts and Science College, Coimbatore 641029, Tamil Nadu, India
* Correspondence: perfume@kangwon.ac.kr (S.K.); sowndhar1982@gmail.com (K.S.); Tel.: +82-33-250-6447 (S.K.); +91-422-2642095 (K.S.)

Received: 27 April 2020; Accepted: 25 May 2020; Published: 28 May 2020

Abstract: Bak-ri-hyang (*Thymus quinquecostatus* Celak.) is an important medicinal and aromatic plant in Korea. *T. quinquecostatus* population and is always mixed with other thyme cultivars during cultivation and marketing. Hence, this study aimed to determine the genetic variability and the essential oil composition of three Korean native thyme, *T. quinquecostatus* cultivars collected from the Wolchul, Jiri, and Odae mountains, in comparison with six commercial thyme cultivars (*T. vulgaris*), to distinguish Bak-ri-hyang from other thyme cultivars. The composition of essential oils obtained from nine individuals was analyzed by gas chromatography–mass spectrometry (GC–MS). The random amplified polymorphic DNA (RAPD) analysis was accomplished using 16 different primers. The GC–MS analysis revealed that Wolchul, creeping, golden, and orange cultivars belong to the geraniol chemotype. Whereas the Odae, lemon, and silver cultivars belong to the thymol chemotype. Further, linalool was the most abundant component in carpet and Jiri cultivars. The RAPD analysis demonstrated that all thyme cultivars showed characteristic RAPD patterns that allowed their identification. In total, 133 bands were obtained using 16 primers, and 124 bands were polymorphic, corresponding to 93.2% polymorphism. Cluster analysis of RAPD markers established the presence of clear separation from nine thyme cultivars. The highest dissimilarity and similarity coefficient of the RAPD markers were 0.58 and 0.98, respectively. According to the RAPD patterns, the nine thyme cultivars could be divided into two major clusters. Among three Korean cultivars, the Wolchul and Odae cultivars were placed into the same cluster, but they did not show identical clustering with their essential oil compositions. The findings of the present study suggest that RAPD analysis can be a useful tool for marker-assisted identification of *T. quinquecostatus* from other *Thymus* species.

Keywords: essential oil; genetic; RAPD; thyme; *Thymus quinquecostatus*; *Thymus vulgaris*

1. Introduction

The genus *Thymus* (Lamiaceae) consists of approximately 300 species of herbaceous perennials and sub-shrubs, distributed throughout the world and predominantly found in the Mediterranean basin [1,2]. They are widely used as spices, herbal tea, and insecticide in addition to flavor and fragrance materials. Among these, *Thymus quinquecostatus* (Bak-ri-hyang) is a scrubby subshrub and an important aromatic plant in Korea. Two varieties of *T. quinquecostatus* such as *T. quinquecostatus* Celak and *T. quinquecostatus* var. *japonica* are found in Korea [3]. In traditional systems of medicine, *T. quinquecostatus* is used for the treatment of cough, inflammation, preventing excessive intestinal gas, and diaphoresis [3–5].

Recent scientific studies reported that *T. quinquecostatus* has antioxidant, antimicrobial, insecticidal, immunological, antidiabetic, and antitumor properties [6–8]. The essential oil of *T. quinquecostatus* is extensively used in cosmetic industries for fragrance purposes. Owing to its medicinal and aromatic properties, it is also broadly used in pharmaceutical and food industries [9–11]. The most abundant components in *T. quinquecostatus* essential oil are thymol, γ-terpinene, and *p*-cymene [8].

Currently, thyme seeds are commercially available in the market. However, most thyme cultivars are not yet chemically or genetically characterized. In particular, *T. quinquecostatus* cultivar is often mixed with other thyme cultivars during cultivation and in nature. Therefore, it is important to validate the methods for the identification of the Korean native thyme cultivar, *T. quinquecostatus*. The essential oil components were commonly used to examine variations between populations [12–14]. It was reported that the genetic constitution and environmental conditions highly influenced the yield and essential oil composition of various plant species [1]. The chemical composition of essential oils might be further altered due to cross hybridization, morphogenesis, polyploidization, extraction methods, drying conditions, stages of harvesting, etc. According to the type, the major components represent 60–95% of the total essential oil. The essential oil composition might vary with the cultivar type. In *T. vulgaris*, seven different chemotypes, such as thymol, carvacrol, geraniol, linalool, thujanol-4, terpinen-4-ol, and 1,8-cineole are described [15,16]. However, the identification of the factors responsible for the chemical polymorphism registered within species is the most challenging aspect of the essential oil analysis.

In addition, the morphological similarity and anatomical features of thyme cultivars create a problem for the correct identification. The genetic variation of plants was also affected by evolution in both inter- and intra-species. Since the distinctness of *Thymus* species from another is always challenging to identify, several characters might need to be considered. DNA-based molecular markers were used for the successful detection of this genetic variation, in the process of evolution, gene flow, and population diversity in many plant species. In recent decades, a number of molecular techniques were used to assess genetic diversity in plants. Among them, the PCR-based random amplified polymorphic DNAs (RAPD) were used for the identification of cultivar and genetic relationships, among and within plant species. Similarity banding pattern was scored for calculation of genetic relatedness [13]. Furthermore, RAPD analysis did not need any previous knowledge regarding the target sequence on the genome of the species [15]. When compared with other molecular markers, RAPD markers can produce a high percentage of polymorphism in plants with very similar genetic characteristics. RAPD analysis has various advantages, including easiness, rapidity, requiring a small amount of genomic DNA as a template, and the possibility for detecting dissimilarity in the coding and noncoding areas of the genome [17–19]. Previously, several authors have reported the use of RAPD markers to study the genetic diversity, phylogenetic relationship, and its combination with the analysis of essential oil composition in various *Thymus* species [1,13,15,20,21]. RAPD data are also important to solve the taxonomic issues within and among plant species. Furthermore, the physiological and morphological variations, essential oil composition, ploidy level, and the relationship between the chemical and genetic evaluations of the *Thymus* species were assessed [15].

In recent times, several companies have commercialized thyme seeds. Hence, Korean native thyme, *T. quinquecostatus* (Bak-ri-hyang) is always mingled with commercial thyme species during cultivation and marketing. In addition, there are no studies on the RAPD evaluation of *T. quinquecostatus* and most studies focus on its essential oil composition. In this context, the present study aimed to evaluate the chemical and genetic variations among *T. quinquecostatus* cultivars from Korea and six commercial *T. vulgaris* cultivars, by using different RAPD markers and essential oil profiles, in order to distinguish Korean native thyme, Bak-ri-hyang from commercial thyme cultivars.

2. Methods

2.1. Plant Materials

First, a survey of the native thyme species (*T. quinquecostatus*) grown in Korea was carried out. We also interviewed experts who had ethnobotanical knowledge on Korean native thyme. According to their ethnobotanical information, we collected *T. quinquecostatus* from three accessions, such as Odae Mt, Wolchul Mt, and Jiri Mt in Korea, during April 2018 (Figures 1 and 2). In addition, fresh plants of six *T. vulgaris* cultivars (lemon, golden lemon (golden), carpet, orange, silver, and creeping) were purchased from Daerim Horticulture, Gwachon, Happy Horticulture, Goyang and Nature Horticulture, Yangju, Republic of Korea (Figure 2).

The plants were authenticated and deposited in the Herbarium, Daejin University, Pocheon, Gyeonggi-do, Republic of Korea, with voucher numbers: lemon—DJU20180713, golden—DJU20180712, carpet—DJU20180717, orange—DJU20180715, silver—DJU20180714, creeping—DJU20180716, Odae Mt.—DJU20180718, Wolchul Mt.—DJU20180719, and Jiri Mt.—DJU20180720. The collected samples were kept at −20 °C for the essential oil analysis and −80 °C for the molecular analysis.

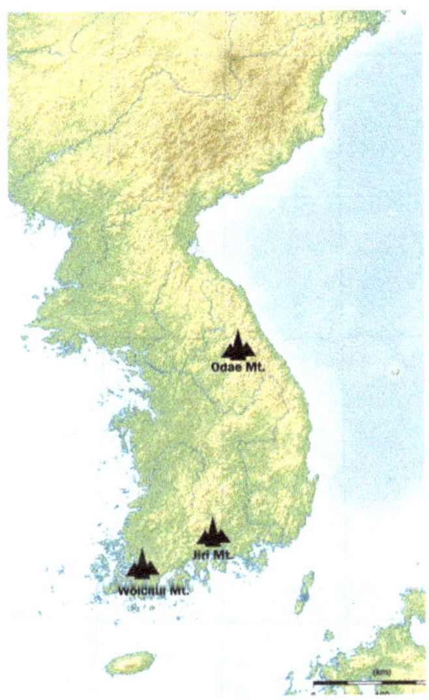

Figure 1. The map showing the collection sites of three accessions (Wolchul, Odae, and Jiri mountains) of *Thymus quinquecostatus* in South Korea.

2.2. Morphological Characteristics

The morphological parameters such as stem type, stem branch, stem color, leaf shape, number of auxiliary leaves, and trichome position were observed for the six commercial and three Korean native thyme cultivars.

2.3. Essential Oil Extraction

The essential oil from nine thyme samples was isolated by steam distillation, using a Clevenger-type apparatus. The steam distillation was performed at 100 °C for 90 min. The essential oil isolation was carried out in triplicates and the yield (%) was calculated as volume (mL) of the isolated oil per 100 g of the fresh plant material. The isolated essential oil was dried using anhydrous sodium sulfate and stored at 4 °C, until tested. The color of essential oils obtained from the three Korean native *T. quinquecostatus* cultivars was measured, using the Chromameter CT-300 (Mintola Camera Co. Ltd.,

Japan). The intensity of the color was expressed in terms of L^* lightness, a^* greenness, and b^* yellowness. The color values of L^*, a^*, and b^* were taken in triplicates for each sample.

Figure 2. The morphology of six commercial *Thymus vulgaris* cultivars and three Korean native *Thymus quinquecostatus* cultivars. (1) Lemon; (2) golden; (3) carpet; (4) orange; (5) silver; (6) creeping; (7) Odae Mt.; (8) Wolchul Mt.; and (9) Jiri Mt.

2.4. Gas Chromatography–Mass Spectrometry (GC–MS) Analysis

The identification of the essential oil components from different thyme cultivars was performed using a Varian CP3800 gas chromatograph coupled with a Varian 1200 L mass detector (Varian, CA, USA). The GC–MS was equipped with a VF-5MS polydimethylsiloxane capillary column (30 m × 0.25 mm × 0.25 μm). The oven temperature was programmed from 50 °C to 250 °C, at a rate of 5 °C/min. The injector temperature was 250 °C and the ionization detector temperature was 200 °C. Helium was the carrier gas (1 mL/min) and the injected volume of the sample was 2 μL, with a split ratio of 10:1. For mass spectra, an electron ionization system with ionization energy of 70 eV was used. The mass range was 50–500 m/z. The determination of the percentage composition of each component was based on the normalization of the GC peak areas. The identification of the essential oil components was based on the comparison of their retention indices (RIs), relative to a homologous series of *n*-alkanes (C_8–C_{22}) and mass spectra from the National Institute of Standards and Technology (NIST, 3.0) library and literature data [22].

2.5. DNA Extraction

The total genomic DNA was isolated from one gram of young leaves of plants, according to the CTAB (cetyl trimethylammonium bromide) extraction method [15]. DNA pellets were dissolved in TE (Tris–EDTA) buffer and RNA was removed by digestion with DNase-free RNase A. The purified total DNA was quantified and its quality was verified using a spectrophotometer, and a diluted solution with the same concentration (10 ng/μL) was prepared by adding TE buffer and was stored at 4 °C.

2.6. Randomly Amplified Polymorphic DNA (RAPD) Analysis

A total of 16 primers (OPA-09, OPA-10, OPA-11, OPA-12, OPA-13, OPA-14, OPA-15, OPA-16, OPA-17, OPA-18, OPA-19, OPA-20, OPB-01, OPB-02, OPB-03, and OPB-04) were used for the RAPD analysis (Table 1). The selection of primers was based on high polymorphisms and good reproducibility of the fragments generated. RAPD amplification was performed in a volume of 25 μL containing 10 ng total DNA, 1× PCR buffer, 3.0 mM $MgCl_2$, 200 μM deoxynucleotide triphosphates (dNTPs), 1 μM primer, 1 μg/mL (w/v) Bovine Serum Albumin (BSA), and 1 unit *Taq* DNA polymerase (Invitrogen). The amplification reactions were performed in a thermocycler and consisted of an initial 5 min denaturation step at 95 °C, followed by 40 cycles of 20 s at 95 °C, 40 s at 35 °C, and 60 s at 72 °C. A final extension of 5 min at 72 °C completed the amplification. The PCR products were separated in 1.2% agarose gels 1× TAE buffer (Tris–Acetate). The gels were stained with ethidium bromide, visualized with a UV transilluminator.

Table 1. The names and sequences of the primers used for random amplified polymorphic DNA (RAPD) analysis.

S. No.	Primer	Sequence
1	OPA-09	GGGTAACGCC
2	OPA-10	GTGATCGCAG
3	OPA-11	CAATCGCCGT
4	OPA-12	TCGGCGATAG
5	OPA-13	CAGCACCCAC
6	OPA-14	TCTGTGCTGG
7	OPA-15	TTCCGAACCC
8	OPA-16	AGCCAGCGAA
9	OPA-17	GACCGCTTGT
10	OPA-18	AGGTGACCGT
11	OPA-19	CAAACGTCGG
12	OPA-20	GTTGCGATCC
13	OPB-01	GTTTCGCTCC
14	OPB-02	TGATCCCTGG
15	OPB-03	CATCCCCCTG
16	OPB-04	GGACTGGAGT

2.7. Statistical Analysis

To calculate RAPD polymorphism, the RAPD markers were scored for the presence (1) or absence (0) of amplified bands for 9 thyme cultivars. Genetic similarity was estimated using the Jaccard's coefficients. Cluster analysis was performed using the unweighted pair group method with an arithmetic mean (UPGMA), and dendrograms were drawn using NTSYS software version 2.02.

3. Results

3.1. Morphological Characteristics of the Thyme Cultivars

The morphological characteristics of six commercial and three Korean native thyme cultivars are presented in Table 2. In these, all three *T. quinquecostatus* cultivars had a creeping type of stem. On the other hand, lemon, golden, orange, and silver cultivars possessed an erect stem type. The length of the stem branch varied among different cultivars. In the creeping stem type, the length of the stem branch ranged from 2 to 8 cm, whereas the length of the stem branch in the erect type ranged from 3 to 11 cm. Carpet cultivar possessed a higher number of stem branches than the other cultivars. The shape of the leaves was mainly oval, followed by oblanceolate. In the case of the Bak-ri-hyang cultivars, the Odae and Jiri cultivars had an oval shape of leaves. The leaf shape of the Wolchul cultivar was oblanceolate. Furthermore, the Bak-ri-hyang cultivars possessed a higher number of auxiliary leaves when compared with the commercial thyme cultivars. The trichome position was mainly observed at the leaf petiole. In the Wolchul cultivar, the trichome position was observed at the leaf margin.

3.2. The Chemical Composition of Essential Oils

The yield and chemical composition of essentials oils obtained from the nine thyme cultivars are presented in Tables 3 and 4. The essential oil components and their concentration produced by thyme cultivars were very diverse. The essential oil yields ranged between 0.12% and 0.43% (v/w) for the *T. quinquecostatus* cultivars. The highest yield was obtained from the Odae cultivar (0.43%). For commercial cultivars of *T. vulgaris*, the essential oil yields ranged from 0.23% to 0.33%. In these, the highest yields were obtained from the lemon and silver cultivars (0.34% and 0.33%, respectively), and the lowest yield was obtained from the carpet cultivar (0.23%). The color profile of the essential oils obtained from the *T. quinquecostatus* cultivars was measured. The L^* value of the essential oils of the Wolchul, Odae, and Jiri cultivars was 92.48, 92.54, and 92.39, respectively. Wolchul and Jiri cultivars possessed similar a^* value (0.18). With regards to the b^* value, the Wolchul cultivar showed the highest value (2.29) and the Odae cultivar showed the lowest value (1.89). The total number of components in the analyzed essential oils ranged between 32 (creeping cultivar) and 43 (lemon cultivar). In these nine samples, twelve compounds were detected in all essential oil samples and these oils were dominated by monoterpenes, accounting for 79.95–92.16% with 0.03–46.47% of monoterpene hydrocarbons and 43.86–88.46% of oxygenated monoterpenes. Whereas sesquiterpenes achieved 6.50–29.17% with 5.83–15.07% of sesquiterpene hydrocarbons and 0.32%–17.23% of oxygenated sesquiterpenes.

Geraniol, geranyl acetate, linalool, phenylethyl alcohol, γ-terpinene, and thymol were detected as the most abundant components, which comprised more than 20% in at least one essential oil (Figure 3). Results of the essential oil composition revealed that all three *T. quinquecostatus* cultivars tested belonged to different chemotypes (Supplementary Figure S1). Cultivars of Wolchul, Odae, and Jiri were mainly composed of geraniol (42.94%), thymol (30.54%), and linalool (47.89%), respectively. In the case of the commercial cultivars of *T. vulgaris*, the lemon and silver cultivars belonged to the thymol chemotype (43.91% and 66.24%, respectively). On the other hand, the creeping, golden, and orange cultivars belonged to the geraniol chemotype (29.57%, 65.99%, and 44.70%, respectively). With regards to the carpet cultivar, linalool (48.16%) was recorded as the most abundant component. Furthermore, geranyl acetate was detected as a major component in the Wolchul (26.49%) and silver (29.86%) cultivars. γ-Terpinene (23.92%) and *p*-cymene (11.13%) were major components in the Odae cultivar. In the creeping cultivar, neral (11.75%) and geranial (18.21%) were also recorded as major components. Other important compounds detected in all essential oils were caryophyllene (2.87%–7.02%), borneol (0.41%–5.91%), β-bisabolene (0.23%–3.86%), and 1-octen-3-ol (0.39%–3.61%). α-Elemol (11.62% and 8.52%, respectively) was also recorded as a major component in the carpet and creeping cultivars.

3.3. RAPD Analysis

The molecular analysis revealed that the RAPD primers produced clear and reproducible polymorphic bands (Figure 4) among 9 thyme cultivars, and generated a total of 133 amplicons from 16 primers. The number of bands per primer varied from 4 (OPA-12, OPA-14, and OPB-04) to 16 (OPA-19), with an average of 8.31 bands per primer. In these, 124 amplicons were polymorphic, corresponding to 93.23% polymorphism (Table 5). Eight primers gave the highest percentage of polymorphism (100%), while the lowest percentage of polymorphism (75%) was obtained by OPA-12 and OPB-04 primers (Table 5).

Table 2. Morphological characteristics of six commercial *Thymus vulgaris* cultivars and three Korean native *Thymus quinquecostatus* cultivars.

Thymus Sp.	Variety	Stem Type	Stem Branch		Stem Color			Leaf Shape	Number of Auxiliary Leaves	Trichome Position		
			Length (cm)	Number	Purple	Pale Purple	Pale Green			Leaf Petiole	Leaf Margin	Leaf Surface
Thymus vulgaris	Lemon	Erect	>3	2	-	-	√	Oval	0	-	-	-
	Golden	Erect	0~8	2	-	√	√	Elliptical	2	√	-	-
	Carpet	Creeping	0~4	4	-	√	-	Oval	0	√	√	√
	Orange	Erect	0~11	[illegible]	-	-	√	Oblanceolate /Oblong	0~2	√	-	-
	Silver	Erect	0~5	0	-	√	-	Oblong	0~2	-	-	-
	Creeping	Creeping	0~2	2	√	-	-	Oval	0~2	√	-	-
Bak-ri-hyang (*Thymus quinquecostatus*)	Odae Mt.	Creeping	>2	0	√	-	-	Oval	2~6	√	-	-
	Wolchul Mt.	Creeping	0~8	2	-	√	√	Oblanceolate	2~8	-	√	-
	Jiri Mt.	Creeping	>2	0	√	-	-	Oval	2~6	√	-	-

Table 3. The yield and color of the essential oils isolated from the six commercial *Thymus vulgaris* cultivars and the three Korean native *Thymus quinquecostatus* cultivars.

Thymus Sp.	Cultivar	Essential Oil Yield (%)	Essential Oil Color
Thymus vulgaris	Lemon	0.34 ± 0.013	Yellow
	Golden	0.29 ± 0.020	Yellow
	Carpet	0.23 ± 0.015	Pale yellow
	Orange	0.29 ± 0.033	Pale yellow
	Silver	0.33 ± 0.042	Yellow
	Creeping	0.24 ± 0.021	Yellow
Thymus quinquecostatus (Bakrihyang)	Odae Mt.	0.43 ± 0.067	Yellow
	Wolchul Mt.	0.28 ± 0.017	White
	Jiri Mt.	0.12 ± 0.014	Yellow

Table 4. The chemical composition of essential oils isolated from the six commercial *Thymus vulgaris* cultivars and the three Korean native *Thymus quinquecostatus* cultivars.

S. No.	Compound Name	RI Lit	RI Cal	Area (%)								
				Lemon	Golden	Carpet	Orange	Silver	Creeping	Odae	Wolchul	Jiri
1	Tricyclene	926	923	-	-	-	-	-	-	0.05 ± 0.01	0.01 ± 0.00	-
2	α-Thujene	930	926	0.17 ± 0.01	-	-	-	0.04 ± 0.00	-	1.91 ± 0.50	0.01 ± 0.00	0.11 ± 0.02
3	α-Pinene	939	934	0.09 ± 0.01	-	-	-	0.02 ± 0.00	-	1.32 ± 0.32	0.26 ± 0.01	0.09 ± 0.02
4	Camphene	954	950	0.11 ± 0.01	-	-	0.01 ± 0.00	0.02 ± 0.00	-	1.69 ± 0.42	0.67 ± 0.01	0.18 ± 0.03
5	β-Pinene	974	978	0.24 ± 0.33	-	-	-	0.19 ± 0.02	-	1.89 ± 0.47	0.17 ± 0.01	0.30 ± 0.03

Table 4. *Cont.*

S. No.	Compound Name	RI Lit	RI Cal	Area (%)								
				Lemon	Golden	Carpet	Orange	Silver	Creeping	Odae	Wolchul	Jiri
6	1-Octen-3-ol	979	985	3.61 ± 0.04	0.70 ± 0.06	0.57 ± 0.06	0.39 ± 0.02	1.46 ± 0.05	0.44 ± 0.03	2.81 ± 0.54	2.47 ± 0.02	3.29 ± 0.30
7	3-Octanone	983	989	0.76 ± 0.03	0.34 ± 0.02	-	0.01 ± 0.00	0.17 ± 0.01	0.0 8 ± 0.00	0.07 ± 0.00	0.82 ± 0.02	1.27 ± 0.20
8	3-Octanol	991	1000	0.38 ± 0.01	0.36 ± 0.02	-	0.06 ±0.01	0.23 ±0.02	0.15 ± 0.01	0.08 ± 0.02	0.70 ± 0.01	1.21 ± 0.19
9	α-Phellandrene	1002	1007	0.11 ± 0.01	-	-		0.04 ± 0.00		0.31 ± 0.09	-	0.05 ± 0.01
10	3-Carene	1011	1009	0.02 ± 0.01	-	-				0.08 ± 0.02	-	0.02 ± 0.01
11	*p*-Cymene	1024	1026	2.84 ± 0.10	0.41 ± 0.03	-	0.01 ± 0.00	2.22 ± 0.04		11.13 ± 0.62	0.05 ± 0.00	2.44 ± 0.15
12	D-Limonene	1029	1031	0.17 ± 0.01	0.01 ± 0.00	-		0.03 ± 0.01		0.53 ± 0.05	0.11 ± 0.01	0.10 ± 0.02
13	Eucalyptol	1032	1034	0.29 ± 0.01	0.09 ± 0.01	-	0.02 ± 0.01	0.41 ± 0.03		0.02 ± 0.01	0.44 ± 0.01	-
14	β-Ocimene	1050	1049	0.25 ± 0.01	-	-				0.09 ± 0.00	0.04 ± 0.00	0.12 ± 0.02
15	γ-Terpinene	1059	1060	8.44 ± 0.01	0.88 ± 0.06	0.05 ± 0.00	0.04 ± 0.00	3.85 ± 0.04	0.03 ± 0.00	23.92 ± 3.30	0.12 ± 0.00	3.43 ± 0.24
16	Sabinene hydrate	1070	1073	0.87 ± 0.03	0.05 ±0.01	-	0.15 ± 0.01	0.74 ± 0.06		3.02 ± 0.51	-	-
17	1-Nonen-3-ol	1078	1083	-	-	-	0.18 ± 0.01	0.30 ± 0.02	-	-	-	-
18	Terpinolene	1088	1087	0.93 ± 0.04	0.06 ± 0.01	-	-	-	-	3.55 ± 0.87	0.88 ± 0.02	0.56 ± 0.05
19	Nonanone	1090	1088	-	-	-				-	0.09 ± 0.01	-
20	Linalool	1096	1103	2.60 ± 0.12	0.47 ± 0.04	48.16 ± 0.67	0.36 ± 0.02	2.08 ± 0.11	3.86 ± 0.09	0.11 ± 0.03	1.49 ± 0.01	47.89 ± 3.11
21	Nonanal	1100	1108	-	-	-	-	-	-	-	0.07 ± 0.00	-
22	1-Octen-3-yl-acetate	1112	1109	-	-	-	-	-	-	-	0.02 ± 0.00	-
23	Chrysanthemal	1124	1120	-	-	-	-	-	0.43 ± 0.01	-	-	-
24	Verbenol	1141	1145	-	0.27 ± 0.01	-	-	-	0.85 ± 0.17	-	0.03 ± 0.00	-
25	Camphor	1146	1151	0.23 ± 0.03	-	0.04 ± 0.01	0.03 ± 0.00	0.62 ± 0.04	-	2.47 ± 0.45	-	0.10 ± 0.01
26	β-Pinene oxide	1159	1166	0.14 ± 0.02	0.02 ± 0.01	-	-	-	0.09 ± 0.02	-	-	-
27	Borneol	1169	1180	2.07 ± 0.07	0.41 ± 0.03	1.57 ± 0.16	3.05 ± 0.21	1.11 ± 0.11	0.73 ± 0.02	2.17 ± 0.49	5.91 ± 0.03	2.17 ± 0.34
28	Terpinen-4-ol	1177	1185	0.40 ± 0.01	-	0.15 ± 0.03	0.17 ± 0.01	0.37 ± 0.05	0.05 ± 0.00	1.49 ± 0.33	0.63 ± 0.03	0.17 ± 0.01
29	α-Terpineol	1188	1200	0.11 ± 0.03	0.05 ± 0.00	0.67 ± 0.01	0.04 ± 0.00	0.17 ± 0.04	0.03 ± 0.02	0.21 ± 0.06	0.09 ± 0.01	0.05 ± 0.01
30	Dihydrocarvone	1192	1202	-	-	-	0.04 ± 0.01	0.06 ± 0.02	-	-	0.13 ± 0.01	0.02 ± 0.01
31	Decanal	1201	1211	0.05 ± 0.01	-	0.06 ± 0.02	0.02 ± 0.00	-	-	0.07 ± 0.03	0.10 ± 0.01	0.05 ± 0.02
32	Nerol	1229	1233	2.66 ± 0.12	1.16 ± 0.05	0.17 ± 0.02	0.47 ± 0.03	1.99 ± 0.05	4.70 ± 0.05	-	1.34 ± 0.06	-
33	Thymol methyl ether	1235	1235	3.00 ± 0.09	0.26 ± 0.03	0.09 ± 0.03	-	2.17 ± 0.10	-	-	-	2.27 ± 0.16
34	Neral	1238	1243	0.99 ± 0.06	3.39 ± 0.21	-	0.26 ± 0.02	0.18 ± 0.05	11.75 ± 0.12	-	0.85 ± 0.03	-
35	2-Isopropyl-4-methylanisole	1244	1242	-	-	-	-	-	-	-	-	1.27 ± 0.18
36	Geraniol	1252	1262	6.03 ± 0.14	65.99 ± 2.30	0.73 ± 0.47	44.70 ± 0.67	3.02 ± 0.26	29.57 ± 0.65	0.35 ± 0.12	42.94 ± 0.32	0.03 ± 0.01
37	Geranial	1276	1264	1.49 ± 0.02	5.42 ± 0.18	0.03 ± 0.01	-	0.31 ± 0.04	18.21 ± 0.83	-	-	-
38	Cyclodecane	1271	1281	-	-	-	-	-	-	-	-	0.08 ± 0.03
39	1-Decanol	1269	1283	-	-	-	-	-	-	-	0.18 ± 0.01	-
40	Bornyl acetate	1285	1292	-	-	0.09 ± 0.00	0.19 ± 0.01	-	0.04 ± 0.01	-	0.53 ± 0.02	-
41	Thymol	1290	1295	43.91 ± 1.64	2.70 ± 0.09	13.36 ± 0.31	8.05 ± 0.03	66.24 ± 1.66	2.17 ± 0.08	30.54 ± 4.37	0.44 ± 0.02	15.98 ± 0.18
42	Carvacrol	1299	1305	-	-	0.78 ± 0.01	-	2.62 ± 0.03	-	-	-	0.27 ± 0.00
43	Methyl geranate	1324	1324	-	0.11 ± 0.01	-	-	-	0.03 ± 0.00	-	-	-
44	Thymol acetate	1352	1349	0.25 ± 0.01	-	-	0.57 ± 0.03	0.42 ± 0.03	-	0.29 ± 0.08	-	0.10 ± 0.02

Table 4. *Cont.*

S. No.	Compound Name	RI Lit	RI Cal	Area (%)								
				Lemon	Golden	Carpet	Orange	Silver	Creeping	Odae	Wolchul	Jiri
45	Geranyl acetate	1381	1386	2.34 ± 0.11	2.89 ± 0.22	-	29.86 ± 0.84	1.09 ± 0.09	6.75 ± 0.08	0.20 ± 0.08	26.49 ± 0.05	-
46	β-Bourbonene	1388	1387	0.04 ± 0.00	0.13 ± 0.02	0.10 ± 0.01	-	-	0.15 ± 0.01	-	-	0.14 ± 0.02
47	β-Cubebene	1388	1390	2.61 ± 0.14	2.20 ± 0.17	5.07 ± 0.02	0.40 ± 0.04	0.81 ± 0.10	2.44 ± 0.16	0.19 ± 0.08	-	3.47 ± 0.27
48	β-Elemene	1390	1394	0.06 ± 0.00	0.07 ± 0.01	0.19 ± 0.02	1.02 ± 0.09	0.08 ± 0.01	0.10 ± 0.02	0.65 ± 0.22	-	0.67 ± 0.06
49	Caryophyllene	1419	1428	3.41 ± 0.16	4.56 ± 0.18	2.62 ± 0.01	4.74 ± 0.66	4.48 ± 0.27	2.87 ± 0.09	4.74 ± 1.04	4.73 ± 0.02	7.02 ± 0.39
50	Neryl propionate	1432	1439	-	0.46 ± 0.22	-	-	-	1.34 ± 0.17	-	-	-
51	α-Bergamotene	1434	1438	-	-	-	-	-	-	-	0.04 ± 0.01	-
52	γ-Elemene	1436	1442	-	0.03 ± 0.01	-	-	-	-	-	-	-
53	Aromadendrene	1441	1438	-	-	0.12 ± 0.02	-	-	-	0.08 ± 0.02	-	-
54	α-Humulene	1454	1463	3.51 ± 0.15	-	0.17 ± 0.01	-	0.18 ± 0.02	-	-	0.66 ± 0.03	-
55	β-Farnesene	1456	1454	-	-	0.04 ± 0.00	0.02 ± 0.01	-	-	-	-	-
56	Germacrene D	1480	1487	-	-	-	-	-	-	-	0.53 ± 0.02	-
57	β-Selinene	1490	1494	-	-	0.03 ± 0.00	-	-	-	-	-	-
58	α-Farnesene	1505	1506	-	-	-	-	-	-	-	-	0.40 ± 0.07
59	β-Bisabolene	1505	1514	2.22 ± 0.10	3.70 ± 0.26	2.47 ± 0.31	3.84 ± 0.45	0.23 ± 0.03	1.81 ± 0.11	1.21 ± 0.34	3.86 ± 0.03	3.00 ± 0.27
60	γ-Cadinene	1513	1518	-	-	-	-	-	-	0.05 ± 0.01	-	-
61	Butylated hydroxytoluene	1515	1518	0.14 ± 0.02	0.14 ± 0.02	0.60 ± 0.09	0.06 ± 0.01	0.18 ± 0.01	0.24 ± 0.04	-	-	-
62	δ-Cadinene	1523	1522	-	-	0.11 ± 0.01	-	-	-	0.12 ± 0.03	-	-
63	β-Sesquiphellandrene	1522	1528	0.12 ± 0.01	0.06 ± 0.01	1.02 ± 0.12	0.08 ± 0.01	0.05 ± 0.01	0.15 ± 0.03	-	0.10 ± 0.01	0.36 ± 0.07
64	α-Elemol	1549	1554	0.31 ± 0.02	0.05 ± 0.01	11.62 ± 0.25	-	0.04 ± 0.01	8.52 ± 0.52	-	-	-
65	Geranyl butyrate	1562	1558	-	1.08 ± 0.10	-	-	-	0.53 ± 0.12	-	-	-
66	cis-3-Hexenyl benzoate	1566	1570	-	-	0.09 ± 0.01	-	-	-	-	-	-
67	Germacrene D-4-ol	1575	1583	-	-	-	-	-	-	0.89 ± 0.38	-	-
68	Spathulenol	1578	1584	-	-	-	0.04 ± 0.00	-	-	-	0.03 ± 0.01	0.06 ± 0.01
69	Caryophyllene oxide	1583	1590	0.08 ± 0.01	0.15 ± 0.01	0.15 ± 0.02	0.20 ± 0.01	0.32 ± 0.06	0.23 ± 0.04	0.22 ± 0.10	0.26 ± 0.01	0.26 ± 0.07
70	Humulene epoxide II	1608	1618	-	-	-	-	-	-	-	0.02 ± 0.00	-
71	γ-Eudesmol	1632	1639	-	-	1.94 ± 0.19	-	-	0.09 ± 0.03	-	-	-
72	T-Muurolol	1646	1651	-	-	-	-	-	-	0.03 ± 0.01	-	-
73	β-Eudesmol	1650	1663	-	-	2.53 ± 0.26	-	-	0.26 ± 0.07	-	-	-
74	α-Cadinol	1654	1664	0.04 ± 0.01	0.02 ± 0.01	-	0.02 ± 0.01	-	-	0.08 ± 0.03	-	-
75	α-Bisabolol	1685	1691	0.04 ± 0.01	-	-	0.01 ± 0.00	-	-	-	-	-
76	Benzyl benzoate	1760	1767	-	-	0.30 ± 0.03	-	-	-	-	-	-
77	1-Hexadecanol	1875	1884	-	-	-	-	0.12 ± 0.02	-	-	-	-
	Total			**98.15 ± 0.51**	**98.67 ± 0.35**	**95.71 ± 1.60**	**99.09 ± 0.30**	**98.66 ± 0.11**	**98.61 ± 0.35**	**98.60 ± 0.56**	**98.32 ± 0.11**	**99.00 ± 0.30**
	Monoterpene hydrocarbons			13.38	1.37	0.05	0.21	7.11	0.03	46.47	2.31	7.39
	Oxygenated monoterpenes			72.17	84.67	66.48	88.46	85.05	79.87	43.86	85.76	76.22
	Sesquiterpene hydrocarbons			11.98	10.74	11.94	10.09	5.83	7.51	7.04	9.93	15.07
	Oxygenated sesquiterpenes			0.62	1.89	17.23	0.33	0.67	11.20	1.23	0.32	0.32

RI Lit, comparison of retention indices with those reported in the literature (Adams, 2007). RI Cal, retention indices relative to n-alkanes (C_3–C_{22}) on the VF-5MS column. Values are the mean of the three replicate determinations ± standard deviation.

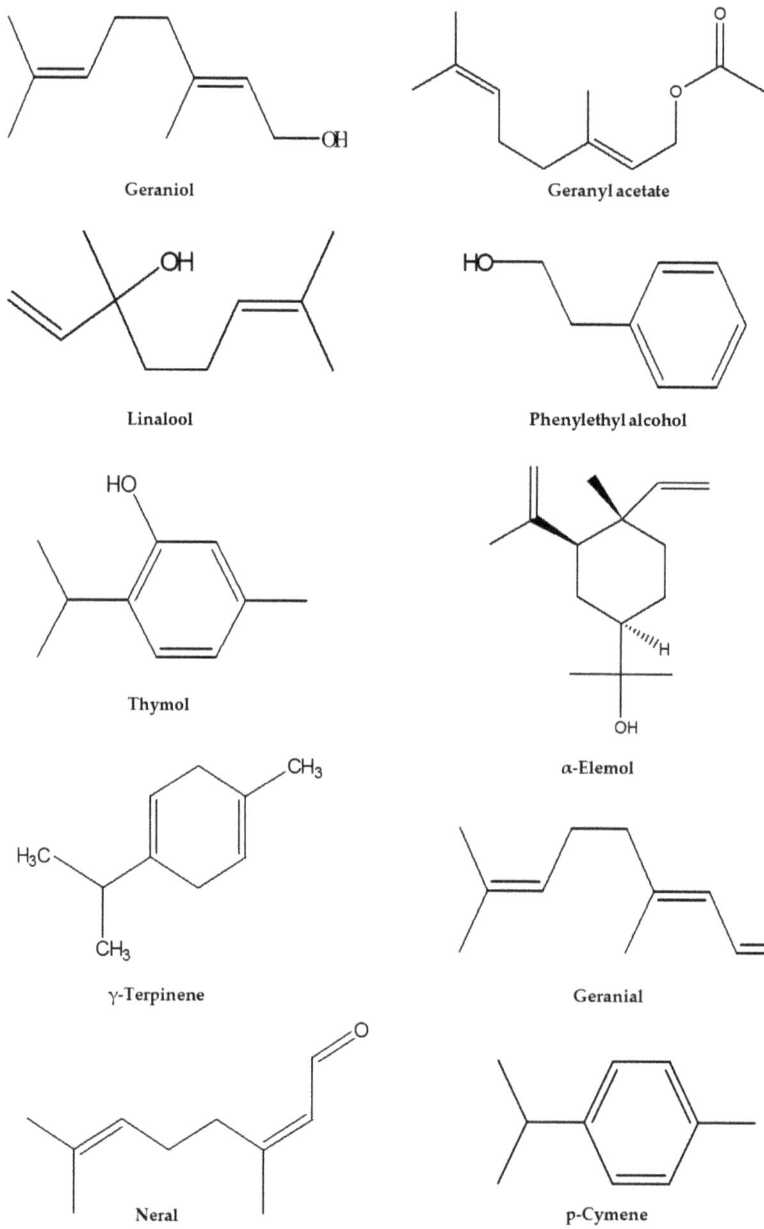

Figure 3. Structure of the major components identified in the essential oils of commercial *Thymus vulgaris* cultivars and Korean native *Thymus quinquecostatus* cultivars. Geraniol, geranyl acetate, linalool, phenylethyl alcohol, γ-terpinene, and thymol were identified as the major components, comprising >20% in at least one of the essential oil obtained from the different cultivars.

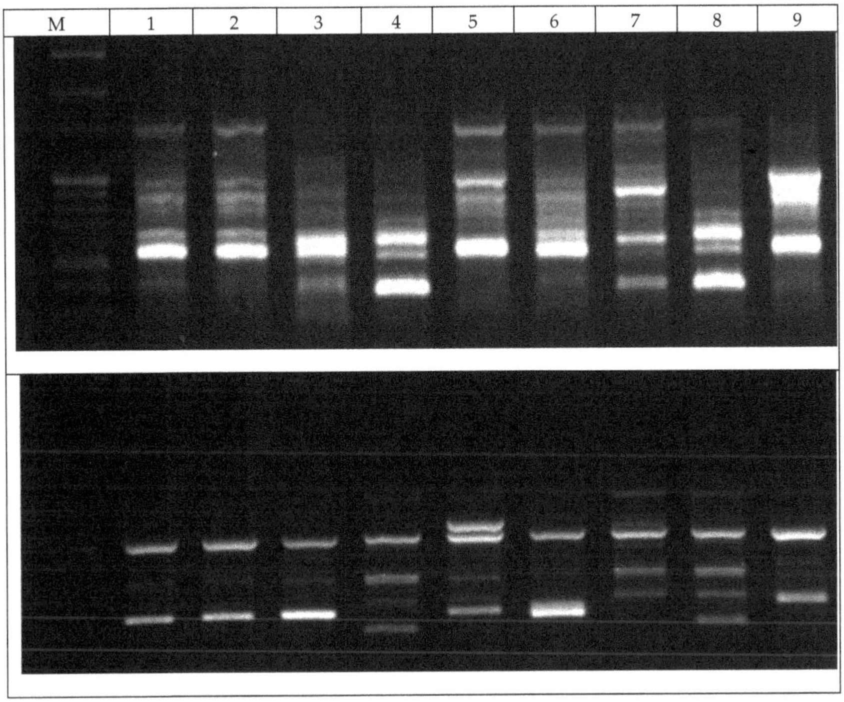

Figure 4. The example of a pattern among the three Korean native *Thymus quinquecostatus* cultivars and the six commercial *Thymus vulgaris* cultivars using OPB-01 (top) and OPA-11 (bottom) primers separated in 1.2% agarose gel electrophoresis. M, PCR marker; 1–6, commercial *Thymus vulgaris* cultivars: (1, lemon; 2, golden; 3, carpet; 4, orange; 5, silver; and 6, creeping); 7–9, Korean native *Thymus quinquecostatus* cultivars (7, Odae Mt; 8, Wolchul; 9, Jiri).

Table 5. Bands and polymorphism revealed by the RAPD primers among the 9 *Thymus* cultivars.

Primer	Total No. of Bands	No. of Polymorphic Bands	Polymorphism (%)
OPA-09	10	10	100
OPA-10	8	8	100
OPA-11	12	11	91.67
OPA-12	4	3	75
OPA-13	10	8	80
OPA-14	4	4	100
OPA-15	8	7	87.50
OPA-16	10	9	90
OPA-17	5	5	100
OPA-18	9	9	100
OPA-19	16	16	100
OPA-20	10	9	90
OPB-01	8	8	100
OPB-02	7	7	100
OPB-03	8	7	87.5
OPB-04	4	3	75
Total	**133**	**124**	
Average	**8.31**	**7.75**	**93.23**

The dendrogram realized from the RAPD markers grouped the 9 thyme cultivars into two major clusters and showed a clear separation (Figure 5). Levels of genetic similarity indices ranged from

0.58 to 0.98. Cluster 1 consisted of lemon, golden, creeping, silver, carpet, and Jiri. Whereas cluster 2 consisted of orange, Wolchul, and Odae.

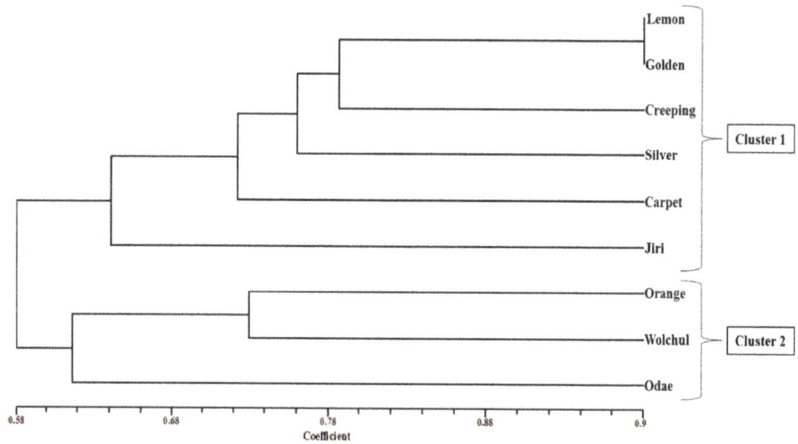

Figure 5. Clustering tree of the three *Thymus quinquecostatus* cultivars and the six commercial *Thymus vulgaris* cultivars, based on the unweighted pair-group method with the arithmetic average (UPGMA), using 16 RAPD markers.

4. Discussion

The identification of the *Thymus* species is extremely difficult because of the high levels of diversity within the genus. This genus contains several commercially important aromatic species. For this purpose, the relationship among the chemical composition of essential oils and molecular analysis was carried out for different *Thymus* species [20,23]. In this context, the essential oil composition and molecular analysis of nine thyme cultivars were investigated in this study, to distinguish between commercial thyme cultivars and Korean native thyme cultivars. In the morphological study, the *T. quinquecostatus* and *T. vulgaris* cultivars exhibited a significant level of variability in recorded parameters. In the qualitative traits, a considerable variability was observed in stem type, stem color, length and number of stem branches, leaf shape, and trichome position, among and within *T. quinquecostatus* and *T. vulgaris* cultivars.

The present study showed a high chemical diversity among nine thyme cultivars. Results revealed that essential oils from Korean cultivars (*T. quinquecostatus*) belonged to the geraniol, thymol, and linalool chemotypes. Essential oils from the commercial thyme cultivars (*T. vulgaris*) such as creeping, golden, and orange belonged to the geraniol chemotype and lemon, and the silver cultivars belonged to the thymol chemotype. Further, carpet cultivar belonged to the linalool chemotype. In particular, these essential oils were dominated by monoterpenes. 1-Octen-3-ol, γ-terpinene, linalool, borneol, α-terpineol, nerol, geraniol, thymol, β-cubebene, β-elemene, caryophyllene, β-bisabolene, butylated hydroxytoluene, β-sesquiphellandrene, and caryophyllene oxide were detected in all six essential oils from the commercial cultivars. With regards to the chemical composition of *T. vugaris* essential oils, seven different chemotypes such as thymol, carvacrol, linalool, geraniol, thujanol-4, terpineol, and 1,8-cineole were identified [15,16]. In the case of *T. quinquecostatus* essential oils, Shin and Kim [8] found that thymol (41.70%), γ-terpinene (16.00%), and *p*-cymene (13.00%) were the most prominent compounds. Similarly, thymol (30.54%), γ-terpinene (23.92%), and *p*-cymene (11.13%) were the major components in the essential oil obtained from the Odae cultivar. However, the major components in the essential oils obtained from Wolchul and Jiri cultivars of *T. quinquecostatus* were not identical. In the Wolchul cultivar, geraniol (42.94%) and geranyl acetate (26.49%) were detected as the major components, whereas linalool (47.89%) and thymol (15.98%) were found to be abundant in the Jiri cultivar.

Hudaib and Aburjai [24] determined variations in the composition of essential oils from cultivated and wild-growing plants of *T. vulgaris* grown in Jordan. Higher oil yields were obtained in plants growing wild, when compared to the cultivated plants. Among the four different samples, thymol (0.8–63.8%) and carvacrol (6.9–86.1%) were the most abundant components in the *T. vulgaris* essential oils. A study indicated that the essential oil composition of *T. vulgaris* highly varied both qualitatively and quantitatively during the vegetative cycle [25]. The variations in the yield and composition of essential oils could be influenced by various factors, such as the geographical region of the plant, plant's maturity, cultivation practices, and weather parameters (temperature, humidity, sunlight duration, and rainfall) [26–28]. In addition, the genetic constitution of the cultivars also played a considerable role in the essential oil composition [1,25].

According to previous reports, it is difficult to distinguish *Thymus* species and cultivars by analyzing the essential oil profile alone. Hence, the combined analysis of chemical composition and molecular techniques was used for the correct identification of the different plant species. In recent decades, the correlation between the chemical composition and molecular analysis of different *Thymus* species were investigated by various researchers [1,13,20,21]. Previous studies showed that both essential oil composition and RAPD analysis could be used to distinguish different thyme cultivars, and especially, to determine their relationships [1]. In addition, RAPD analysis revealed high polymorphisms even when using closely related genotypes. Even though the essential oil composition of plants was different from one another, RAPD analysis clustered these plants together, owing to their similar genetic background [15].

In the present study, 16 primers were used to amplify segments of DNA of the genome of three Korean thyme cultivars and six commercial thyme cultivars, to investigate the genetic variations. A total of 133 bands were obtained and the average percentage of the polymorphic bands was 93.23%. Based on the RAPD data, the similarity of the cultivars, estimated by the Jaccard's coefficient, is depicted in Figure 5. The nine cultivars of thyme fell into two clusters. Cluster 1 was formed by six cultivars (lemon, golden, creeping, silver, carpet, and Jiri) and cluster 2 by three cultivars (orange, Wolchul, and Odae). This emphasized the obvious variation between the Korean cultivars (except Jiri cultivar) and the commercial cultivars. The dendrogram indicated a clear separation of *T. quinquecostatus* from *T. vulgaris*, with the exception of the Jiri cultivar. According to the RAPD similarity matrix, it was observed that the Wolchul and Odae cultivars were closely related. Nevertheless, there was no significant relationship between the essential oil composition and RAPD data. The ability to discriminate all studied cultivars using RAPD bands indicated that RAPD analysis can provide a rapid and inexpensive technique to identify phenotypically similar thyme cultivars.

Based on previous reports, a high correlation between genetic and chemical relationships was attained in several plants. These data indicated that the composition of the essential oil is regulated by a number of genes that are extensively distributed throughout the plant genome [1,29,30]. Khalil et al. [31] used RAPD analysis to determine the genetic relationship between *T. vulgaris* populations collected in Syria. In their study, 13 individuals were analyzed using 27 primers, which generated 180 polymorphic bands from 198 bands. The authors found a significant correlation between *T. vulgaris* populations and their geographic areas. The present study also proved that the geographic distribution had a significant influence on genetic variation. Comparing the groups formed by the cluster analysis based on RAPD data (Figure 5) and chemotype, based on essential oil composition, we can observe that the groups formed in both cases were not identical.

In another study, the composition of essential oils and genetic relationships between six commercial cultivars of *T. vulgaris* were analyzed. A total of 104 were polymorphic RAPD bands (63.8%) were obtained using 15 primers. Among 15 primers, the highest percentage of polymorphism was obtained by the OPA-05 primer (90.9%). Similar to the essential oil composition, the six *T. vulgaris* cultivars fell into two major clusters, according to the RAPD patterns, with a correlation coefficient of −0.779 [1].

The chemical and genetic variations of 20 taxa from four Hungarian *Thymus* species (*T. glabrescens*, *T. pannonicus*, *T. praecox*, and *T. pulegioides*) were studied by Pluhár et al. [23]. In the molecular analysis, 114 polymorphic RAPD bands (80.8%) were obtained using 13 primers. The results revealed that partial correlation was found between the essential oil and RAPD analyses. The essential oil composition and genetic variation in six micropropagated genotypes (in vitro and in vivo) of *T. saturejoides* were investigated by Nordine et al. [32]. RAPD results and the essential oil composition grouped these six genotypes into three clusters exhibiting significant intraspecific chemical and genetic differences. Furthermore, a significant correlation was observed between RAPD and essential oil composition obtained from the in vitro genotypes.

Similar to our report, several studies also reported that the combined use of RAPD and essential oil analyses were not significantly correlated. For example, the genetic and chemical relationships among 31 individuals of *T. caespititius* collected from the islands of Pico, Sao Jorge, and Terceira (Azores) were determined. In the RAPD analysis, 187 polymorphic bands were obtained using 17 primers. However, there was no close relationship between the collection site, the essential oil composition, and RAPD analysis [15]. Rustaiee et al. [20] also studied the essential oil composition and genetic variability between some *Thymus* species such as *T. daenensis* (two populations), *T. fallax*, *T. fedtschenkoi*, *T. migricus*, and *T. vulgaris*, using GC-MS and RAPD. Although the RAPD markers allowed a perfect distinction among different *Thymus* species according to their characteristic genetic background, there was no identical clustering with the essential oil composition. In addition, Masi et al. [33] found that the essential oil compositions did not match with the results achieved from agronomic and genetic analyses in *Ocimum basilicum*. In another study, there was no correlation between RAPD and the essential oil obtained from the in vivo genotypes of *T. saturejoides* [32]. Based on the previous and present studies, marker-assisted RAPD technique had a high advantage for the assessment of the genetic differences of plant species without prior molecular knowledge.

Results of the present study revealed that there was a significant correlation between the genetic and geographic distances of the Korean thyme cultivars (Wolchul and Odae cultivars), compared to the commercial thyme cultivars. However, the chemical polymorphism of these thyme cultivars is not well-understood. Hence, other molecular techniques should be investigated in order to understand this question in *T. quinquecostatus* and other *Thymus* cultivars.

5. Conclusions

The present study emphasized that RAPD analysis allowed a perfect distinction between the Korean thyme cultivars (Wolchul and Odae) and commercial thyme cultivars, based on their unique genetic background. However, the chemical composition of the Wolchul and Odae cultivars was not identical. Furthermore, there was no significant relationship between the RAPD data and essential oil composition of both *T. quinquecostatus* and *T. vulgaris* cultivars. The chemical composition and molecular data obtained in this study delivered a good starting point for future investigations. It could be concluded that the RAPD markers proved to be an effective tool for discriminating different *Thymus* species. The sample collection must be done from different geographical regions in Korea to understand the genetic and chemical variability of the *T. quinquecostatus* cultivars.

Supplementary Materials: The following are available online at http://www.mdpi.com/2079-6382/9/6/289/s1, Figure S1: GC-MS chromatograms of essential oils from three Korean native *Thymus quinquecostatus* cultivars. A, Odae cultivar; B, Wolchul cultivar; C, Jiri cultivar.

Author Contributions: M.K. performed the experiment; K.S. compiled the data and wrote the manuscript; J.-C.M. analyzed the data; and S.K. designed the experiment and revised the manuscript. All authors have read and agreed to the published version of the manuscript.

Funding: This research received no external funding.

Acknowledgments: This study was carried out with the support of "Cooperative Research Program for Agriculture Science and Technology Development (Project No. PJ014506012020)", Rural Development Administration, Republic of Korea.

Conflicts of Interest: The authors declare no conflict of interest.

References

1. Echeverrigaray, S.; Agostini, G.; Atti-Serfini, L.; Paroul, N.; Pauletti, G.F.; dos Santos, A.C. Correlation between the chemical and genetic relationships among commercial thyme cultivars. *J. Agric. Food Chem.* **2001**, *49*, 4220–4223. [CrossRef]
2. Maissa, B.J.; Walid, H. Antifungal activity of chemically different essential oils from wild Tunisian *Thymus* spp. *Nat. Prod. Res.* **2015**, *29*, 869–873. [CrossRef] [PubMed]
3. Kim, Y.S.; Lee, S.J.; Hwang, J.W.; Kim, E.K.; Kim, S.E.; Kim, E.H.; Moon, S.H.; Jeon, B.T.; Park, P.J. In vitro protective effects of *Thymus quinquecostatus* Celak extracts on t-BHP-induced cell damage through antioxidant activity. *Food Chem. Toxicol.* **2012**, *50*, 4191–4198. [CrossRef] [PubMed]
4. Jia, P.; Gao, T.; Xin, H. Changes in structure and histochemistry of glandular trichomes of *Thymus quinquecostatus* Celak. *Sci. World J.* **2012**, *2012*, 187261. [CrossRef] [PubMed]
5. Li, X.; He, T.; Wang, X.; Shen, M.; Yan, X.; Fan, S.; Wang, L.; Wang, X.; Xu, X.; Sui, H.; et al. Trditional uses, chemical constituents and biological activities of plants from the genus *Thymus*. *Chem. Biodivers.* **2019**, *17*, e1900254.
6. Hyun, T.K.; Kim, H.C.; Kim, J.S. Antioxidant and antidiabetic activity of *Thymus quinquecostatus* Celak. *Ind. Crops Prod.* **2014**, *52*, 611–616. [CrossRef]
7. Jia, P.; Liu, H.; Gao, T.; Xin, H. Glandular trichomes and essential oil of *Thymus quinquecostatus*. *Sci. World J.* **2013**, *2013*, 387952.
8. Shin, S.; Kim, J.H. Antifungal activities of essential oils from *Thymus quinquecostatus* and *T. magnus*. *Planta Med.* **2004**, *70*, 1090–1092. [CrossRef]
9. Kykkidou, S.; Giatrakou, V.; Papavergou, A.; Kontominas, M.G.; Savvaidis, I.N. Effect of thyme essential oil and packaging treatments on fresh Mediterranean swordfish fillets during storage at 4 °C. *Food Chem.* **2009**, *115*, 169–175. [CrossRef]
10. Chang, Y.L.; Shen, M.; Ren, X.Y.; He, T.; Wang, L.; Fan, S.S.; Wang, X.H.; Li, X.; Wang, X.P.; Chen, X.Y.; et al. Multi-response extraction optimization based on anti-oxidative activity and quality evaluation by main indicator ingredients coupled with chemometric analysis on *Thymus quinquecostatus* Celak. *Molecules* **2018**, *23*, 957. [CrossRef]
11. Solomakos, N.; Govaris, A.; Koidis, P.; Botsoglou, N. The antimicrobial effect of thyme essential oil, nisin, and their combination against *Listeria monocytogenes* in minced beef during refrigerated storage. *Food Microbiol.* **2008**, *25*, 120–127. [CrossRef] [PubMed]
12. Adams, R.P.; Socorro, M.; Elizondo, G.; Elizondo, M.G.; Slinkman, E. A new subalpine variety of *Juniperus blancoi* Martinez (Cupressaceae) from Durango, Mexico. *Biochem. Syst. Ecol.* **2006**, *34*, 205–211. [CrossRef]
13. Younsi, F.; Rahali, N.; Mehdi, S.; Boussaid, M.; Messaoud, C. Relationship between chemotypic and genetic diversity of natural populations of *Artemisia herba-alba* Asso growing wild in Tunisia. *Phytochemistry* **2018**, *148*, 48–56. [CrossRef] [PubMed]
14. Boulila, A.; Bejaoui, A.; Messaoud, C.; Boussaid, M. Variation of volatiles in Tunisian populations of *Teucrium polium* L. (Lamiaceae). *Chem. Biodivers.* **2008**, *7*, 1389–1400. [CrossRef] [PubMed]
15. Trindade, H.; Costa, M.M.; Sofia, B.L.A.; Pedro, L.G.; Figueiredo, A.C.; Barroso, J.G. Genetic diversity and chemical polymorphism of *Thymus caespititius* from Pico, Sao Jorge and Terceira islands (Azores). *Biochem. Syst. Ecol.* **2008**, *36*, 790–797. [CrossRef]
16. Torras, J.; Dolors Grau, M.; Lopez, J.F.; de Las Heras, F.X.C. Analysis of essential oils from chemotypes of *Thymus vulgaris* in Catalonia. *J. Sci. Food Agric.* **2007**, *87*, 2327–2333. [CrossRef]
17. Gajera, B.B.; Kumar, N.; Singh, A.S.; Punvar, B.S.; Ravikiran, R.; Subhash, N.; Jadeja, G.C. Assessment of genetic diversity in castor (*Ricinus communis* L.) using RAPD and ISSR markers. *Ind. Crops Prod.* **2010**, *32*, 491–498. [CrossRef]
18. Mahajan, V.; Chouhan, R.; Kitchlu, S.; Bindu, K.; Koul, S.; Singh, B.; Bedi, Y.S.; Gandhi, S.G. Assessment of chemical and genetic variability in *Tanacetum gracile* accessions collected from cold desert of Western Himalaya. *3 Biotech* **2018**, *8*, 284. [CrossRef]

19. Chowdhury, T.; Mandal, A.; Roy, S.C.; Sarker, D.D. Diversity of the genus *Ocimum* (Lamiaceae) through morpho-molecular (RAPD) and chemical (GC–MS) analysis. *J. Genet. Eng. Biotechnol.* **2017**, *15*, 275–286. [CrossRef]
20. Rustaiee, A.R.; Yavari, A.; Nazeri, V.; Shokrpour, M.; Sefidkon, F.; Rasouli, M. Genetic diversity and chemical polymorphism of some *Thymus* species. *Chem. Biodivers.* **2013**, *10*, 1088–1098. [CrossRef]
21. Trindade, H.; Costa, M.M.; Sofia, B.L.A.; Pedro, L.G.; Figueiredo, A.C.; Barroso, J.G. A combined approach using RAPD, ISSR and volatile analysis for the characterization of *Thymus caespititius* from Flores, Corvo and Graciosa islands (Azores, Portugal). *Biochem. Syst. Ecol.* **2009**, *37*, 670–677. [CrossRef]
22. Adams, R.P. *Identification of Essential Oil Components by gAs Chromatography/Mass Spectrometry*; Allured Publishing Co.: Carol Stream, IL, USA, 2007.
23. Pluhár, Z.; Kocsis, M.; Kuczmog, A.; Csete, S.; Simkó, H.; Sárosi, S.; Molnár, P.; Horváth, G. Essential oil composition and preliminary molecular study of four Hungarian *Thymus* species. *Acta Biol. Hung.* **2012**, *63*, 81–96. [CrossRef] [PubMed]
24. Hudaib, M.; Aburjai, T. Volatile components of *Thymus vulgaris* L. from wild-growing and cultivated plants in Jordan. *Flavour Frag. J.* **2007**, *22*, 322–327. [CrossRef]
25. Hudaib, M.; Speroni, E.; Di Pietra, A.M.; Cavrini, V. GC/MS evaluation of thyme (*Thymus vulgaris* L.) oil composition and variations during the vegetative cycle. *J. Pharm. Biomed. Anal.* **2002**, *29*, 691–700. [CrossRef]
26. Dhouioui, M.; Boulila, A.; Chaabane, H.; Zina, M.S.; Casabianca, H. Seasonal changes in essential oil composition of *Aristolochia longa* L. ssp. *paucinervis Batt. (Aristolochiaceae) roots and its antimicrobial activity. Ind. Crops Prod.* **2016**, *83*, 301–306.
27. Mewes, S.; Kruger, H.; Pank, F. Physiological, morphological, chemical and genomic diversities of different origins of thyme (*Thymus vulgaris* L.). *Genet. Resour. Crop Evol.* **2008**, *55*, 1303–1311. [CrossRef]
28. Rajabi, Z.; Ebrahimi, M.; Farajpour, M.; Mirza, M.; Ramshini, H. Compositions and yield variation of essential oils among and within nine *Salvia* species from various areas of Iran. *Ind. Crops Prod.* **2014**, *61*, 233–239. [CrossRef]
29. Vieira, R.F.; Grayer, R.J.; Paton, A.; Simon, J.E. Genetic diversity of *Ocimum gratissimum* L. based on volatile oil constituents, flavonoids and RAPD markers. *Biochem. Syst. Ecol.* **2001**, *29*, 287–304. [CrossRef]
30. Keskitalo, M.; Pehu, E.; Simon, J.E. Variation in volatile compounds from tansy (*Tanacetum vulgare* L.) related to genetic and morphological differences of genotypes. *Biochem. Syst. Ecol.* **2001**, *29*, 267–285. [CrossRef]
31. Khalil, R.; Khalil, R.; Li, Z. Determination of genetic variation and relationship in *Thymus vulgaris* populations in Syria by random RAPD markers. *Plant Biosyst.* **2012**, *146*, 217–225. [CrossRef]
32. Nordine, A.; Udupa, S.M.; Iraqi, D.; Meksem, K.; Hmamouchi, M.; ElMeskaoui, A. Correlation between the chemical and genetic relationships among *Thymus saturejoides* genotypes cultured under *in vitro* and *in vivo* environments. *Chem. Biodivers.* **2016**, *13*, 387–394. [CrossRef] [PubMed]
33. Masi, L.; Siviero, P.; Esposito, C.; Castaldo, D.; Siano, F.; Laratta, B. Assessment of agronomic, chemical and genetic variability in common basil (*Ocimum basilicum* L.). *Eur. Food Res. Technol.* **2006**, *223*, 273–281. [CrossRef]

antibiotics

MDPI

Article

Natural Oregano Essential Oil May Replace Antibiotics in Lamb Diets: Effects on Meat Quality

Ivan A. Garcia-Galicia [1], Jose A. Arras-Acosta [1], Mariana Huerta-Jimenez [2], Ana L. Rentería-Monterrubio [1], Jose L. Loya-Olguin [3], Luis M. Carrillo-Lopez [2], Juan M. Tirado-Gallegos [1] and Alma D. Alarcon-Rojo [1,*]

[1] Facultad de Zootecnia y Ecología, Universidad Autónoma de Chihuahua, Chihuahua 31453, Mexico; igarciag@uach.mx (I.A.G.-G.); ing.jarras@gmail.com (J.A.A.-A.); arenteria@uach.mx (A.L.R.-M.); jtirado@uach.mx (J.M.T.-G.)

[2] Catedrático CONACYT-UACH, Facultad de Zootecnia y Ecología, Universidad Autónoma de Chihuahua, Chihuahua 31453, Mexico; mhuertaj@uach.mx (M.H.-J.); lmcarrillo@uach.mx (L.M.C.-L.)

[3] Posgrado en Ciencias Biológico Agropecuarias/Unidad Académica de Medicina Veterinaria y Zootecnia, Universidad Autónoma de Nayarit, Compostela 63700, Mexico; joselenin28@hotmail.com

* Correspondence: aalarcon@uach.mx; Tel.: +52-6142-168099

Received: 4 April 2020; Accepted: 3 May 2020; Published: 12 May 2020

Abstract: A study was conducted to investigate the effect of oregano essential oil (OEO) and monensin sodium on the oxidative stability, colour, texture, and the fatty acid profile of lamb meat (m. *Longissimus lumborum*). Twenty Dorper x Pelibuey lambs were randomly divided into five treatments; control (CON), monensin sodium (SM, Rumensin 200® 33 mg/kg), a low level of OEO (LO, 0.2 g/kg dry matter (DM)), a medium level of OEO (MO, 0.3 g/kg DM), and a high level of OEO (HO, 0.4 g/kg DM). Dietary supplementation of OEO at any concentration lowered the compression strength in comparison with CON and SM. MO had the highest a* values (7.99) and fatty acid concentration (C16:1n7, C18:1n9c, C18:1n6c, C20:1n9, and C18:2n6c) during storage for 7 d at 3 °C. Lipid oxidation was not promoted ($p > 0.05$) by the moderated supplementation of oregano essential oil; however, OEO at 0.3 g/kg DM showed a slight lipid pro-oxidant effect. Dietary supplementation of MO and SM had the same effect on colour, tenderness, and the fatty acid profile of lamb (*L. lumborum*). It was demonstrated that oregano essential oil was beneficial for lambs feeding, and it could be a natural alternative to replace monensin in lamb diets with improvements in the quality of the meat.

Keywords: lamb; carvacrol; monensin; meat tenderness; TBARS

1. Introduction

Feeding systems in animal production are using additives that improve growth. These additives modify the weight daily gain, composition and yield of carcass, nutritional value, stability, and shelf life of the meat. Nowadays, the consumers are demanding naturally raised, trusted, and organic meat and meat products. Hence, there is a need to look for alternatives to the use of synthetic growth promoters [1,2].

Monensin sodium (SM) is an antimicrobial ionophore produced by *Streptomyces cinamonensis* and is used in ruminants to improve feed efficiency [3]. Even though antimicrobial growth promoters in animals intended for human consumption are strongly limited in some countries [4,5], they are still used in the U.S.A. and Mexico.

Phytochemicals have properties that benefit not only the animals fed with them but indirectly might impact the consumers of the meat [6,7]. Oregano essential oil (OEO) has antioxidant and antimicrobial properties, due to its phenolic components, mainly thymol and carvacrol [8–10]. The active components (carvacrol and thymol) of OEO are potent antimicrobials affecting populations such as *E. coli*, *Staphylococcus aureus*, *Salmonella typhimurium*, protozoa, fungi, *Ruminococcus fibrisolvens*, and *Fibrobacter succinogenes*; this change in populations modifies ruminal fermentation, which is fundamental in the conversion of dietary nutrients to muscle tissue [11–13]. Specifically in sheep, there is evidence that carvacrol potentially decreases acetate concentrations, and it increases propionate and butyrate. Both are volatile fatty acids precursors of muscle and fat components in the animal [14].

Specific nutritional components in the lamb diet might directly affect the quality of the meat. Ruminant meat research has drawn considerable attention because ruminant meat contains some bioactive lipids, including n-3 long chain polyunsaturated fatty acids (n-3 LC-PUFA), and the fatty acid profiles of ruminant meat can be enhanced through dietary supplementation [15,16]. Polyunsaturated fatty acids (PUFAs) are not only essential nutrients for humans, but also significant in providing protection from cardiovascular diseases, inflammatory diseases, diabetes, some cancers and behavioral disorders [17–19]. Therefore, the consumption of adequate n-3 LC-PUFAs is crucial to maintaining a healthy body and for the prevention of chronic diseases [20]. Humans can obtain n-3 LC-PUFA or their C18 PUFA precursors from various sources including aquatic, farm livestock products, oilseeds, fruits, herbs, cyanobacteria, and microorganisms [21].

The formation of large amounts of saturated fatty acids (SFAs) in the rumen is a result of the biohydrogenation process when bacteria convert unsaturated fatty acids to SFA. So, the fatty acids (FAs) occurring in the rumen are highly saturated and take part in the absorption as well as deposition of the fat in muscles. The decrease in SFA and increase of health-beneficial fatty acids (PUFAs) content have been an important objective in ruminant meat studies. According to the Food and Agriculture Organization of the United Nations [22], the ratio of PUFA and SFA is a significant indicator for the nutritional evaluation of fat, with a recommendation of about 0.40. It is essential to increase the consumption of eicosapentaenoic (EPA) and docosahexaenoic acid (DHA) in human diets, because the synthesis of these FAs from dietary α-linolenic acid (ALA, C18:3 n-3) is very restricted [23].

Dietary supplementation of OEO in pigs, chickens, and cattle rendered meat with lower microbial concentration and higher antioxidant capacity [24–30]. Essential oils have shown to decrease methane production [31] and biohydrogenation [32]. The FAs profile in meat may be affected by the form of the provided lipid. However, the exact influence of oregano essential oil on the FA profile of lamb meat is limited.

Limited information exists on the benefits of phytochemicals in the productive performance and carcass and meat quality of ovine and other small ruminants. Moreover, information is available in terms of OEO and oxidative stability, but little is known about the effects on the fatty acid profile of lamb meat. Here, the aim of this study was to investigate the effect of the dietary supplementation of three levels of oregano essential oil and sodium monensin on the oxidative stability, colour, texture, and the fatty acid profile of lamb meat (m. *Longissimus lumborum*).

2. Results

2.1. Lipid Oxidation

Oregano essential oil (OEO) and monensin (SM) did not have advantages ($p = 0.28$) on the lipid oxidative stability of lamb (Figure 1). However, meat from male lambs fed with a high level of OEO (HO) had a 46% higher ($p < 0.05$) malonaldehyde (MDA) formation (1.15 ± 1.25 mg MDA/kg) when compared to SM (0.79 ± 1.25 mg MDA/kg).

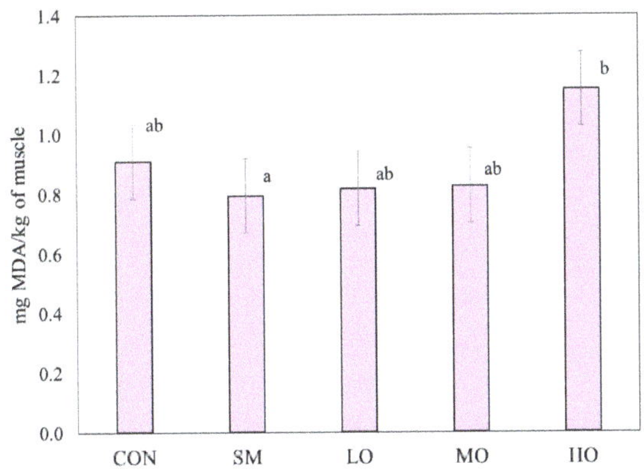

Figure 1. Means (± S.E.) of thiobarbituric acid reactive substances (TBARS) values (mg malonaldehyde (MDA)/kg of muscle) in m. *Longissimus lumborum* of male lambs unsupplemented (CON) or supplemented with monensin (SM), or three different levels of oregano essential oils: low (LO = 0.2 g/kg of dry matter (DM)); medium (MO = 0.3 g/kg of DM) or high (HO = 0.4 g/kg of DM). TBARS analysis was performed after 7 d of simulated retail display in modified atmosphere packaging (O_2:CO_2, 75:25%). [a,b] Different superscripts mean significant difference ($p < 0.05$) among treatments within the variables.

2.2. Compression Strength

OEO dietary supplementation reduced the compression force (Figure 2) of lamb ($p = 0.03$). The meat with the medium level of OEO (MO) and the control (CON) meat had the lowest and highest compression strength values, 12.51 ± 1.13 and 17.98 ± 1.32 N/cm^3, respectively. At day 7, the meat with a high level of OEO (HO) presented the lowest compression strength and the highest oxidation rate. Therefore, a greater PUFA deposition ($p < 0.05$) could be assumed.

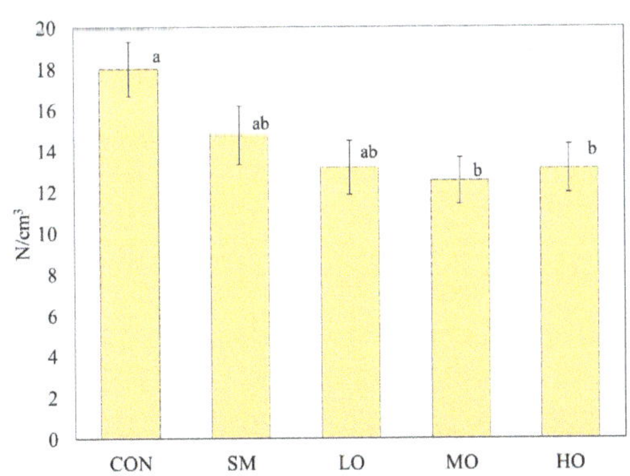

Figure 2. Means (± S.E.) of compression strength (N/cm^3) in m. *Longissimus lumborum* of male lambs unsupplemented (CON) or supplemented with monensin (SM), or three different levels of oregano essential oils (LO = 0.2 g/kg of DM; MO = 0.3 g/kg of DM or HO = 0.4 g/kg of DM). Compression strength analyses were performed after 7 d of simulated retail display in modified atmosphere packaging (O_2:CO_2, 75:25%). [a, b] Different superscripts mean significant difference ($p < 0.05$) among treatments within variable.

2.3. Colour

The interaction between treatment and day of shelf life was not significant ($p > 0.05$). Treatment and storage time affected the colour of *L. lumborum* ($p < 0.001$). Seven d of storage in 75% O_2 packs negatively affected ($p < 0.05$) a *, b * and c *, decreasing the values of the three colour coordinates (Figure 3).

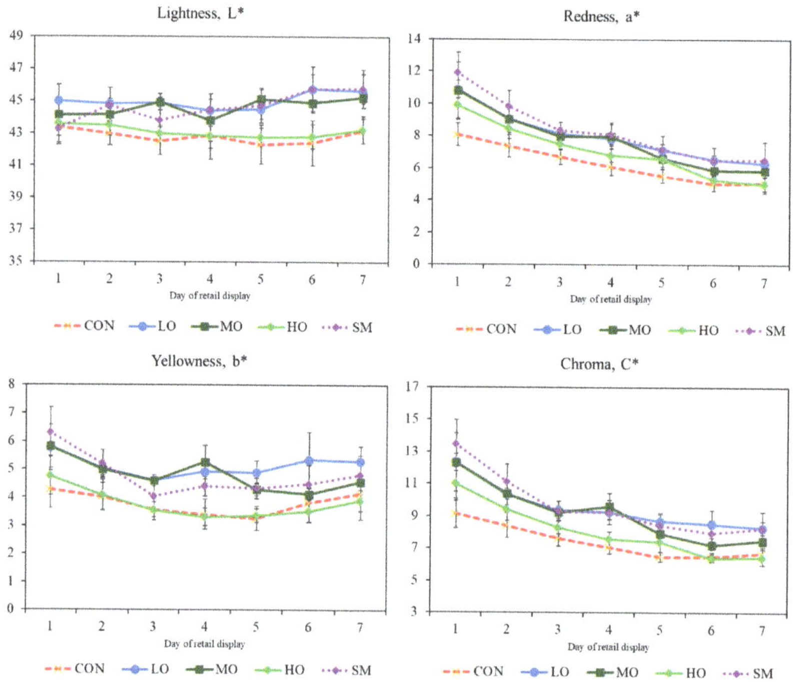

Figure 3. Colour coordinates (L* = lightness, a* = redness, b* = yellowness, and C* = Chroma. ± S.E.) in m. *Longissimus lumborum* of male lambs unsupplemented (CON) or supplemented with three different levels of oregano essential oils (LO = 0.2 g/kg of DM; MO = 0.3 g/kg of DM; or HO = 0.4 g/kg of DM) or monensin (SM), during 7 d of simulated retail display in modified atmosphere packaging (O2:CO2, 75:25%).

The value of a* differed among treatments ($p < 0.001$). SM, MO, and the meat with a low level of OEO (LO) had the highest values during storages (Figure 3, $p < 0.001$). SM, MO, and LO had also the higher b* values throughout storage compared to CON and HO (Figure 3). Treatment also affected C*. In this case, SM, MO, and LO were the highest ($p < 0.001$); meanwhile, CON and HO were similar ($p = 0.03$).

These statements can be confirmed with the means of the colour coordinates by treatment (Figure 4). It can be established that the dietary addition of 0.2–0.3 g/kg dry matter (DM) OEO and SM (commercial dosage) reduces colour loss after 7 d under MAP.

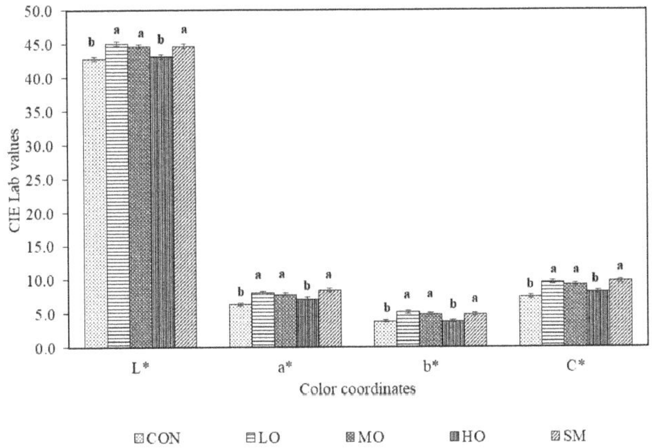

Figure 4. Means (± S.E.) of colour coordinates (L*, lightness; a*, redness; b*, yellowness; C*, saturation) in m. *Longissimus lumborum* of male lambs unsupplemented (CON) or supplemented with three different levels of oregano essential oils (LO = 0.2 g/kg of DM; MO = 0.3 g/kg of DM; or HO = 0.4 g/kg of DM) or monensin (SM), with 7 d of simulated retail display in modified atmosphere packaging (O_2:CO_2, 75:25%). [a,b] Different superscripts mean significant difference ($p < 0.05$) among treatments.

2.4. Fatty Acid Profile

In general, the inclusion of OEO or SM affected the fatty acid profile ($p > 0.05$). Eighteen FAs were analyzed, and their percentage was calculated relative to all FAs (Table 1). Differences were observed ($p < 0.05$) among treatments in a single FA. The treatment modified the percentage of C16:1n7, C20:1n9, and C22:6n3. Furthermore, it also modified the content (mg/kg of muscle) of C14:0, C16:0, C18:0, C16:1n7, C18:1n9c, C18:1n6c, C20:1n9, and C18:2n6c (Table 2).

Table 1. Percentage of fatty acids in the total fat of *Longissimus lumborum* of male lambs unsupplemented (CON), supplemented with three different levels (HO = 0.4 g/kg DM; MO = 0.3 g/kg DM; LO = 0.2 g/kg DM) of oregano essential oils or supplemented with monensin (SM) at commercial doses.

Fatty Acid	Treatments					
	CON	SM	HO	MO	LO	*p*-Value
C14:0	1.95	1.68	1.99	2.00	1.78	0.60
C14:1	0.11	0.10	0.09	0.10	0.10	0.99
C16:0	25.00	23.69	23.26	25.64	25.07	0.49
C16:1n7	1.00^{ab}	1.35^{a}	0.87^{b}	0.91^{ab}	1.06^{ab}	0.04
C18:0	14.69	15.07	15.04	15.87	14.90	0.94
C18:1n9t	2.31	2.22	1.63	2.84	2.20	0.19
C18:1n9c	44.17	46.28	45.83	44.89	46.24	0.52
C18:2n6t	0.25	0.24	0.24	0.22	0.21	0.97
C18:2n6c	3.10	3.32	3.28	2.69	2.85	0.60
$C18:3n^6$	0.09	0.06	0.07	0.05	0.03	0.78
$C18:3n^3$	0.06	0.04	0.04	0.04	0.04	0.17
C20:0	0.38	0.27	0.18	0.25	0.32	0.10
$C20:1n^9$	0.72^{ab}	0.62^{b}	0.69^{ab}	0.82^{a}	0.58^{b}	0.01
$C20:3n^3$	0.05	0.05	0.08	0.04	0.04	0.35
$C20:4n^6$	0.14	0.13	0.08	0.05	0.08	0.46
C20:5n3	1.68	1.48	1.81	1.25	1.12	0.13
$C22:5n^3$	0.86	0.39	0.53	0.33	0.46	0.27
C22:6n3	0.30^{ab}	0.23^{ab}	0.86^{a}	0.21^{b}	0.23^{ab}	0.03

[a,b] Different superscripts mean significant difference ($p < 0.05$) among treatments (columns).

Table 2. Concentration of fatty acids (mg/kg of muscle) in total fat of *Longissimus lumborum* of male lambs affected by supplementation of three different levels (HO = 0.4 g/kg DM; MO = 0.3 g/kg DM; LO = 0.2 g/kg DM) of oregano essential oils or monensin (SM) at commercial doses *vs.* a control diet (CON).

Fatty Acid	Treatments					
	CON	SM	HO	MO	LO	*p*-Value
Saturated						
C14:0	13.99^{b}	14.47^{ab}	9.67^{b}	23.50^{a}	13.18^{b}	0.042
C16:0	172.7^{b}	200.1^{ab}	120.88^{b}	289.77^{a}	183.74^{ab}	0.031
C18:0	94.93^{b}	121.7^{ab}	71.94^{b}	164.17^{a}	103.58^{ab}	0.043
Mono-unsaturated						
C16:1n7	6.60^{ab}	10.68^{a}	4.31^{b}	9.51^{a}	7.14^{ab}	0.024
C18:1n9c	287.41^{b}	369.6^{ab}	221.58^{b}	472.97^{a}	321.59^{ab}	0.043
C18:1n9t	14.67	16.82	14.39	8.12	30.19	0.007
C20:1n9	2.90^{b}	3.54^{b}	4.21^{b}	7.88^{a}	4.46^{b}	0.001
PUFA's						
C18:2n6t	18.74^{ab}	24.88^{ab}	14.04^{b}	27.10^{a}	18.66^{ab}	0.043

a, b Different superscripts mean significant difference ($p < 0.05$) among treatments (column) in the same fatty acid.

HO decreased the concentration of C14, C16, and C18 ($p < 0.05$); meanwhile, MO increased ($p < 0.05$) them. LO and SM did not alter the concentrations of these acids when compared to CON ($p > 0.05$, Table 2). Higher concentrations of C14, C16, C18, C16:1n7, C18:1n, C20:1n9, and C18:2n6t in the total fat of *Longissimus lumborum* were observed ($p < 0.05$) in MO meat compared to the control group.

Concentrations (Table 2) of monounsaturated fatty acids (MUFAs); C16:1n7, C18:1n9c and C20:1n9 were different among treatments ($p < 0.05$). MO increased the concentration of these MUFAs ($p < 0.05$), while HO decreased the concentrations ($p < 0.05$). HO, LO, and CON had similar low levels, whereas SM and MO had higher values ($p < 0.05$).

The treatment ($p < 0.05$) affected the concentration of linoleic acid (C18:2n6t) in total fat. MO increased concentrations of C18:2n6t ($p < 0.05$, Table 2). The lowest concentrations of C18:2 were in HO meat.

3. Discussion

3.1. Lipid Oxidation

In the present study, the incorporation of OEO at levels of 0.2 and 0.3 g/kg DM or monensin in the lamb diet had a similar effect on the lipid oxidation of lamb. The control treatment was slightly higher in malonaldehyde formation in meat. However, this difference was not statistically different from the control treatment. Notwithstanding, the result is still promising as far as the application of OEO concerns, since replacing OEO for monensin in lamb diets shows to be a beneficial choice.

However, meat from male lambs fed with a high level of OEO had significantly higher malonaldehyde (MDA) formation when compared to the control treatment.

The effectiveness of essential oil in preventing oxidation in lamb meat has also been reported by Nieto et al. [33]. They tested distilled dietary rosemary leaf (DRL, 0%, 10% and 20%) to prevent lipid oxidation and the sensory deterioration of cooked lamb, under retail display conditions. Cooked lamb fillets were stored at 0, 2, or 4 d (4 °C) in a display cabinet and then reheated, simulating catering practices. The cooked lamb suffered rapid lipid oxidation and odour and flavour spoilage associated with slight rancidity and warmed-over flavour. DRL feeding delayed lipid oxidation (thiobarbituric acid reactive substances, or TBARS) and volatile compounds more effectively in the first two d of storage. Percentages of 10% and 20% of DRL provided equal antioxidant capacity.

These positive effects of essential oil have also been found in bovine meat. Rivaroli et al. [34] fed crossbred young bulls with different doses of an essential oil blend (oregano, garlic, lemon, rosemary, thyme, eucalyptus, and sweet orange). They found that a dose of 3.5 g/animal/d decreases lipid oxidation. However, higher doses could have a pro-oxidant effect, and they are not recommended in feedlot animals.

Antioxidants that interact with reactive oxidant species (ROS) might become pro-oxidants, causing lipid and protein oxidation [35,36]. Low concentrations of essential oils might prevent this, and antioxidant activity is kept as observed in the present study, where low and medium oregano oil doses (0.2 and 0.3 g/kg DM diet) resulted in lower TBARS values, and the high doses produce an increase of lipid oxidation.

It is important to mention that in the present study, the lipid oxidation is considered still low (TBARS values lower than 2.0), which is in agreement with the report of Campo et al. [37]. They revealed that the TBARS value of 2.0 (2 mg MDA/kg meat) could be considered the threshold where the rancid flavour overpowers beef flavour. Therefore, it is considered as the maximum level for the positive sensory perception of beef. These authors indicated that from that point onwards, it can expected for beef to be rejected due to a strong sensory perception of lipid oxidation.

In physiological conditions, mammals constantly produce reactive oxygen species (ROS). Low concentrations of ROS are essential for several physiological processes, including protein phosphorylation, apoptosis, and cellular defence against microorganisms [38]. Oxidative stress refers to a lack of balance between the production of ROS and the level of antioxidants. Domestic animals are frequently exposed to oxidative stress, especially under intensive breeding systems [39]. Oxidative stress is responsible for numerous disease processes in animals. Many secondary metabolites formed by plants serve as defence agents against physiological and environmental stressors, and pathogenic microorganisms [40]. The main molecules responsible for the antioxidative properties of herbs and spices are phenolic substances. In particular, *Origanum vulgare* is an herb rich in phenolics [41].

Essential oils are rich sources of natural antioxidants, such as the phenolic compounds, and due to their high redox properties and chemical structure, they affect lipid metabolism in animal tissues by exerting beneficial effects on the antioxidant enzyme activity. Furthermore, phenolic compounds also prevent the production of reactive oxygen species and the off-flavors that are formed from the oxidation of polyunsaturated fatty acids [42]. Dietary supplementation with EOs is a simple and convenient strategy to uniformly introduce natural antioxidants into phospholipid membranes, where they may effectively inhibit the oxidative reactions by preventing the formation of radicals, and it appears to be a more effective way of slowing down hte lipid oxidation of animal products compared to post-mortem addition [43–45].

Other benefits of OEO have been stated in the literature. OEO modifies ruminal microflora, which also modifies the concentration of ruminal volatile fatty acid. Fat deposition (mainly unsaturated fatty acids, UFAs) is promoted when the concentration of propionic acid decreases and the acetic acid increases. Under some circumstances, UFAs are more susceptible to oxidation [46–48], and they may promote the formation of MDA in absence of antioxidants, as observed in this study (Figure 1).

As it has been previously pointed out, the structure of some lipid components from the essential oils changes as they transit through the digestive tract, and if they are absorbed in the intestine, the lipid profile and the oxidative stability of the meat might be modified [46–49]. In the present study, monensin has a similar effect to that of OEO in terms of lipid oxidation. This indicates that OEO could safely replace monensin in lamb diets, with the advantage of being a natural additive that promotes other positive changes in lamb, such as colour and shelf life preservation. OEO supplementation demonstrated lipid antioxidant activity in fresh lamb meat. OEO improves the antioxidant activity, which has an influence on retarding the lipid meat oxidation during refrigerated and long-term frozen storage. This process could be explained by carvacrol and thymol action on the permeability of cell membranes, and by the transformation of lipid and hydroxyl radicals into stable products [29]. This effect was supported in the present study.

The antioxidant effect of dietary OEO supplementation has also been demonstrated in poultry [44,50,51]. Moreover, OEO has been studied as an ingredient in meat formulations. In lamb burgers, the addition of 24 mL/kg of oregano extract is recommended as a natural antioxidant in replacement of sodium erythorbate, and the product has good acceptability [52].

3.2. Compression Strength

The tenderness of meat has been associated with intramuscular fat (IMF) content [53], and the increase of monounsaturated fatty acids (MUFAs) and PUFAs concentration in IMF could reduce the compression force of meat, thus producing more tender meat and, in this way, improving the quality.

Some of the intrinsec main factors that influence meat texture are the content and solubility of collagen, sarcomere diameter, intramuscular fat content, and proteolysis by calpains during ageing, among others [54]. The dietary inclusion of OEO decreases the concentration of acetic acid and increases propionic acid in rumen, which favours fat deposition [55] and improves meat tenderness. An increased quantity of subcutaneous fat and intramuscular fat decreases the rate of temperature decline, enhances the activity of autolytic enzymes in the muscle, lessens the myofibrillar shortening, and thereby increases the tenderness of cooked meat [56]. In the present study, it can be assumed that differences in tenderness between CON and MO are related to intramuscular fat deposition, since MUFA and PUFA are oilier in texture than saturated fatty acids. Apparently, the MO inclusion promoted a greater amount of MUFA and PUFA in the meat.

There are no other studies showing an improvement of lamb tenderness when animals were fed OEO. In the study of Simitzis et al. [57], the dietary oregano essential oil supplementation on lamb did not influence the tenderness of *Longissimus thoracis* muscle. Demirel et al. [58] reported that the effect of oregano oil was not significant on carcass and lamb meat quality attributes.

Contrasting effects of OEO on the tenderness and shear force of meat from other species are reported. Cheng et al. [59] observed that dietary OEO enhanced the tenderness and overall acceptance of pork. Forte et al. [60] showed that dietary oregano essential oil increased the meat tenderness, but it did not modify other pork quality traits, such as the pH, colour, drip loss, and cooking loss. However, OEO improved consumer perceptions of the meat quality, such as consistency and overall liking. In contrast, Ranucci et al. [61] evaluated a plant extract mix (chestnut and oregano essential oil) in a pig diet and evaluated the pig performance and meat quality. The fresh meat colour, pH, and WB shear force was not affected by OEO supplementation. Simitzis et al. [29] did not find any change in the meat shear force and sensory traits of meat from pigs supplemented OEO. As well, Rossi et al. [62] reported an enhancement of sensory attributes in meat from pigs supplemented plant extract (*Lippia* spp.) but did not find any tenderness improvement in the meat.

When adding essential oils to meat products, it has been pointed out that protein oxidation reduces meat tenderness, but the essential oils of oregano and rosemary can protect the thiols in pork patties and reduce the disulphide crosslinks of the myosin heavy chains, avoiding the tenderness reduction of meat [63].

Finally, Lei et al. [64] demonstrated that the addition of essential oil-cobalt had a significant effect on the meat quality of tested goats. Similarly, Velasco et al. [65] found that the incorporation of dietary dry oregano at 1% and 5% in the diet of Boer goats did not affect the meat quality characteristics.

3.3. Colour

The addition of the OEO and/or monensin in the lamb diet influences the colour (L*, a*, b*, and C*) parameters of the meat. According to recent reports by Payne et al. [66], the colour values in finishing lambs (240 d old) are L* = 34.3, a* = 5.7 and b* = 16.9. The L* values in the present study are higher (L* = 40), meaning a lighter meat. The high L* value could be attractive to consumers that prefer lighter meat [34]. A positive result could be that yellowness (b*) was relatively lower compared to Payne et al. [66], since consumers do not expect to find high b* in fresh meat. Lightness (L*) was higher in the meat from oregano and monensin treatments compared to the control. As noted by Rivaroli et al. [34], in feedlot-finished young bulls that were fed with essential oils, L* values were superior to the other literature data of cattle finished in feedlot.

Colour is one of the most important quality characteristics to determine the consumer decision for purchasing meat. The natural colour of meat is produced by the myoglobin and hemoglobin pigments. These three components that define the colour of meat are all highly susceptible to oxidation [67,68]. An unattractive brown colour can result from the oxidation of red oxymyoglobin to metmyoglobin. The mechanisms that modify pigment distribution in animal tissues could be activated by lowering hemoglobin oxidation by dietary OEO supplementation [57]. Antioxidants have the ability to retard meat colour deterioration by extending the red colour and delaying metmyoblobin formation. Simitzis et al. [57] included 1 mL OEO/kg in lambs diet and found higher a* and b* values. In lamb meat, Nieto et al. [33] indicates that lambs fed with 3.7% and 7.5% of oregano leaves produced significant differences regarding the colour values. In this study, as the storage period was prolonged, the L* and b* values increased and the a* value decreased. Similarly, Simitzis et al. [29] pointed out that supplementing lamb diets with OEO resulted in significant effects on meat colour (L*, a*, and b*).

Different results have been found in other animal species. The colour of pork patties was investigated by Carpenter et al. [69]. They did not find significant changes in colour parameters by the addition of grape seed and bearberry extracts to the diet. Similar results were obtained for fresh chicken breast meat [25], whereas the incorporation of rosemary and oregano extracts in pig rations resulted in significant differences in the luminosity of meat.

Similar results have been reported by Camo et al. [24], who reported that the packaging of lamb meat using rosemary and oregano extracts resulted in the difference in meat redness of the treated animals compared to the controls. Intrinsic characteristics of the animals have also an effect on meat colour. Lamb meat colour changes by body weight, sex, and breed [70]. In this way, Hopkins and Fogatry [71] found that the colour of the m. *Longissimus thoracis* varied with breed. Based on the findings obtained in this study, the effect of OEO on meat colour parameters was found to be in agreement with the literature and within the reference ranges.

Possibly, components from OEO accumulate on the meat, as it has been reported in non-ruminants [72]. Essential oils have an antioxidant activity when used directly on the meat or supplemented ante-mortem [46,73,74], which may protect meat pigment from oxidation throughout storage. If dietary OEO are accumulated in meat, it might mean that they passed the rumen without being degraded. Alternatively, colour might remain stable, as carvacrol supports the activity of glutathione peroxidase and superoxide dismutase, which are two of the most important antioxidant enzymatic complexes in mammals [75].

It is important to highlight that dietary MO not only maintained a higher and more stable redness, yellowness, and saturation during storage, but it also reduced the compression force, and, although not significant ($p > 0.05$), lower TBARS were observed. Improvement of the oxidative stability of MO meat was shown by the stable colour during storage. Dietary antioxidants such as tocopherol deposited in meat may avoid rancidity or the oxidation of tissue components [76]. Carvacrol has a high antioxidant activity [65]. It is possible that the antioxidant activity of OEO is more related to protein protection (pigments) than to lipid components. Some spices and their extracts such us oregano have a high antioxidant activity due to their phenolic compound content, which improves the nutritive value and the quality of meat, because they prevent meat oxidation [65].

Moura et al. [77] evaluated dietary monensin (SM) and incrementing levels of copaiba (*Copaifera* spp.) essential oil (CO) on nutrient intake, time spent eating and ruminating, performance, carcass traits, and the meat quality of feedlot lambs. They observed that the addition of CO at 1.5 g/kg increased Warner Bratzler shear force and decreased L* intensity in *Semimembranosus* meat in comparison to SM.

3.4. Fatty Acid Profile

The supplementation treatments, SM and MO, modified the fatty acid profile compared to the other treatments, whereas HO treatment modified the fatty acid profile undesirably. The similarity between SM and MO might imply that as they modify the rumen environment, the growth rate of rumen microflora changes, resulting in changes in the fermentation profile [49,55,78]. These changes impact the fatty acid profile [79], as it has been reported that monensin was at least partially effective to inhibit the biohydrogenation of unsaturated FAs in the rumen. This consequently increased the percentage of n-6 and n-3 PUFAs and conjugated linoleic acid in milk.

SM (Rumensin®) in ruminant diets [55] and essential oils have bactericidal or bacteriostatic effects [13]. The antibacterial effect is more evident in Gram-positive bacteria, where the cell membrane acts directly with hydrophobic components [80]. SM and some compounds in essential oils are lipophilic; hence, they do not penetrate the membrane of Gram-negative bacteria [81,82]. However, Gram-negative bacteria are not completely resistant to the lipophilic compounds in essential oils, because low molecular weight molecules can interact with the cellular lipid bilayer [82]. Thymol and carvacrol can also disintegrate the external membrane of Gram-negative bacteria [83]. Hence, SM and essential oils affect equally Gram-positive and Gram-negative bacteria, but they use different pathways. The levels of essential oil inclusion are fundamental, because it has been reported that low levels are not enough to modify the ruminal microflora and high levels reduced significantly the bacterial counts, while neither of them change the ruminal fermentation rate [78].

In this regard, several authors have already shown the mechanism of SM inducing ruminal environmental changes. It has been pointed out that SM modifies the ruminal and intestinal microflora, which causes a higher nitrogen and carbon retention in the animal [3]. Additionally, SM promotes the growth of propionic acid-producing microorganisms. Therefore, the concentration of propionic and butyric acids increase, while acetate decreases in ruminal fluid. This leads to an acetate:propionate ratio decline [3,84–86], which in turn favours the recovery of energy used by the animal [79]. Additionally, SM reduces the formation of methane and lactic acid produced by other microorganisms [87,88].

In the present study, most of the FAs that were statistically different are saturated or monounsaturated. This might indicate that triglycerides are accumulating in the intramuscular adipocytes within the neutral lipid fraction. Nevertheless, phospholipidic variations may take place, considering that this fraction is easily altered with the diet [89,90]. An advantage of monensin is that it does not only change the microbial populations in the rumen to such levels that the fatty acid profile is modified, but it also changes the digestibility of nutrients and the utilisation of proteins [3]. Ionophores such as SM alter the fat deposition in beef, particularly arachidonic (C20:4) and linolenic (C18:3n3) acids [91,92]. Furthermore, in bovine milk, SM also changes the amount of fat and increases C18:2 [93]. However, an outstanding characteristic of OEO is that its active components (carvacrol and thymol) of OEO are potent antimicrobials affecting populations such as *E. coli*, *Staphylococcus aureus*, *Salmonella typhimurium*, protozoa, fungi, *Ruminococcus fibrisolvens* and *Fibrobacter succinogenes*, which modifies ruminal fermentation and is fundamental in the conversion of dietary nutrients to muscle tissue [11–13]. Specifically in sheep, there is evidence that carvacrol decreases acetate concentrations and increases propionate and butyrate. Both are volatile fatty acid precursors of muscle and fat components in the animal [14].

Other essential oils have also been studied in lamb nutrition, and their results are promising. Parvar et al. [94] investigated the effects of *Ferulago angulata* (chavil) essential oil (FAE) dietary supplementation on growth performance, meat quality characteristics, and the fatty acid composition of *longissimus* muscle (LM) in fattening lambs. They found that the supplements increased the concentrations of PUFA and decreased SFA contents in meat. Lambs that used diets containing FAE had a lower n-6:n-3 fatty acid ratio compared to the control treatment. They concluded that FAE (up to 750 mL/kg DM) can be used in diets without adverse effects on physical parameters or the chemical composition of meat, and it enhanced the anti-oxidative status of lamb's meat. On the other hand, negative effects of monensin in sheep have been observed. A study of lamb supplementation with monensin (zilpaterol hydrochloride, ZH; 0 or 10 mg/lamb daily) showed a decrease in the content of C20:5n3 (eicosapentaenoic acid), C22:6n-3 (docosahexaenoic acid), and total omega-3 fatty acids, compared with the zero ZH group [95].

In monogastric animals such as chickens, the inclusion of carvacrol and thymol fat increases PUFA and decreases SFA in breasts [75]. In this study, the PUFAS concentration was not different ($p > 0.05$) between the control and monensin treatment. However, the PUFAs concentration was higher in MO.

Promising results of OEO have also been reported in pork. Cheng et al. [59] reported that dietary OEO enhanced the sensory attributes and anti-oxidative status of pork meat by improving IMF and n-3 PUFA proportion and antioxidant capacity.

4. Materials and Methods

4.1. Animal Handling and Treatment Description

Twenty male lambs (Dorper x Pelibuey. Initial body weight, 26.2 ± 3.9 kg) were randomly assigned to one of five treatments (*n* = 4 per treatment); CON: control, basal diet; SM: basal diet + 33 mg/kg monensin sodium (Rumensin 200®); LO: basal diet + 0.2 g/kg DM (dry matter) of OEO; MO: basal diet + 0.3 g/kg DM of OEO, and HO: basal diet + 0.4 g/kg DM of OEO. The basal diet was formulated for 27 kg-lambs to gain 250 g daily [96], and it consisted of 20% alfalfa hay and 80% concentrate DM basis (Table 3). OEO (62.7% carvacrol concentration) was extracted from the leaves (*Lippia* S. *berlandieri*) by steam distillation and obtained from Natural Solution™ Jimenez, México. The diet adaptation period lasted 15 d, and the experimental period was 70 d.

Table 3. Ingredients and chemical composition (DM basis) of diets of finishing hair lambs supplemented with Carvacrol.

Ingredients (%)	Treatment				
	Control	Monensin	Low Oil	Medium Oil	High Oil
Rolled sorghum	36.32	36.32	36.32	36.32	36.32
Soybean meal	34.79	34.79	34.79	34.79	34.79
Alfalfa hay, full bloom	20	20	20	20	20
Molasses cane	5	5	5	5	5
Corn gluten (60% CP)	2	2	2	2	2
Calcium carbonate	0.883	0.883	0.883	0.883	0.883
Salt	0.5	0.5	0.5	0.5	0.5
Mineral premix 1	0.5	0.5	0.5	0.5	0.5
Essential oil (Carvacrol, g/Kg MS)	0	0	0.2	0.3	0.4
Monensin (pmm)	0	33	0	0	0
Calculated chemical composition (% DM basis)					
CP	23.68	23.68	23.68	23.68	23.68
ME (Mcal/kg)	2.739	2.739	2.739	2.739	2.739
Ca	0.791	0.791	0.791	0.791	0.791
P	0.453	0.453	0.453	0.453	0.453

CON: 0 g/Kg MS Carvacrol; M: 33 ppm/Kg MS Monensin; Low: 0.2 g/Kg MS Carvacrol; Medium: 0.3 g/Kg MS Carvacrol; High: 0.4 g/Kg MS Carvacrol. [1] P 12%; Ca 11.5%; Mg 0.6%; Mn 2160 ppm; Zn 2850 ppm; Fe 580 ppm; Cu 1100 ppm; I 102 ppm; Co 13 ppm; Se 9 ppm; Vitamins: A 22,000 UI/Kg; E 24,500 UI/Kg.

Animals were housed in individual pens (equipped with feeding and drinking troughs) and fed twice daily (8:00 and 16:00 h). All experimental procedures with the animals complied with the institutional Bioethics code and Animal Welfare Guidelines, fulfilling the Official Mexican Norms. The protocol was approved by the Institutional Bioethics and Animal Welfare Committee on September 21, 2007, with official number P/302/2017. Description and declarations in this document also followed the Animal Research: Reporting of In Vivo Experiments (ARRIVE) guidelines.

At day 71 (average body weight 45.9 ± 2.82 kg) and after a 12-h fasting period, the animals were electrically stunned and slaughtered by exsanguination using conventional methods in the Meat Science Complex, Department of Animal Science and Ecology (UACH). The dressed carcasses were hung by Achilles tendon suspension in a chiller at 2 °C. At 48 h post-mortem, a portion of *Longissimus lumborum* (LL, 15 cm approx.) was excised longitudinally (after the 12th rib) from the right half of the carcass. The portions of *LL* were vacuum packed (Easy-pack, Rhino, Germany) in 7 μm thickness pouches and fast-frozen (−20 °C) until further analyses.

After two weeks, frozen muscles were sliced transversally to obtain 16 steaks (2 cm thickness) transversal to the muscle fibre direction. Steaks were analysed as follows: the two cranial steaks were tested for lipid oxidation, the next 2 were tested for colour, the following three were tested for texture, and the last 2 were tested for the fatty acid profile.

Samples for lipid oxidation, colour, and texture were thawed at 3 °C for 24 h. Then, to simulate commercial retail display, samples were packed under a modified atmosphere (MAP, 75% O_2-25% CO_2, PVC Cryovac® trays, PET-PVDC-PE top lidding. Rhino 4 sealed packer. USA) and placed in a chilled storage (3 °C, artificial led light, 12 h/d, 700 lx). Samples for fatty acids were re-packed in vacuum bags and kept frozen for 7 d more until fatty acid analysis. All measurements were performed in triplicate except for the compression test, for which measurements were taken at least six times.

4.2. Lipid Oxidation

Lipid oxidation was determined after 7 d of simulated commercial retail display by TBARS (thiobarbituric acid reactive substances), and the values were expressed as mg malonaldehyde (MDA) per kg of meat, according to the distillation method [97].

4.3. Compression Strenght Analysis

To determine the compression strength of the samples, the methodology of the American Meat Science Association Guidelines [98] was followed. MAP chilled stored samples were placed in sealed plastic bags and cooked in a water bath (Fisher Scientific® mod. Isotemp 215, Waltham, MA, USA) until an internal temperature of 72 ± 1 °C was reached at the geometric centre. Temperature was monitored with a thermocouple wire, which was attached to an infrared digital thermometer (Fisherbrand™ Traceable™ Infrared Thermometer with Trigger Grip). Subsequently, cooked steaks (2 cm thickness) were stored at 1 °C for 24 h. Compression strengt was determined by the "punch and die" method [99], which was modified by Jones et al. [100]. Steaks (2 cm thickness) were perforated transversally (parallel to muscle fibers) at least 6 times, and compression values were averaged per every steak (3 steaks per animal). Compression strength analysis was performed with a TA.XT2*i* texture analyser (Stable Micro Systems, Surrey, UK), attached to a 30 kg load cell, and set with a 20 mm cylindrical probe (Crosshead speed of 100 mm min^{-1} at 3 cm of distance). Data are expressed in Newtons/cm^3.

4.4. Colour

Colour (L *, a *, b *, and c *) was evaluated daily for 7 d. Measurements were taken thrice in MAP raw meat with a colorimeter (Minolta® Konica Minolta Camera, Tokyo, Japan), 8 mm Illuminant C. Standard observer, C: Y = 94.2, x = 0.3130 and y = 0.3190following the methodology of the Commission Internationale l'Eclairage with the CIELAB scale [101].

4.5. Fatty Acids (FA) Profile Analysis

Fatty acid profile analysis was carried out in the steaks without simulated retail display. Lipid extraction was carried out following Bligh and Dyer [102,103]. FA derivatization was obtained by saponification, methylation, and esterification [104]. FA were analysed by gas chromatography (Claurus 400 Perkin Elmer, Waltham, MA, USA), with a polar column (100 m × 0.25 mm × 0.20 μm; Sigma, Bellfonte, PA, USA). Peak identification was achieved by comparing the retention times of the unknowns with the standard SupelcoTM 37 Component FAME mix (Sigma).

4.6. Analysis of Data

Variables measured only once (TBARs values, compression values, fatty acid, and concentrations) were analysed with a one-way ANOVA, where diet (5 treatments) was the independent variable. If the treatments were significantly different ($p < 0.05$), means were compared with a Tukey test. The effect of diet on colour (L *, a *, b*, an c *) was analysed with a general linear mixed model, considering the day of the storage as random variable and diet (5 treatments) as a fixed effect. All analyses were performed using the 'R' statistical software version 3.6.0 [105].

5. Conclusions

This study demonstrated that the dietary inclusion of 0.3 g (per kg of DM) of OEO increases lamb tenderness and, similarly to monensin and low oregano oil (0.1 g/kg DM) supplementation, it preserves the meat colour after 7 d of storage. Thus, oregano oil could be an alternative to monensin in lamb diets. The supplementation of oregano essential oil also significantly affected the fatty acid profile, increasing the content of C16:1n7, C18:1n9c, C20:1n9, and C18:2n6 in meat. This study demonstrated that oregano essential oil was beneficial for lamb meat quality. Overall, this study confirmed the remarkable beneficial effect of oregano essential oil on the colour, tenderness, lipid oxidation, and fatty acid profile of lamb meat, which is of significant importance, aiming to evaluate the benefits of phytochemicals to replace monensin and obtaining advantages in lamb meat quality. Further research is needed to identify the main metabolic pathway of oregano essential oil as well as the crucial active components that favourably alter the quality characteristics in lamb meat.

Author Contributions: Conceptualization, I.A.G.-G. and A.D.A.-R.; methodology, I.A.G.-G., J.A.A.-A., A.L.R.-M.; formal analysis, I.A.G.-G., M.H.-J., and L.M.C.-L.; investigation, J.A.A.-A. and J.L.L.-O.; resources, A.D.A.-R.; writing—original draft preparation, I.A.G.-G., A.L.R.-M., J.L.L.-O. and A.D.A.-R.; writing—review and editing, M.H.-J., and J.M.T.-G.; project administration, I.A.G.-G.; funding acquisition, I.A.G.-G. All authors have read and agreed to the published version of the manuscript.

Funding: This research was funded by PROMEP, SEP; clave UACH-PTC-216.

Acknowledgments: The authors would like to thank Guillermo Villalobos for providing the meat and the information from his research on oregano essential oils supplementation on performance and digestibility of sheep.

Conflicts of Interest: The authors declare no conflict of interest. The funders had no role in the design of the study; in the collection, analyses, or interpretation of data; in the writing of the manuscript, or in the decision to publish the results.

References

1. Resurreccion, A.V. Sensory aspects of consumer choices for meat and meat products. *Meat Sci.* **2004**, *66*, 11–20. [CrossRef]
2. Kinnucan, H.W.; Xiao, H.; Hsia, C.-J.; Jackson, J.D. Effects of Health Information and Generic Advertising on U.S. Meat Demand. *Am. J. Agric. Econ.* **1997**, *79*, 13–23. [CrossRef]
3. Schelling, G.T. Monensin mode of action in the rumen. *J. Anim. Sci.* **1984**, *58*, 1518–1527. [CrossRef]
4. Millet, S.; Maertens, L. The European ban on antibiotic growth promoters in animal feed: From challenges to opportunities. *Vet. J.* **2011**, *187*, 143–144. [CrossRef] [PubMed]
5. Hughes, P.; Heritage, J. Antibiotic growth-promoters in food animals. In assessing quality and safety of animal feeds. In *Assesing Quality and Safety of Animal Feeds*; Jutzi, S., Ed.; Food and agriculture organization of the United Nations: Rome, Italy, 2004; pp. 129–153.
6. Kumar, Y.; Yadav, D.N.; Ahmad, T.; Narsaiah, K. Recent Trends in the Use of Natural Antioxidants for Meat and Meat Products. *Compr. Rev. Food Sci. Food Saf.* **2015**, *14*, 796–812. [CrossRef]
7. Shah, M.A.; Bosco, S.J.D.; Mir, S.A. Plant extracts as natural antioxidants in meat and meat products. *Meat Sci.* **2014**, *98*, 21–33. [CrossRef] [PubMed]
8. Xu, J.; Zhou, F.; Ji, B.P.; Pei, R.S.; Xu, N. The antibacterial mechanism of carvacrol and thymol against Escherichia coli. *Lett. Appl. Microbiol.* **2008**, *47*, 174–179. [CrossRef]
9. Aguirre, A.; Borneo, R.; León, A.E. Antimicrobial, mechanical and barrier properties of triticale protein films incorporated with oregano essential oil. *Food Biosci.* **2013**, *1*, 2–9. [CrossRef]
10. Piñon, M.I.; Alarcon-Rojo, A.D.; Renteria, A.L.; Mendez, G.; Janacua-Vidales, H. Reduction of microorganisms in marinated poultry breast using oregano essential oil and power ultrasound. *Acta Aliment.* **2015**, *44*, 527–533. [CrossRef]
11. Gündüz, G.T.; Gönül, Ş.A.; Karapınar, M. Efficacy of oregano oil in the inactivation of *Salmonella typhimurium* on lettuce. *Food Control* **2010**, *21*, 513–517. [CrossRef]
12. Valenzuela-Grijalva, N.V.; Pinelli-Saavedra, A.; Muhlia-Almazan, A.; Domínguez-Díaz, D.; González-Ríos, H. Dietary inclusion effects of phytochemicals as growth promoters in animal production. *J. Anim. Sci. Technol.* **2017**, *59*, 1–17. [CrossRef] [PubMed]

13. Walsh, S.E.; Maillard, J.Y.; Russel, A.D.; Catrenich, C.E.; Charbonneau, A.L.; Bartolo, R.G. Activity and mechanisms of action of selected biocidal agents on Gram-positive and -negative bacteria. *J. Appl. Microbiol.* **2003**, *94*, 240–247. [CrossRef] [PubMed]
14. Koyuncu, M.; Canbolat, O. Effect of carvacrol on intake, rumen fermentation, growth performance and carcass characteristics of growing lambs. *J. Appl. Anim. Res.* **2010**, *38*, 245–248. [CrossRef]
15. Chikwanha, O.C.; Vahmani, P.; Muchenje, V.; Dugan, M.E.R.; Mapiye, C. Nutritional enhancement of sheep meat fatty acid profile for human health and wellbeing. *Food Res. Int.* **2018**, *104*, 25–38. [CrossRef]
16. Vargas-Bello-Pérez, E.; Larraín, R.E. Impacts of fat fromruminants' meat on cardiovascular health and possible strategies to alter its lipid composition. *J. Sci. Food Agric.* **2017**, *97*, 1969–1978. [CrossRef]
17. Xu, R. Important bioactive properties of omega-3 fatty acids. *Ital. J. Food Sci.* **2015**, *27*, 129–135.
18. Chen, G.C.; Yang, J.; Eggersdorfer, M.; Zhang, W.; Qin, L.Q. N-3 long-chain polyunsaturated fatty acids and risk of all-cause mortality among general populations: A meta-analysis. *Sci. Rep.* **2016**, *6*, 1–9. [CrossRef]
19. Zárate, R.; el Jaber-Vazdekis, N.; Tejera, N.; Pérez, J.A.; Rodríguez, C. Significance of long chain polyunsaturated fatty acids in human health. *Clin. Transl. Med.* **2017**, *6*. [CrossRef]
20. Van Le, H.; Nguyen, D.V.; Vu Nguyen, Q.; Malau-Aduli, B.S.; Nichols, P.D.; Malau-Aduli, A.E.O. Fatty acid profiles of muscle, liver, heart and kidney of Australian prime lambs fed different polyunsaturated fatty acids enriched pellets in a feedlot system. *Sci. Rep.* **2019**, *9*, 1–11. [CrossRef]
21. Abedi, E.; Sahari, M.A. Long-chain polyunsaturated fatty acid sources and evaluation of their nutritional and functional properties. *Food Sci. Nutr.* **2014**, *2*, 443–463. [CrossRef]
22. *World Health Organization Diet, Nutrition and the Prevention of Chronic Diseases*; WHO Technical Report Series 916: Geneva, Switzerland, 2003.
23. Miltko, R.; Majewska, M.P.; Bełżecki, G.; Kula, K.; Kowalik, B. Growth performance, carcass and meat quality of lambs supplemented different vegetable oils. *Asian-Australas J. Anim. Sci.* **2019**, *32*, 767–775. [CrossRef] [PubMed]
24. Camo, J.; Lorés, A.; Djenane, D.; Beltrán, J.A.; Roncalés, P. Display life of beef packaged with an antioxidant active film as a function of the concentration of oregano extract. *Meat Sci.* **2011**, *88*, 174–178. [CrossRef] [PubMed]
25. Chouliara, E.; Karatapanis, A.; Savvaidis, I.N.; Kontominas, M.G. Combined effect of oregano essential oil and modified atmosphere packaging on shelf-life extension of fresh chicken breast meat, stored at 4 °C. *Food Microbiol.* **2007**, *24*, 607–617. [CrossRef]
26. Gallo, M.; Ferracane, R.; Naviglio, D. Antioxidant addition to prevent lipid and protein oxidation in chicken meat mixed with supercritical extracts of Echinacea angustifolia. *J. Supercrit. Fluids* **2012**, *72*, 198–204. [CrossRef]
27. Karabagias, I.; Badeka, a.; Kontominas, M.G. Shelf life extension of lamb meat using thyme or oregano essential oils and modified atmosphere packaging. *Meat Sci.* **2011**, *88*, 109–116. [CrossRef]
28. Petrou, S.; Tsiraki, M.; Giatrakou, V.; Savvaidis, I.N. Chitosan dipping or oregano oil treatments, singly or combined on modified atmosphere packaged chicken breast meat. *Int. J. Food Microbiol.* **2012**, *156*, 264–271. [CrossRef]
29. Simitzis, P.E.; Symeon, G.K.; Charismiadou, M.A.; Bizelis, J.A.; Deligeorgis, S.G. The effects of dietary oregano oil supplementation on pig meat characteristics. *Meat Sci.* **2010**, *84*, 670–676. [CrossRef]
30. Giannenas, I.; Tzora, A.; Bonos, E.; Sarakatsianos, I.; Karamoutsios, A.; Anastasiou, I.; Skoufos, I. Effects of dietary oregano essential oil, laurel essential oil and attapulgite on chemical composition, oxidative stability, fatty acid profile and mineral content of chicken breast and thigh meat. *Eur. Poult. Sci.* **2016**, *80*, 1–18.
31. Cobellis, G.; Trabalza-Marinucci, M.; Yu, Z. Critical evaluation of essential oils as rumen modifiers in ruminant nutrition: A review. *Sci. Total Environ.* **2016**, *545–546*, 556–568. [CrossRef]
32. Lourenço, M.; Cardozo, P.W.; Calsamiglia, S.; Fievez, V. Effects of saponins, quercetin, eugenol, and cinnamaldehyde on fatty acid biohydrogenation of forage polyunsaturated fatty acids in dual-flow continuous culture fermenters1. *J. Anim. Sci.* **2008**, *86*, 3045–3053. [CrossRef]
33. Nieto, G.; Bañón, S.; Garrido, M.D. Effect of supplementing ewes' diet with thyme (*Thymus zygis* ssp. *gracilis*) leaves on the lipid oxidation of cooked lamb meat. *Food Chem.* **2011**, *125*, 1147–1152.
34. Rivaroli, D.C.; Guerrero, A.; Velandia Valero, M.; Zawadzki, F.; Eiras, C.E.; del Mar Campo, M.; Sañudo, C.; Mendes Jorge, A.; Nunes do Prado, I. Effect of essential oils on meat and fat qualities of crossbred young bulls finished in feedlots. *Meat Sci.* **2016**, *121*, 278–284. [CrossRef] [PubMed]

35. Galati, G.; O'Brien, P.J. Potential toxicity of flavonoids and other dietary phenolics: Significance for their chemopreventive and anticancer properties. *Free Radic. Biol. Med.* **2004**, *37*, 287–303. [CrossRef] [PubMed]
36. Nemeikaite-Čeniene, A.; Imbrasaite, A.; Sergediene, E.; Čenas, N. Quantitative structure-activity relationships in prooxidant cytotoxicity of polyphenols: Role of potential of phenoxyl radical/phenol redox couple. *Arch. Biochem. Biophys.* **2005**, *441*, 182–190. [CrossRef]
37. Campo, M.M.; Nute, G.R.; Hughes, S.I.; Enser, M.; Wood, J.D.; Richardson, R.I. Flavour perception of oxidation in beef. *Meat Sci.* **2006**, *72*, 303–311. [CrossRef]
38. Miller, J.K.K.; Brzezinska-Slebodzinska, E.; Madsen, F.C.C. Oxidative stress, antioxidants, and animal function. *J. Dairy Sci.* **1993**, *76*, 2812–2823. [CrossRef]
39. Aurousseau, B. Oxygen radicals in farm animals. Physiological effects and consequences on animal products. *INRA Prod. Anim.* **2002**, *15*, 67–82.
40. Bhatt, N. Herbs and Herbal Supplements, a Novel Nutritional Approach in Animal Nutrition. *Iran. J. Appl. Anim. Sci.* **2015**, *5*, 497–516.
41. Fasseas, M.K.; Mountzouris, K.C.; Tarantilis, P.A.; Polissiou, M.; Zervas, G. Antioxidant activity in meat treated with oregano and sage essential oils. *Food Chem.* **2008**, *106*, 1188–1194. [CrossRef]
42. Rice-evans, C.A.; Miller, N.J.; Bolwell, P.G.; Bramley, P.M.; Pridham, J.B. The Relative Antioxidant Activities of Plant-Derived Polyphenolic Flavonoids. *Free Radic. Res.* **1995**, *22*, 375–383. [CrossRef]
43. Decker, E.A.; Park, Y. Healthier meat products as functional foods. *Meat Sci.* **2010**, *86*, 49–55. [CrossRef] [PubMed]
44. Govaris, A.; Botsoglou, N.; Papageorgiou, G.; Botsoglou, E.; Ambrosiadis, I. Dietary versus post-mortem use of oregano oil and/or α-tocopherol in turkeys to inhibit development of lipid oxidation in meat during refrigerated storage. *Int. J. Food Sci. Nutr.* **2004**, *55*, 115–123. [CrossRef] [PubMed]
45. Simitzis, P.E. Enrichment of Animal Diets with Essential Oils—A Great Perspective on Improving Animal Performance and Quality Characteristics of the Derived Products. *Medicines* **2017**, *4*, 35. [CrossRef] [PubMed]
46. Hashemipour, H.; Khaksar, V.; Rubio, L.A.; Veldkamp, T.; van Krimpen, M.M. Effect of feed supplementation with a thymol plus carvacrol mixture, in combination or not with an NSP-degrading enzyme, on productive and physiological parameters of broilers fed on wheat-based diets. *Anim. Feed Sci. Technol.* **2016**, *211*, 117–131. [CrossRef]
47. Chaves, A.V.; Stanford, K.; Gibson, L.L.; McAllister, T.A.; Benchaar, C. Effects of carvacrol and cinnamaldehyde on intake, rumen fermentation, growth performance, and carcass characteristics of growing lambs. *Anim. Feed Sci. Technol.* **2008**, *145*, 396–408. [CrossRef]
48. Cardozo, P.W.; Calsamiglia, S.; Ferret, A.; Kamel, C. Screening for the effects of natural plant extracts at different pH on in vitro rumen microbial fermentation of a high-concentrate diet for beef cattle1. *J. Anim. Sci.* **2005**, *83*, 2572–2579. [CrossRef]
49. Benchaar, C.; Calsamiglia, S.; Chaves, A.V.; Fraser, G.R.; McAllister, T.A.; Beauchemin, K.A. A review of plant-derived essential oils in ruminant nutrition and production. *Anim. Feed Sci. Technol.* **2008**, *145*, 209–228. [CrossRef]
50. Botsoglou, N.A.; Grigoropoulou, S.H.; Botsoglou, E.; Govaris, A.; Papageorgiou, G. The effects of dietary oregano essential oil and α-tocopheryl acetate on lipid oxidation in raw and cooked Turkey during refrigerated storage. *Meat Sci.* **2003**, *65*, 1193–1200. [CrossRef]
51. Botsoglou, N.A.; Christaki, E.; Fletouris, D.J.; Florou-Paneri, P.; Spais, A.B. The effect of dietary oregano essential oil on lipid oxidation in raw and cooked chicken during refrigerated storage. *Meat Sci.* **2002**, *62*, 259–265. [CrossRef]
52. Fernandes, R.P.P.; Trindade, M.A.; Tonin, F.G.; Pugine, S.M.P.; Lima, C.G.; Lorenzo, J.M.; de Melo, M.P. Evaluation of oxidative stability of lamb burger with *Origanum vulgare* extract. *Food Chem.* **2017**, *233*, 101–109. [CrossRef]
53. Watkins, P.J.; Frank, D.; Singh, T.K.; Young, O.A.; Warner, R.D. Sheepmeat Flavor and the Effect of Different Feeding Systems: A Review. *J. Agric. Food Chem.* **2013**, *61*, 3561–3579. [CrossRef]
54. Wheeler, T.L.; Shackelford, S.D.; Koohmaraie, M. Variation in proteolysis, sarcomere length, collagen content, and tenderness among major pork muscles. *J. Anim. Sci.* **2000**, *78*, 958–965. [CrossRef] [PubMed]
55. Castillejos, L.; Calsamiglia, S.; Ferret, A. Effect of essential oil active compounds on rumen microbial fermentation and nutrient flow in in vitro systems. *J. Dairy Sci.* **2006**, *89*, 2649–2658. [CrossRef]

56. Smith, G.C.; Dutson, T.R.; Hostetler, R.L.; Carpenter, Z.L. Fatness, rate of chilling and tenderness of lamb. *J. Food Sci.* **1976**, *41*, 748–756. [CrossRef]
57. Simitzis, P.E.; Deligeorgis, S.G.; Bizelis, J.A.; Dardamani, A.; Theodosiou, I.; Fegeros, K. Effect of dietary oregano oil supplementation on lamb meat characteristics. *Meat Sci.* **2008**, *79*, 217–223. [CrossRef]
58. Demirel, G.; Ekiz, B.; Unal, A.; Yalcintan, H.; Kocabagli, N.; Yilmaz, A. Effects of dietary oregano essential oil on carcass and meat quality of kivircik lambs. *J. Anim. Vet. Adv.* **2013**, *12*, 991–995.
59. Cheng, C.; Liu, Z.; Zhou, Y.; Wei, H.; Zhang, X.; Xia, M.; Deng, Z.; Zou, Y.; Jiang, S.; Peng, J. Effect of oregano essential oil supplementation to a reduced-protein, amino acid-supplemented diet on meat quality, fatty acid composition, and oxidative stability of *Longissimus thoracis* muscle in growing-finishing pigs. *Meat Sci.* **2017**, *133*, 103–109. [CrossRef]
60. Forte, C.; Ranucci, D.; Beghelli, D.; Branciari, R.; Acuti, G.; Todini, L.; Cavallucci, C.; Trabalza-Marinucci, M. Dietary integration with oregano (*Origanum vulgare* L.) essential oil improves growth rate and oxidative status in outdoor-reared, but not indoor-reared, pigs. *J. Anim. Physiol. Anim. Nutr.* **2017**, *101*, e352–e361. [CrossRef]
61. Ranucci, D.; Beghelli, D.; Trabalza-Marinucci, M.; Branciari, R.; Forte, C.; Olivieri, O.; Badillo Pazmay, G.V.; Cavallucci, C.; Acuti, G. Dietary effects of a mix derived from oregano (*Origanum vulgare* L.) essential oil and sweet chestnut (*Castanea sativa* Mill.) wood extract on pig performance, oxidative status and pork quality traits. *Meat Sci.* **2015**, *100*, 319–326. [CrossRef]
62. Rossi, R.; Pastorelli, G.; Cannata, S.; Tavaniello, S.; Maiorano, G.; Corino, C. Effect of long term dietary supplementation with plant extract on carcass characteristics meat quality and oxidative stability in pork. *Meat Sci.* **2013**, *95*, 542–548. [CrossRef]
63. Nieto, G.; Jongberg, S.; Andersen, M.L.; Skibsted, L.H. Thiol oxidation and protein cross-link formation during chill storage of pork patties added essential oil of oregano, rosemary, or garlic. *Meat Sci.* **2013**, *95*, 177–184. [CrossRef] [PubMed]
64. Lei, Z.; Zhang, K.; Li, C.; Wu, J.; Davis, D.; Casper, D.; Jiang, H.; Jiao, T.; Wang, X.; Wang, J. Dietary supplementation with Essential-oils-cobalt for improving growth performance, meat quality and skin cell capacity of goats. *Sci. Rep.* **2018**, *8*. [CrossRef] [PubMed]
65. Velasco, V.; Orellana, C.; Williams, P.; Campos, J.; Melín, P. Meat Quality Characteristics of Goats Fed with Dietary Dry Oregano (*Origanum vulgare* L.). *Chil. J. Agric. Anim. Sci.* **2011**, *27*, 95–104.
66. Payne, C.E.; Pannier, L.; Anderson, F.; Pethick, D.W.; Gardner, G.E. Lamb age has little impact on eating quality. *Foods* **2020**, *9*, 187. [CrossRef] [PubMed]
67. Barbut, S. Principles of meat processing. In *The Science of Poultry and Meat Processing*; Barbut, S., Ed.; University of Guelp: Guelp, ON, Canada, 2015.
68. Mancini, R.A.; Hunt, M.C. Current research in meat color. *Meat Sci.* **2005**, *71*, 100–121. [CrossRef] [PubMed]
69. Carpenter, R.; O'Grady, M.N.; O'Callaghan, Y.C.; O'Brien, N.M.; Kerry, J.P. Evaluation of the antioxidant potential of grape seed and bearberry extracts in raw and cooked pork. *Meat Sci.* **2007**, *76*, 604–610. [CrossRef]
70. Teixeira, A.; Batista, S.; Delfa, R.; Cadavez, V. Lamb meat quality of two breeds with protected origin designation. Influence of breed, sex and live weight. *Meat Sci.* **2005**, *71*, 530–536. [CrossRef]
71. Hopkins, D.L.; Fogarty, N.M. Diverse lamb genotypes -2. Meat pH, colour and tenderness. *Meat Sci.* **1998**, *49*, 477–488. [CrossRef]
72. Ramos, F.A.; Martínez, A.P.; Montes, E.S.; Gaytán, C.N.; Selene, A.; Cázarez, H.; Tovar, J.C.; Sánchez, J.G.; Rodríguez, C. Oregano oil use in broiler diet increases accumulation of carvacrol and thymol in breast meat Broiler production. *Acta Univ.* **2017**, *27*, 34–39.
73. Boroski, M.; Giroux, H.J.; Sabik, H.; Petit, H.V.; Visentainer, J.V.; Matumoto-Pintro, P.T.; Britten, M. Use of oregano extract and oregano essential oil as antioxidants in functional dairy beverage formulations. *Food Sci. Technol.* **2012**, *47*, 167–174. [CrossRef]
74. Hernández-Hernández, E.; Ponce-Alquicira, E.; Jaramillo-Flores, M.E.; Guerrero Legarreta, I. Antioxidant effect rosemary (*Rosmarinus officinalis* L.) and oregano (*Origanum vulgare* L.) extracts on TBARS and colour of model raw pork batters. *Meat Sci.* **2009**, *81*, 410–417. [CrossRef] [PubMed]
75. Hashemipour, H.; Kermanshahi, H.; Golian, A.; Veldkamp, T. Effect of thymol and carvacrol feed supplementation on performance, antioxidant enzyme activities, fatty acid composition, digestive enzyme activities, and immune response in broiler chickens. *Poult. Sci.* **2013**, *8*, 2059–2069. [CrossRef] [PubMed]

76. Petron, M.J.; Raes, K.; Claeys, E.; Lourenco, M.; Fremaut, D.; De Smet, S. Effect of grazing pastures of different botanical composition on antioxidant enzyme activities and oxidative stability of lamb meat. *Meat Sci.* **2007**, *75*, 737–745. [CrossRef] [PubMed]
77. Moura, L.V.; Oliveira, E.R.; Fernandes, A.R.M.; Gabriel, A.M.A.; Silva, L.H.X.; Takiya, C.S.; Cônsolo, N.R.B.; Rodrigues, G.C.G.; Lemos, T.; Gandra, J.R. Feed efficiency and carcass traits of feedlot lambs supplemented either monensin or increasing doses of copaiba (*Copaifera* spp.) essential oil. *Anim. Feed Sci. Technol.* **2017**, *232*, 110–118. [CrossRef]
78. Calsamiglia, S.; Busquet, M.; Cardozo, P.W.; Castillejos, L.; Ferret, A. Invited Review: Essential Oils as Modifiers of Rumen Microbial Fermentation. *J. Dairy Sci.* **2007**, *90*, 2580–2595. [CrossRef]
79. Ogunade, I.; Schweickart, H.; Andries, K.; Lay, J.; Adeyemi, J. Monensin Alters the Functional and Metabolomic Profile of Rumen Microbiota in Beef Cattle. *Animals* **2018**, *8*, 211. [CrossRef] [PubMed]
80. Chao, S.C.; Young, D.G.; Oberg, C.J. Screening for Inhibitory Activity of Essential Oils on Selected Bacteria, Fungi and Viruses. *J. Essent. Oil Res.* **2000**, *12*, 639–649. [CrossRef]
81. Cox, S.D.; Mann, C.M.; Markham, J.L.; Gustafson, J.E.; Warmington, J.R.; Wyllie, S.G. Determining the antimicrobial actions of tea tree oil. *Molecules* **2001**, *6*, 87–91. [CrossRef]
82. Nazzaro, F.; Fratianni, F.; De Martino, L.; Coppola, R.; De Feo, V. Effect of essential oils on pathogenic bacteria. *Pharmaceuticals* **2013**, *6*, 1451–1474. [CrossRef]
83. Helander, I.M.; Alakomi, H.-L.; Latva-Kala, K.; Mattila-Sandholm, T.; Pol, I.; Smid, E.J.; Gorris, L.G.M.; von Wright, A. Characterization of the Action of Selected Essential Oil Components on Gram-Negative Bacteria. *J. Agric. Food Chem.* **1998**, *46*, 3590–3595. [CrossRef]
84. Castro-Montoya, J.M.; Makkar, H.P.S.; Becker, K. Effects of monensin on the chemical composition of the liquid associated microbial fraction in an in vitro rumen fermentation system. *Livest. Sci.* **2012**, *150*, 414–418. [CrossRef]
85. Boucqué, C.V.; Fiems, L.O.; Cottyn, B.G.; Casteels, M.; Buysse, F.X. Monensin-sodium as a performance-promoting additive for fattening bulls and its impact on carcass and meat quality characteristics. *Anim. Feed Sci. Technol.* **1982**, *7*, 401–410. [CrossRef]
86. Smith, S.B.; Kawachi, H.; Choi, C.B.; Choi, C.W.; Wu, G.; Sawyer, J.E. Cellular regulation of bovine intramuscular adipose tissue development and composition. *J. Anim. Sci.* **2009**, *87*, 72–82. [CrossRef] [PubMed]
87. Russell, J.B.; Strobel, H.J. Effect of ionophores on ruminal fermentation. *Appl. Environ. Microbiol.* **1989**, *55*, 1–6. [CrossRef]
88. Isabel Osorio-Teran, A.; David Mendoza-Martínez, G.; Alberto Miranda-Romero, L.; Martínez-Gomez, D.; Abel Hernández-García, P.; Antonio Martínez-García, J.; Brasileira de Zootecnia, R. Effect of calcium propionate and monensin on in vitro digestibility and gas production. *Brazilian J. Anim. Sci.* **2017**, *46*, 348–353. [CrossRef]
89. Wood, J.D.; Enser, M.; Fisher, A.V.; Nute, G.R.; Sheard, P.R.; Richardson, R.I.; Hughes, S.I.; Whittington, F.M. Fat deposition, fatty acid composition and meat quality: A review. *Meat Sci.* **2008**, *78*, 343–358. [CrossRef]
90. Wood, J.D.; Nute, G.R.; Richardson, R.I.; Whittington, F.M.; Southwood, O.; Plastow, G.; Mansbridge, R.; Da Costa, N.; Chang, K.C. Effects of breed, diet and muscle on fat deposition and eating quality in pigs. *Meat Sci.* **2004**, *67*, 651–667. [CrossRef]
91. Ladeira, M.M.; Santarosa, L.C.; Chizzotti, M.L.; Ramos, E.M.; Machado Neto, O.R.; Oliveira, D.M.; Carvalho, J.R.R.; Lopes, L.S.; Ribeiro, J.S. Fatty acid profile, color and lipid oxidation of meat from young bulls fed ground soybean or rumen protected fat with or without monensin. *Meat Sci.* **2014**, *96*, 597–605. [CrossRef]
92. Daley, C.; Abbott, A.; Doyle, P.; Nader, G.; Larson, S. A review of fatty acid profiles and antioxidant content in grass-fed and grain-fed beef. *Nutr. J.* **2010**, *9*. [CrossRef]
93. Silva, D.C.; Santos, G.T.; Branco, A.F.; Damasceno, J.C.; Kazama, R.; Matsushita, M.; Horst, J.A.; Santos, W.B.R.; Petit, H.V. Production performance and milk composition of dairy cows fed whole or ground flaxseed with or without monensin. *J. Dairy Sci.* **2007**, *90*, 2928–2936. [CrossRef]
94. Parvar, R.; Ghoorchi, T.; Kashfi, H.; Parvar, K. Effect of Ferulago angulata (Chavil) essential oil supplementation on lamb growth performance and meat quality characteristics. *Small Rumin. Res.* **2018**, *167*, 48–54. [CrossRef]
95. Dávila-Ramírez, J.L.; Avendaño-Reyes, L.; Macías-Cruz, U.; Peña-Ramos, E.A.; Islava-Lagarda, T.Y.; Zamorano-García, L.; Valenzuela-Melendres, M.; Camou, J.P.; González-Ríos, H. Fatty acid composition

and physicochemical and sensory characteristics of meat from ewe lambs supplemented with zilpaterol hydrochloride and soybean oil. *Anim. Prod. Sci.* **2017**, *57*, 767–777. [CrossRef]
96. Committee on the Nutrient Requirements of Small Ruminants, National Research Council. *Nutrient Requirements of Small Ruminants: Sheep, Goats, Cervids, and New World Camelids*; The National Academies Press: Washington, DC, USA, 2007.
97. Tarladgis, B.; Watts, B.; Younathan, M.; Dugan, L. A distillation method for the quantitative determination of malonaldehyde in rancid foods. *J. Am. Oil Chem. Soc.* **1960**, *37*, 44–48. [CrossRef]
98. American Meat Science Association. *Research Guidelines for Cookery, Sensory Evaluation, and Instrumental Tenderness Measurements of Meat*; AMSC: Champaign, IL, USA, 2016.
99. Segars, R.A.; Hamel, R.G.; Kapsalis, J.G.; Klutter, R.A. A punch and die test cell for determining the textural qualities of meat. *J. Texture Stud.* **1975**, *6*, 211–225. [CrossRef]
100. Jones, R.C.; Dransfield, E.; Robinson, J.M.; Crosland, A.R. Correlation of mechanical properties, composition and perceived texture of beefburgers. *J. Texture Stud.* **1985**, *16*, 241–262. [CrossRef]
101. Roufs, J.A.J. *Light as a True Visual Quantity: Principles of Measurement*; Commission Internationale de l'Éclairage: Paris, France, 1978.
102. Bligh, E.G.; Dyer, W.J. A rapid method of total lipid extraction and purification. *Can. J. Biochem. Physiol.* **1959**, *37*, 911–917. [CrossRef] [PubMed]
103. Jensen, S.K. Improved Bligh and Dyer extraction procedure. *Lipid Technol.* **2008**, *20*, 280–281. [CrossRef]
104. *AOAC Official Methods of Analysis*; Association of Official Analytical Chemists: Washington, DC, USA, 1995.
105. R Team Core. *R: A Language and Environment for Statistical Computing*; R Foundation for Statistical Computing: Vienna, Austria, 2018.

antibiotics

MDPI

Article

In Vitro Antimicrobial Activity of Frankincense Oils from *Boswellia sacra* Grown in Different Locations of the Dhofar Region (Oman)

Vita Di Stefano [1], Domenico Schillaci [1,*], Maria Grazia Cusimano [1], Mohammed Rishan [2] and Luay Rashan [2]

1 Department of Science and Technology Biological, Chemical and Pharmaceutical (STEBICEF) Università degli Studi di Palermo, Via Archirafi, 32, 90123 Palermo, Italy; vita.distefano@unipa.it (V.D.S.); mariagrazia.cusimano@unipa.it (M.G.C.)

2 Frankincense and Biodiversity Lab, Dhofar University, 2509 Salalah, Oman; mrishan@du.edu.om (M.R.); lrashan@du.edu.om (L.R.)

* Correspondence: domenico.schillaci@unipa.it

Received: 2 April 2020; Accepted: 17 April 2020; Published: 20 April 2020

Abstract: Frankincense essential oils from *Boswellia sacra* have been commonly used to treat microbial infections from as early as the 11th century. The main feature of the plant is its gum resin, from which it is possible to obtain essential oils. In the present study, we focused on the comparative study of the oils extracted from the resins of three different *Boswellia sacra* cultivars (Najdi, Sahli and Houjri). From each of frankincense resin three successive essential oil samples (Grade 1, Grade 2, Grade 3) were obtained. Houjri gum resin gave the lowest percentage (5%) of total essential oil content but showed the maximum number of volatile components in all three grades. Najdi Grade 2 essential oil showed a minimum inhibitory concentration (MIC) of 52 mg/mL toward relevant pathogens *Staphylococcus aureus* and *Pseudomonas aeruginosa*, and samples from Grade 2 of Sahily and Houjiri were particularly active against a dermatological strain *Propionibacterium acnes*, displaying MIC values of 0.264 and 0.66 mg/mL, respectively. Data obtained from in vitro studies showed that all essential oils had a significant antifungal effect against *Candida albicans* and *Malassezia furfur*, showing MIC values ranging from 54.56 to 0.246 mg/mL. This work aims to increase the number of substances available in the fight against pathogens and to combat the phenomenon of antibiotic resistance, encouraging the use of alternative resources, especially in non-clinical settings (farms, food processing, etc.).

Keywords: *Boswellia sacra*; frankincense essential oil; GC/MS analysis; antimicrobial activity; *Staphylococcus aureus*; *Pseudomonas aeruginosa*; *Propionibacterium acnes*; *Candida albicans*; *Malassezia furfur*

1. Introduction

The *Boswellia* genus of plants includes about twenty species distributed in northeastern coastal areas of Africa, the Arabic peninsula and the Indian subcontinent. *Boswellia sacra*, a small tree of up to 5 m of height with papery peeling bark and densely tangled branches with leaves clustered at the ends, is indigenous to the southern parts of Oman in the Dhofar region, and is also cultivated in other parts of Oman.

Boswellia sacra is well known for its oleo-gum resin named frankincense or olibanum, which is usually harvested from deep incisions made into the tree trunk. Generally, in almost all of the cultures where frankincense is traded, it is used for fragrance and fumigating objects used specifically for a religious purpose. In traditional medicine, frankincense has a unique place amongst remedies for the treatment of many disorders (dermatological, gastric, hepatic, rheumatoid arthritis, etc.) [1].

Investigation of the pharmacological activities of essential oils of different *Boswellia* ssp. resins reveals a wide range of medicinal uses. In particular, *B. sacra* has been used as an analgesic, antioxidant, cardio protective substance and an anti-inflammatory [2–4]. Frankincense essential oils have been commonly used to treat microbial infections from as early as the 11th century, and some authors have studied the effect of frankincense on urinary tract infections [5]. Resin essential oils have an antimicrobial activity against important human pathogens, both bacterial and fungal organisms, such as *Staphylococcus aureus*, *Escherichia coli*, *Proteus vulgaris* and *Candida albicans* [6,7]. *B. sacra* oil also shows significant activity on *Aspergillus parasiticus* and *Aspergillus flavus* growth, and consequently suppresses aflatoxin production [8].

There are various compounds detected in *Boswellia* species based on several factors, which include color, purity, aroma, clump size, age of the tree, season of harvest and the geographical location of the plant source. The essential oil of *B. sacra* contains a high proportion of monoterpenes (97.3%). The common compounds include α- and β-pinene, limonene, myrcene, linalool and others.

Başer [9] claimed that octyl acetate (39.9%) was the main constituent, followed by 1-octanol (11.9%). Al-Harrasi [10] found limonene (33.5%) and (*E*)-β-ocimene (32.2%) to be the predominant compounds in olibanum from *B. sacra*. Camarda et al. [6] also reported limonene to be the dominant substance, albeit at half the abundance found by Al-Harrasi (18.2%). Furthermore, Camarda identified α-pinene as the second most abundant substance (15.1%). Al-Saidi identified α-pinene as the main volatile component, followed by octyl acetate (13.4%) [11].

Omani resin is commercially available in *B. sacra* different accessions, under the local names Houjri, Najdi, and Sahli or Shaebi (Figure 1), based on different geographic locations in Dhofar from where the resins are harvested. Houjri, the first grade resin, has the lightest color and a larger clump size, is collected from trees growing in the north of the Samhan mountains, and is the most expensive. Najdi, the second resin, has a pale yellow color and is collected from the plateau behind the Dhofar mountains. Finally, Sahli or Shaebi, which also has a darker color, is collected from the valleys and is the cheaper one.

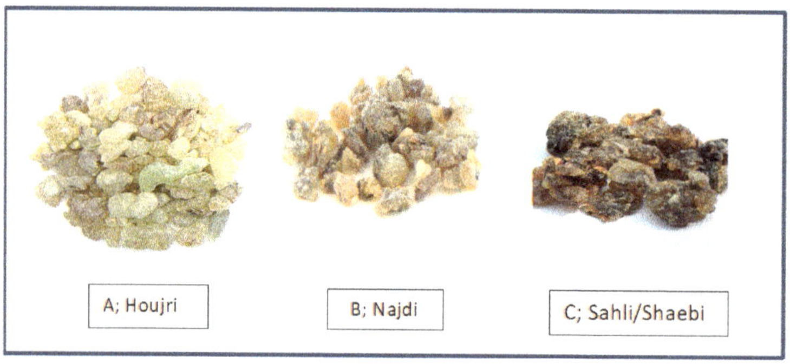

Figure 1. Omani commercially available *B. sacra* resin samples: (**A**) Houjri, (**B**) Najdi, and (**C**) Sahli/Shaebi.

In the present investigation, we report on the comparative study of the constituents of frankincense essential oil from three different cultivars grown in various agro-climatic conditions in the Dhofar region in Southern Oman, on the eastern border with Yemen. In addition, the authors compared in vitro antibacterial and antifungal activities of *B. sacra* resin essential oils, with the goal to find new strategies in the struggle against antibiotic-resistant microorganisms.

2. Results

In Table 1, the essential oil yields from three different frankincense gum cultivars, called Najdi, Sahli and Houjri, are shown. The samples were subjected to three successive hydrodistillations (2, 4, 6 h) to obtain three corresponding essential oils that were called Grade 1, Grade 2 and Grade 3. The essential oils from *B. sacra* oleogum resins were obtained in yields of 8.92–12% (*w*/*w*) in Grade 1. Regarding Grade 2 essential oils, Najdi recorded the highest percentage of 2.20% (*w*/*w*), followed by

Sahli, of 2.08% (*w*/*w*). Grade 3 showed the lowest yield percentages of 0.40–0.72% (*w*/*w*) of essential oils (Table 1). However, Grade 1 essential oil samples recorded much higher yield values in comparison to values mentioned in the literature [11]. The essential yields of hydrodistilled *B. sacra* oleo gum resin are comparable to a previous study [10].

Table 1. Percentage yields of essential oils obtained from three different frankincense gum cultivars called Najdi, Sahli and Houjri, subjected to three successive hydrodistillations.

Hydrodistillation Extract (Time)	Percentage Yield of Essential Oils (% *w*/*w*)		
	Najdi	Sahli	Houjri
Grade 1 (2 h)	9.32	12.0	8.92
Grade 2 (4 h)	2.20	2.08	1.48
Grade 3 (6 h)	0.68	0.40	0.72
Total	12.20	14.48	11.12

All the essential oil samples were investigated and compared for the presence of volatile components, see Table 2. The data obtained from gas chromatography–mass spectrometry (GC/MS) analysis of Grade 1 essential oil indicate the presence of more than 30 components. These chemical components were grouped into three categories, namely monoterpenes, diterpenes and sesquiterpenes.

Table 2. Chemical constituents identified in Grade 1, 2 and 3 (respectively 2, 4 and 6 h hydrodistillation time) oleogum resin essential oils. RI = retention index relative to C8–C36 *n*-alkanes on ZB-WAX column, MS = NIST and Wiley library and the literature, tr. = trace (<0.01%).

Components of Essential Oils			Grade 1 (2 h)			Grade 2 (4 h)			Grade 3 (6 h)		
			Percentage Essential Oil Composition%								
Compound	RT (min.)	RI	Najdi	Sahli	Houjri	Najdi	Sahli	Houjri	Najdi	Sahli	Houjri
n-nonane	6.05	900	–	–	–	–	–	–	–	–	–
tricyclene	6.65	918	0.10	0.16	0.20	0.10	0.21	0.08	0.08	0.11	0.20
α-thujene	6.82	923	0.10	0.37	0.43	0.21	0.93	0.58	0.22	1.28	0.81
α-pinene	7.08	930	79.59	78.69	71.09	77.21	73.31	63.11	64.45	61.82	62.57
camphene	7.61	946	3.23	2.66	3.00	2.95	2.06	1.62	1.58	1.79	1.48
thujadiene	7.74	950	0.08	010	0.24	0.20	0.37	0.08	0.21	0.43	0.21
sabinene	8.41	970	0.71	7.78	7.63	1.05	3.99	2.40	0.42	2.95	2.48
β-pinene	8.58	975	2.39	2.25	2.17	1.62	2.31	1.53	1.58	1.96	1.43
β-mircene	9.02	988	0.35	0.16	0.32	0.23	0.38	0.36	0.36	0.28	0.14
n-decane	9.45	1000		–		–	–	–	–	–	–
δ-3-carene	9.74	1007	9.94	0.41	2.16	5.39	0.60	0.83	5.88	0.50	1.82
p-cymene	10.38	1023	1.55	0.67	0.90	1.56	0.87	1.49	1.09	1.21	2.60
limonene	10.56	1027	1.23	2.31	0.83	1.35	2.88	1.47	1.84	3.53	2.21
eucalyptole	10.68	1030	0.04	0.06	0.14	0.02	0.01	0.16	0.01	tr.	0.04
cis-sabinene hydrate	12.37	1070	tr.	tr.	tr.	tr.	tr.	0.01	0.02	0.04	0.10
terpinolene	12.92	1084	0.07	tr.	tr.	0.05	0.08	0.02	0.30	0.18	tr.
p-cymenene	13.13	1089	tr.	tr.	tr.	tr.	tr.	tr.	tr.	tr.	tr.
linalool	13.48	1097	0.01	0.17	0.19	tr.	0.02	0.12	tr.	0.02	0.12
n-undecane	13.61	1100		–		–	–	–	–	–	–
fenchone	13.93	1108	tr.	0.13	0.17	tr.	tr.	0.33	tr.	tr.	0.17
α-campholenol	14.71	1125	tr.	0.21	0.35	tr.	0.15	0.45	tr.	0.24	0.15
trans-pinocarveol	15.29	1139	0.02	0.15	0.71	0.09	0.70	1.06	0.12	0.45	0.62
cis-verbenol	15.52	1144	tr.	0.80	1.85	0.26	0.61	1.73	0.10	0.50	0.84
pinocarvone	16.25	1160	tr.	tr.	0.02	tr.	tr.	tr.	tr.	tr.	0.04
cis-sabinol	16.69	1170	tr.	tr.	tr.	0.01	0.53	0.08	0.37	1.22	0.06
4-terpineol	17.08	1180	tr.	0.28	0.76	0.01	1.70	1.50	0.16	2.43	2.27
p-cymen-8-ol	17.41	1187	tr.	tr.	0.21	tr.	0.24	0.88	0.01	0.24	0.53
α-terpineol	17.74	1195	0.01	0.18	0.60	0.97	0.90	1.42	1.84	1.09	1.34
n-dodecane	18.07	1200		–		–	–	–	–	–	–
verbenone	18.24	1204	tr.	0.28	0.92	tr.	0.53	2.06	0.15	0.69	1.35

Table 2. *Cont.*

Components of Essential Oils			Grade 1 (2 h)			Grade 2 (4 h)			Grade 3 (6 h)		
			Percentage Essential Oil Composition%								
Compound	**RT (min.)**	**RI**	**Najdi**	**Sahli**	**Houjri**	**Najdi**	**Sahli**	**Houjri**	**Najdi**	**Sahli**	**Houjri**
trans-carveol	18.84	1218	tr.	tr.	tr.	tr.	tr.	0.23	0.01	tr.	0.01
bornyl acetate	21.66	1281	0.31	0.49	0.93	0.72	1.01	1.49	1.87	1.50	1.11
thymol	22.24	1294	tr.	tr.	0.48	tr.	tr.	0.63	tr.	tr.	0.18
n-tridecane	22.51	1300				–	–	–	–	–	–
carvacrol	22.77	1306	tr.	tr.	0.11	tr.	tr.	0.19	tr.	tr.	0.07
δ-elemene	24.39	1344	tr.	tr.	0.12	0.17	0.19	0.74	0.32	0.60	0.32
α-copamene	25.580	1371	0.02	0.04	0.08	0.37	0.11	0.22	0.50	0.21	0.25
β-bourbonene	25.90	1379	0.06	0.24	0.09	0.46	0.48	0.87	0.96	1.00	0.33
β-elemene	26.18	1385	0.01	0.44	0.87	1.40	1.18	3.72	4.68	2.79	4.55
n-tetradecane	26.82	1400		–		–	–	–	–	–	–
β-caryophyllene	27.38	1414	0.07	0.28	0.28	0.59	0.76	0.73	0.75	1.48	0.64
α-humulene	28.87	1450	tr.	0.10	0.11	0.41	0.23	0.28	0.54	0.46	0.35
allo-aromadendrene	29.04	1454	tr.	tr.	0.03	tr.	0.10	0.14	0.21	0.16	0.12
γ-muurolene	29.63	1470	tr.	tr.	tr.	0.01	tr.	0.01	0.01	tr.	0.04
β-eudesmene	30.24	1483	0.06	0.39	0.78	1.01	1.18	2.50	3.52	3.07	3.46
azulene	30.53	1490	0.01	0.15	0.30	0.83	0.55	0.90	1.87	1.35	1.19
n-pentadecane	30.94	1500		–		–	–	–	–	–	–
γ-cadinene	31.45	1513	tr.	0.03	tr.	0.62	0.50	0.05	2.71	1.39	0.17
caryophyllene oxide	33,10	1527	tr.	tr.	0.19	tr.	tr.	1.01	0.09	0.21	1.14
n-hexadecane	34,01	1600		–		–	–	–	–	–	–
viridiflorol	35,15	1633	tr.	tr.	tr.	0.12	tr.	0.38	0.40	0.10	0.46
τ-cadinol	36,77	1638	tr.	tr.	0.02	tr.	tr.	tr.	tr.	tr.	0.12
α-eudesmol	37,19	1641	tr.	tr.	0.03	tr.	tr.	0.80	tr.	tr.	0.91

2.1. Composition of Resin Essential Oils

In the case of the cultivar Najdi, a total of 23 active components were identified in the Grade 1 essential oil by GC/MS analysis. The highest percentage was reported for α-pinene, followed by δ-3-carene, camphene and β-pinene (79.59%, 9.94%, 3.23%, and 2.39%, respectively). In the Grade 1 essential oil of Sahli, GC/MS data showed that 29 constituents were present in the essential oil. The major components were α-pinene (78.69%), sabinene (7.78%), camphene (2.66%), limonene (2.31%) and β-pinene (2.66%). In the case of Houjri, GC/MS reports 37 volatile components, which is the highest number compared to Najdi and Sahli with α-pinene, followed by sabinene, camphene, β-pinene, δ-3-carene and cis-verbenol (71.09%, 7.63%, 3.00%, 2.17%, 2.16% and 1.85%, respectively). Data are shown in Table 2.

In Grade 2 essential oil, a sensible reduction of monoterpenes was reported in comparison to Grade 1 essential oil. In all cultivars the major components α-pinene, camphene, β-pinene and δ-3-carene decrease, and the sesquiterpens β-elemene and β-eudesmene increase with a successive 4 h of hydrodistillation.

Grade 3 essential oils show a% reduction in monoterpens (α-pinene, camphene, β-pinene and δ-3-carene) in three cultivars, while at the same time the content in % of sesquiterpenes (β-elemene, β-eudesmene, γ-cadinene) is increased. The highest percentage was recorded for α-pinene. Table 2 shows the results.

2.2. Antibacterial Activity of Essential Oils

All the essential oils from the different frankincense cultivars were tested against two relevant pathogens, *S. aureus* and *P. aeruginosa*, included in the global priority list of antibiotic-resistant bacteria from WHO/OMS [12], and against some reference strains of dermatological interest: *Staphylococcus epidermidis* ATCC 12228, *Staphylococcus hominis* ATCC 27,844 and *Propionibacterium acnes* ATCC 11827. The results are expressed in terms of minimal inhibitory concentration (MIC) in percentage *v*/*v* or in mg/mL and are reported in Table 3. The results indicate that Grade 2 essential oil of Najdi showed an MIC of 52 mg/mL for *S. aureus* ATCC 6538, *P. aeruginosa* ATCC 9027 and *P. aeruginosa* ATCC15442.

The essential oil of Grade 2 Sahli was active against all reference strains of *S.aureus* and *P. aeruginosa*, with MIC values ranging from 440 to 110 mg/mL. It showed the best MIC values, ranging from 6.16 to 0.264 mg/mL, against bacterial strains of dermatological interest *S. epidermidis*, *S. hominis* and *P. acnes*.

Table 3. Antibacterial activity of essential oils expressed in terms of minimum inhibitory concentration (MIC) in %*v*/*v* or mg/mL (between brackets).

Cultivar	Essential Oil	*S. aureus* ATCC 25923	*S. aureus* ATCC 6538	*P. aeruginosa* ATCC 9027	*P. aeruginosa* ATCC 15442	*S. epidermidis* ATCC 12228	*S. hominis* ATCC 27844	*P. acnes* ATCC 11827
Najdi	Grade 1	25	25	>25 (>210)	>25 (>210)	6.2 (52)	25 (210)	>25 (>210)
	Grade 2	25 (210)	6.2 (52)	6.2 (52)	6.2 (52)	50 (420)	>25 (>210)	>25 (>210)
	Grade 3	12.5 (112.5)	12.5 (112.5)	12.5 (112.5)	12.5 (112.5)	25 (225)	>25 (>225)	>25 (>225)
Sahli	Grade 1	25 (210)	12.5 (105)	25 (210)	25 (210)	0.7 (5.88)	25 (210)	25 (210)
	Grade 2	12.5 (110)	25 (220)	50 (440)	12.5 (110)	0.3 (2.64)	0.7 (6.16)	0.03 (0.264)
	Grade 3	25 (210)	25 (210)	>25 (>210)	>25 (>210)	25 (210)	25 (210)	>25 (>210)
Houjri	Grade 1	25 (200)	25 (200)	25 (>200)	25 (>200)	50 (400)	>25 (>200)	>25 (>200)
	Grade 2	25 (237.5)	50 (475)	25 (237.5)	25 (237.5)	1.5 (14.25)	0.3 (2.85)	0.07 (0.66)
	Grade 3	25 (200)	25 (200)	25 (200)	25 (200)	25 (200)	25 (200)	>25 (>200)

2.3. Antifungal Activity of Essential Oils

The antifungal activity was determined against *C. albicans* ATCC 10,231 and *Malassezia furfur* ATCC 14521. All tested samples showed an interesting antifungal activity against *C. albicans* and *M. furfur*, with MIC values ranging from 54.56 to 0.240 mg/mL. In particular, Grade 2 essential oil from Najdi and Grade 1 essential oil from Sahli showed MIC values at the lowest tested concentration corresponding to a percentage *v*/*v* of 0.03 (≤0.252 mg/mL) against both fungal pathogens. The data were shown in the Table 4.

Table 4. Antifungal activity MIC values in %*v*/*v* and in mg/mL between brackets.

Essential oil	*Candida albicans* ATCC 10231	*Malassezia furfur* ATCC 14521
Grade 1 Najdi	≤0.03 (≤ 0.252)	≤0.03 (≤ 0.252)
Grade 2 Najdi	≤0.03 (≤ 0.252)	≤ 0.03 (≤0.252)
Grade 3 Najdi	≤ 0.03 (≤0.270)	≤0.03 (≤0.270)
Grade 1 Sahli	≤ 0.03 (0.252)	≤0.03 (≤0.252)
Grade 2 Sahli	3.10 (27.28)	≤0.03 (≤ 0.264)
Grade 3 Sahli	1.50 (54.56)	1.50 (54.56)
Grade 1 Houjri	≤0.03 (≤0.240)	≤0.03 (≤0.240)
Grade 2 Houjri	1.50 (14.25)	0.15 (1.425)
Grade 3 Houjri	1.50 (12.60)	0.07 (0.588)

3. Discussion

Data presented in Table 2 clearly indicate that there are major differences in the composition of the essential oils of three studied cultivars. All three grades of resin sample gave the highest value for α-pinene, similar findings reported by already mentioned authors [11] and a comparable range for α-pinene was noted in different fractions of the essential oil of *B. sacra* [13]. The percentage of monoterpenes and sesquiterpenes reported for the Grade 1 essential oil of Najdi showed similar values as reported for *Boswellia serrata* [14]. However, there is a major difference in the values recorded for α-pinene in this study, which was 79.59%, while only 5.3% was reported for *B. serrata* [15].

The increased time of extraction through hydrodistillation in three cultivars has resulted in a reduction of monoterpenes, while at the same time the content in % of sesquiterpenes is increased. Moreover, on the basis of data collected, it can be stated that the same species of plants behave differently in terms of the yield of active components of essential oils under different environments. The Najdi sample showed a higher percentage for monoterpenes in all three grades compared to the other two frankincense samples.

Concerning the antimicrobial activity, it was interesting to note that both *S. aureus* and *P. aeruginosa* tested strains were susceptible to most of the essential oils obtained from the three different cultivars. As there are not many antimicrobials concurrently active against the above-mentioned pathogens, these results are relevant because there is a pressing demand for effective antimicrobials towards these microorganisms present on the WHO/OMS priority bacteria list. Therefore, tested essential oils could be a source for new anti-staphylococcal and anti-pseudomonal molecules. Grade 2 of Sahli showed a MIC at 0.264 mg/mL for *P. acnes*, an important pathogen of dermatologic relevance. The activity could be explained by a higher concentration of sesquiterpenes in the extracts obtained by subsequent hydrodistillation of 4 h.

A previous article on four essential oils from different *Boswellia* species showed a significant activity against *Candida tropicalis* and *C. albicans* [16]. In our study we found that all tested samples showed an interesting antifungal activity against *C. albicans* and *M. furfur* with MIC values ranging from 27.28 to 0.240 mg/mL. A high percentage of α-pinene in all Grade 1 essential oils (and in general in monoterpenes) could explain the activity against tested fungal pathogens. Our findings could enable the use of frankincense oil blends in many fields, such as starting products for dermatological bacterial and fungal infections treatments.

In conclusion, GC-MS analysis of *B. sacra* essential oil indicates α-pinene as its primary component. α-Pinene has been found to be one of the integral components of many other antimicrobial essential oils, and thus can be inferred as one of the main players in the antimicrobial activity of *B. sacra* essential oils. Indeed, many of the monoterpenoids putatively identified in our study have been previously reported to have a potent broad antimicrobial spectrum activity. Moreover, the occurrence of drug resistance towards an essential oil in a microorganism is very rare because of its multicomponent nature that necessitates modification of numerous targets [17].

4. Materials and Methods

4.1. Sample Collection

The different *B. sacra* gum resins were purchased from local market, collected over a period from March to May 2016, and compared with authenticated samples preserved in the Herbarium Center of Nizwa University, Oman, under Voucher number UC29.

The sample named Najdi was obtained from a plateau behind the Dhofar Mountains, where a rapid decline in rainfall and moisture was noted in this region. At the same time, temperature variances rise and desert weather changes the climate of the plain, including the southern slopes. Though annual rainfall in the Jebel region ranges from 500–750 mm, precipitation in the Nejd is recorded only in traces. The sample named Sahli was obtained from valleys (Shabi) of the Dhofar region. The normal annual precipitation is around 110 mm but can be from 70 to 360 mm. July to August is usually the rainy period. The sample Houjri is from Jebel Samhan of the Dhofar region. The Jebel hill range forms a distinct agro-climatic region. Thick fogs are detained back through ranges from inner desert areas. Precipitation is predominantly high and it can be from 600 to 700 mm, the highest in the country, which supports an enduring flora cover.

4.2. Extraction of Essential Oils

The essential oils from different cultivars of *B. sacra* were obtained through a hydrodistillation method using Clevenger-type apparatus. A total of 250 g of finely grounded frankincense oleogum resin was added to 2500 mL distilled water in a 5000 mL bottom flask placed in a heating mantle. Hydrodistillation was performed under atmospheric pressure at boiling temperature (about 100 °C) using a closed steam jacket. Following two hours of hydrodistillation, a Grade 1 essential oil of each oleogum resin was collected, whereas Grade 2 and Grade 3 essential oils were collected after 4 and 6 h of successive distillation of the same resin sample. Therefore, a total of nine gum resin essential oils were obtained. The resulted essential oils were collected, dried with anhydrous sodium sulfate, and kept in a dark glass bottle at 4° C until analyzed by gas chromatography–mass spectrometry (GC-MS).

4.3. GC/MS Analysis

GC/MS analysis of obtained essential oils was carried out by using a Thermo Scientific DSQ II single quadrupole system of an EI (electron ionization) type, employed in full scan. The temperatures of the injector and ion source were 265 °C and 260 °C, respectively. The capillary column was a ZB-WAX (30 m x 0.25 mm i.e., film breadth 0.25 μm, (Phenomenex, Italy)). The temperature of the oven was set and the column temperature started at 50 °C, was raised at 3 °C/min to 250 °C, and after that a second gradient was used to 300 °C at 40 °C per minute, which was then seized for 3 min under isothermal conditions. The flow rate of helium, the carrier gas, was 1 mL/min. A sample of 1μL was inoculated with a split ratio of 1:100. The temperature of the ion source was 260 °C, the temperature of the MSn transfer line was 265 °C and the temperature of the injector was 270 °C. Ionization voltage was 70 eV and the mass range scanned was 35–550 *m*/*z*. Each oil component was identified on the basis of its retention index, relative to C8-C36 n-alkanes, from MS library searches using the NIST and Wiley GC-MS databases, and by comparison with mass spectral data in the literature.

The percent composition of the essential oil was computed by a standardization process from the GC peak areas, supposing identical mass response factor for all components. Triplicate analyses were prepared for each oil sample.

4.4. Antimicrobial Activity, Minimum Inhibitory Concentrations (MICs) Determination

The MICs of essential oils for seven reference bacterial strains (two strains of *Staphylococcus aureus* (ATCC 25923; ATCC 6538), two reference strains of *Pseudomonas aeruginosa* (ATCC 15442; ATCC 9027) and *Staphylococcus epidermidis* ATCC 12228, *Staphylococcus hominis* ATCC 27844, and *Propionibacterium acnes* ATCC 11827) were defined by means of an already described technique for a dilution antimicrobial susceptibility test for bacteria that grow aerobically (CLSI), with few amendments [18]. Concisely, 0.1 mL of essential oil was mixed with 0.1mL of culture medium (Mueller Hinton broth, Sigma-Aldrich, Milan, Italy) present in a well of a sterile 96-wells plate, and a 1:2 dilution series with broth medium was performed. A growth control and a negative sterile well control was included for each plate. A bacterial suspension (10 μL) from a suspension in NaCl 0.9% of isolated clusters, preferred from a 18 to 24 h agar plate and attuned to get a turbidity equivalent to a 0.5 McFarland standard, was added into the wells and the final test concentration of bacteria was approximately 5×10^6 CFU/mL bacterial. The 96-well plates were incubated at 37 °C for 24 h and MICs were read by a microplate spectrophotometer (GloMax®-Multi Detection System, Promega Italia s.r.l, Milan, Italy) as the lowest concentration of the sample that fully suppresses bacterial growth in the microdilution wells and whose optical density (OD) at 570 nm is comparable to OD values of negative control wells (only medium, without inoculum).

Antifungal activity against *Candida albicans* (ATCC 10231) and *Malassezia furfur* (ATCC 14521) were evaluated in terms of MICs by using a similar micro-method but by employing Sabouraud broth (Sigma-Aldrich, Milan, Italy) or Sabouraud broth with added 2% olive oil in the case of *M. furfur*, as growth media, and with an incubation time of 48 h.

5. Conclusions

This work aims to increase the number of substances available in the fight against infectious diseases and to combat the phenomenon of antibiotic resistance by promoting the use of alternative molecules in non-clinical settings (farms, food processing, airborne decontamination, etc.), preserving conventional antibiotics for clinical application in humans [17]. Worthy of further investigation is also a combination of approaches, combining *B. sacra* essential oils and conventional antibiotics against the global priority pathogens list of antibiotic-resistant bacteria.

Author Contributions: Conceptualization, L.R., V.D.S., D.S.; investigation, V.D.S., M.G.C., M.R.; data curation, L.R., V.D.S., D.S.; writing—original draft preparation, L.R.; writing—review and editing, V.D.S., D.S; funding acquisition, L.R. All authors have read and agreed to the published version of the manuscript.

Funding: This research was funded by Dhofar University, Sultanate of Oman, grant number DU/AY-2015-2016 FLAB-P002.2

Acknowledgments: The Board of Director and the Vice-Chancellor of Dhofar University are acknowledged for administrative support.

Conflicts of Interest: The authors declare no conflict of interest.

References

1. Al-Yasiry, A.R.M.; Kiczorowska, B. Frankincense—Therapeutic properties. *Postepy Hig. Med. Dosw. Online* **2016**, *70*, 380–391. [CrossRef] [PubMed]
2. Verhoff, M.; Seitz, S.; Paul, M.; Noha, S.M.; Jauch, J.; Schuster, D.; Werz, O. Tetra- and Pentacyclic Triterpene Acids from the Ancient Anti-inflammatory Remedy Frankincense as Inhibitors of Microsomal Prostaglandin E2 Synthase-1. *J. Nat. Prod.* **2014**, *77*, 1445–1451. [CrossRef] [PubMed]
3. Iram, F.; Khan, S.A.; Husain, A. Phytochemistry and potential therapeutic actions of Boswellic acids: A mini-review. *Asian Pac. J. Trop. Biomed.* **2017**, *7*, 513–523. [CrossRef]
4. Michie, C.A.; Cooper, E. Frankincense and myrrh as remedies in children. *J. R. Soc. Med.* **1991**, *84*, 602–605. [PubMed]
5. Vuuren, S.F.V.; Kamatou, G.P.P.; Viljoen, A.M. Volatile composition and antimicrobial activity of twenty commercial frankincense essential oil samples. *S. Afr. J. Bot.* **2010**, *76*, 686–691. [CrossRef]
6. Camarda, L.; Dayton, T.; Di Stefano, V.; Pitonzo, R.; Schillaci, D. Chemical composition and antimicrobial activity of some oleogum resin essential oils from *Boswellia* spp. (*Burseraceae*). *Ann. Chim.* **2007**, *97*, 837–844. [CrossRef] [PubMed]
7. Ljaljević Grbić, M.; Unković, N.; Dimkić, I.; Janaćković, P.; Gavrilović, M.; Stanojević, O.; Stupar, M.; Vujisić, L.; Jelikić, A.; Stanković, S.; et al. Frankincense and myrrh essential oils and burn incense fume against micro-inhabitants of sacral ambients. Wisdom of the ancients? *J. Ethnopharmacol.* **2018**, *219*, 1–14. [CrossRef] [PubMed]
8. El-Nagerabi, S.A.F.; Elshafie, A.E.; AlKhanjari, S.S.; Al-Bahry, S.N.; Elamin, M.R. Biological activities of *Boswellia sacra* extracts on the growth and aflatoxins secretion of two aflatoxigenic species of *Aspergillus* species. *Food Control* **2013**, *34*, 763–769. [CrossRef]
9. Başer, K.H.C.; Demirci, B.; Dekebo, A.; Dagne, E. Essential oils of some Boswellia spp., Myrrh and Opopanax. *Flavour Fragr. J.* **2003**, *18*, 153–156. [CrossRef]
10. Al-Harrasi, A.; Al-Saidi, S. Phytochemical analysis of the essential oil from botanically certified oleogum resin of *Boswellia sacra* (Omani Luban). *Mol. Basel Switz.* **2008**, *13*, 2181–2189. [CrossRef] [PubMed]
11. Al-Saidi, S.; Rameshkumar, K.B.; Hisham, A.; Sivakumar, N.; Al-Kindy, S. Composition and antibacterial activity of the essential oils of four commercial grades of Omani luban, the oleo-gum resin of Boswellia sacra FLUECK. *Chem. Biodivers.* **2012**, *9*, 615–624. [CrossRef] [PubMed]
12. WHO Publishes List of Bacteria for Which New Antibiotics Are Urgently Needed. Available online: https://www.who.int/news-room/detail/27-02-2017-who-publishes-list-of-bacteria-for-which-new-antibiotics-are-urgently-needed (accessed on 27 February 2020).
13. Ni, X.; Suhail, M.M.; Yang, Q.; Cao, A.; Fung, K.-M.; Postier, R.G.; Woolley, C.; Young, G.; Zhang, J.; Lin, H.-K. Frankincense essential oil prepared from hydrodistillation of *Boswellia sacra* gum resins induces human pancreatic cancer cell death in cultures and in a xenograft murine model. *BMC Complement. Altern. Med.* **2012**, *12*, 253. [CrossRef] [PubMed]
14. Woolley, C.L.; Suhail, M.M.; Smith, B.L.; Boren, K.E.; Taylor, L.C.; Schreuder, M.F.; Chai, J.K.; Casabianca, H.; Haq, S.; Lin, H.-K.; et al. Chemical differentiation of *Boswellia sacra* and *Boswellia carterii* essential oils by gas chromatography and chiral gas chromatography-mass spectrometry. *J. Chromatogr. A* **2012**, *1261*, 158–163. [CrossRef] [PubMed]
15. Aisha, A.A.; Tabassum, K.; Siddiqui, A. Kundur (*Boswellia serrata* Roxb)—A boon of nature in the world of Unani system of medicine. *Int. J. Adv. Res. Dev.* **2019**, *4*, 10–14.

16. Schillaci, D.; Arizza, V.; Dayton, T.; Camarda, L.; Di Stefano, V. In vitro anti-biofilm activity of Boswellia spp. oleogum resin essential oils. *Lett. Appl. Microbiol.* **2008**, *47*, 433–438. [CrossRef] [PubMed]
17. Yap, P.S.X.; Yiap, B.C.; Ping, H.C.; Lim, S.H.E. Essential oils, a new horizon in combating bacterial antibiotic resistance. *Open Microbiol. J.* **2014**, *8*, 6–14. [CrossRef] [PubMed]
18. Mauro, N.; Schillaci, D.; Varvarà, P.; Cusimano, M.G.; Geraci, D.M.; Giuffrè, M.; Cavallaro, G.; Maida, C.M.; Giammona, G. Branched High Molecular Weight Glycopolypeptide With Broad-Spectrum Antimicrobial Activity for the Treatment of Biofilm Related Infections. *ACS Appl. Mater. Interfaces* **2018**, *10*, 318–331. [CrossRef] [PubMed]

antibiotics

MDPI

Article

Inhibition of Glucosyltransferase Activity and Glucan Production as an Antibiofilm Mechanism of Lemongrass Essential Oil against *Escherichia coli* O157:H7

Luis A. Ortega-Ramirez, M. Melissa Gutiérrez-Pacheco, Irasema Vargas-Arispuro, Gustavo A. González-Aguilar, Miguel A. Martínez-Téllez and J. Fernando Ayala-Zavala *

Centro de Investigación en Alimentación y Desarrollo, A. C. Carretera Gustavo Enrique Astiazarán Rosas, No. 46. La Victoria, C. P. 83304. Hermosillo 83000, Sonora, Mexico; ingeluis_100@hotmail.com (L.A.O.-R.); melissa.gtzpacheco@gmail.com (M.M.G.-P.); iris@ciad.mx (I.V.-A.); gustavo@ciad.mx (G.A.G.-A.); norawa@ciad.mx (M.A.M.-T.)

* Correspondence: jayala@ciad.mx; Tel.: +52-(662)-289-2400 (ext. 527); Fax: +52-(662)-280-0422

Received: 2 February 2020; Accepted: 27 February 2020; Published: 29 February 2020

Abstract: The resistance of *Escherichia coli* O157:H7 to disinfection is associated with its ability to form biofilms, mainly constituted by glucans produced by glucosyltransferases. Citral and geraniol, terpenes found in the essential oil of *Cymbopogon citratus* (EO), have proven antibacterial activity against planktonic *E. coli*; however, no information was found about their efficacy and mode of action against *E. coli* biofilms. Therefore, the inhibitory effect of *C. citratus* EO, citral, and geraniol on glucans production and glucosyltransferase activity as anti-biofilm mechanism against *E. coli* was evaluated. EO, citral, and geraniol inhibited the planktonic growth of *E. coli* (minimal inhibitory concentration or MIC= 2.2, 1.0, and 3.0 mg/mL, respectively) and the bacterial adhesion (2.0, 2.0, and 4.0 mg/mL, respectively) on stainless steel. All compounds decreased the glucans production; citral and geraniol acted as uncompetitive inhibitors of glucosyltransferase activity (The half maximal inhibitory concentrations or IC_{50} were 8.5 and 6.5 μM, respectively). The evidence collected by docking analysis indicated that both terpenes could interact with the helix finger of the glucosyltransferase responsible for the polymer production. In conclusion, *C. citratus* EO, citral, and geraniol inhibited glucosyltransferase activity, glucans production, and the consequent biofilm formation of *E. coli* O157:H7.

Keywords: extracellular polymeric substance matrix; cellulose synthesis; enzyme inhibition; essential oils

1. Introduction

The persistence and the resistance of *Escherichia coli* O157:H7 to disinfection are associated with its ability to form biofilms on food contact surfaces. Biofilms are communities of microorganisms embedded in an aqueous matrix of extracellular polymeric substances (EPS) produced by the attached cells; EPS are mainly composed by polysaccharides, proteins, lipids, and nucleic acids, which can vary in composition among strains and environmental conditions [1]. The adhesion and the biofilm formation of *E. coli* on food contact surfaces causes cross-contamination, and its consequences are observed on continuous outbreaks every year [2]. It has been reported that *E. coli* O157:H7 biofilms on stainless steel can lead to the release of embedded cells to contaminate other surfaces [3]. This information highlights the importance of studying the characteristics of *E. coli* biofilms to assure effective disinfection procedures.

Exopolysaccharides are secreted during *E. coli* O157:H7 biofilm development, and some of them include cellulose, colanic acid, and the adhesin poly-β-1,6-N-acetyl-glucosamine, and these polymers are involved in the maintenance of biofilm structure and cellular protection against disinfectants [4]. It has been reported that cellulose is the major EPS component of *E. coli* biofilms, and it is essential for its structure and strength, creating cell–cell and cell–surface interactions, retaining water, and avoiding the effect of disinfectants [5]. Previously it was demonstrated that degradation of the EPS matrix of *E. coli* O157:H7 biofilms (mainly composed by glucans) increased their susceptibility to disinfectants. The synthesis and the secretion of glucans are carried out by the enzyme glucosyltransferase, consisting of three transmembrane proteins (BcsA, BcsB, and BcsC) [6]. BcsA is the catalytically active subunit located within the cell, and it is responsible for the uridine diphosphate glucose (UDP-glucose) condensation, then the product is transferred to BcsB and BcsC subunits for processing and extracellular secretion [6]. Thus, blocking this enzymatic process could lead to the inhibition of biofilm production, leaving planktonic *E. coli* more susceptible to disinfectants.

The essential oil (EO) of lemongrass (*Cymbopogon citratus*) is rich in terpenes such as citral (85%) and geraniol (1.5%). *C. citratus* EO has been effective in inhibiting the planktonic growth of *E. coli* O157:H7 with a minimal inhibitory concentration (MIC) of 0.63 mg/mL [7], while Singh et al. [8] reported an MIC value of 0.008 mg/mL. On the other hand, citral and geraniol also showed antibacterial activity against *E. coli* as well as anti-quorum sensing activity at concentrations of 0.01 and 0.06 mg/mL, respectively [9]. On the other hand, *C. citratus* EO in combination with *Allium cepa* EO reduced the presence of *E. coli* in lettuce and spinach [10]. However, its antibacterial activity on planktonic cells could differ from the expected response against biofilms. In addition, *C. citratus* EO was able to inhibit *Staphylococcus aureus* and *Streptococcus mutans* biofilms [11,12]. Previous evidence described the ability of citral and geraniol-like terpenes to traverse the bacterial membrane and interact with vital metabolic enzymes [13]. Previous studies also evidenced the potential of citral to inactivate several enzymes [14,15]. Therefore, the objective of this study was to explore the effect of *C. citratus* EO, citral, and geraniol on the glucans production, glucosyltransferase activity, and biofilm formation of *E. coli* O157:H7.

2. Results

2.1. Susceptibility of Planktonic and Biofilm *E. coli* O157:H7 Cells to *C. citratus* EO, Citral, and Geraniol

Citral was effective inhibiting the growth of planktonic cells (minimal inhibitory concentration or MIC = 1.0 mg/mL) compared to biofilm (minimal biofilm inhibitory concentration or MBIC = 2.0 mg/mL), followed by *C. citratus* EO [MIC = 2.2 mg/mL and MBIC = 2.0 mg/mL] and geraniol (MIC = 3.0 mg/mL and MBIC = 4.0 mg/mL). In general, higher concentrations were needed to inhibit the cell adhesion and the biofilm formation compared to those needed to inhibit the growth of planktonic cells. Lower concentrations than the MIC and the MBIC values of the treatments (*C. citratus* EO = 0.5 mg/mL, citral = 0.5 mg/mL, geraniol = 0.25 mg/mL) were selected to avoid interference of the loss of viability on the biofilm formation and glucans production responses (Figure 1).

Figure 2A shows the *E. coli* O157:H7 biofilm cells on stainless steel coupons in the presence of the treatments. It can be observed that the viable cell in the control bacteria increased steadily as a function of the incubation time, reaching a maximum growth at 12 h at 37 °C. On the other hand, *C. citratus* EO, citral, and geraniol significantly reduced biofilm cells on stainless steel surfaces. *C. citratus* EO reduced 1.64 log CFU/cm^2 the cell adhesion at 12 h compared to the control bacteria, whereas citral and geraniol completely inhibited cell adhesion at the end of incubation time. Figure 2B shows the microphotographs of *E. coli* biofilm development at different incubation times in the absence and presence of the compounds. A significant increase in bacterial aggregation was observed in the control (a) as the incubation time increased, being at 10 h a complete surface colonization. In the case of *C. citratus* EO (b), we observed a significant reduction in aggregation compared to the control, keeping constant at 10 and 12 h, whereas for citral (c), a significant reduction was observed after 8 h. Geraniol

(d) completely inhibited *E. coli* biofilm formation, since no bacterial aggregation was observed after 2 h of incubation at 37 °C.

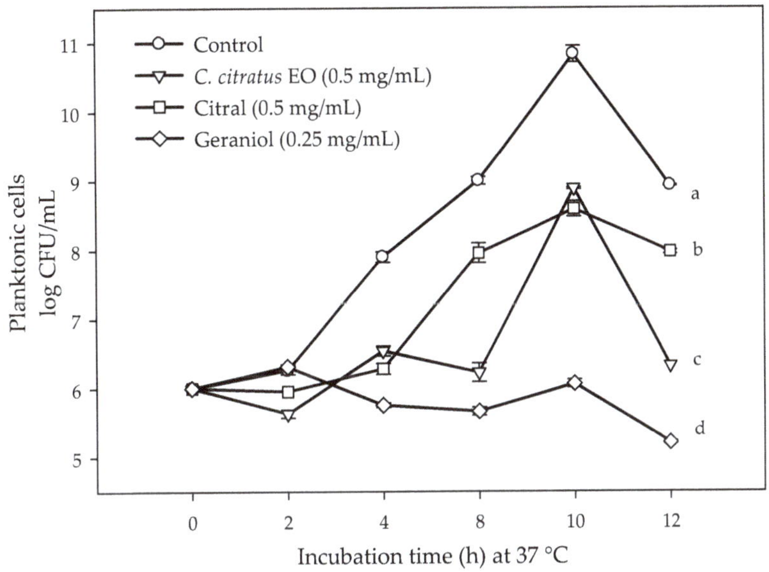

Figure 1. Viability changes of planktonic *E. coli* O157:H7 exposed to non-lethal concentrations of *C. citratus* essential oil (EO), citral, and geraniol. Different letters among treatments indicated significant differences among them ($p < 0.05$). The values are means ± SD, $n = 3$.

2.2. Effect of *C. citratus* EO, Citral, and Geraniol on the Glucans Content in *E. coli* O157:H7 Biofilms

Figure 3 shows the glucan content in *E. coli* biofilms exposed to *C. citratus* EO (0.5 mg/mL), citral (0.5 mg/mL), and geraniol (0.25 mg/mL). It can be observed that the glucan content of control increased exponentially during the incubation time, whereas in those treated with *C. citratus* EO, citral, and geraniol, the glucans production during the biofilm formation was significantly reduced. The stainless steel coupons exposed to citral and geraniol had a lower glucan content compared to the control and the *C. citratus* EO treated bacteria. The relationship between the secreted glucans and the biofilm cells on stainless steel surfaces showed a Pearson correlation coefficient of 0.768 with a probability of 0.0000119.

2.3. Inhibition of Glucosyltransferase Activity by Citral and Geraniol

The activity of pure glucosyltransferase was affected by the presence of citral and geraniol showing IC_{50} values of 8.5 and 6.5 μM, respectively (Figure 4). The reaction pattern of the tested glucosyltransferase showed a Michaelis–Menten kinetic (Figure 5A,B). Table 1 shows the calculated kinetic constants, where both K_m and V_{max} decreased with increasing citral and geraniol concentrations (Figure 5C,D). On the other hand, low K_i values indicated that both inhibitors showed affinity towards the enzyme–substrate complex, this being higher in the case of geraniol. The steric arrangements that could explain the interference of terpenes were proposed by the computational docking analysis. Docking analysis showed that the most probable interactions among citral or geraniol and the enzyme occurred within the hydrophobic pocket located below the gating loop and next to the helix finger of the glucosyltransferase enzyme (Figure 6). The affinity energy obtained for the citral–enzyme-substrate complex was -5.8 kcal/mol with a root-mean-square deviation of atomic positions or RMSD 1.382 Å (Figure 6B), while for the geraniol–enzyme-substrate complex, it was -6.1 kcal/mol with RMSD 1.649 Å (Figure 6C).

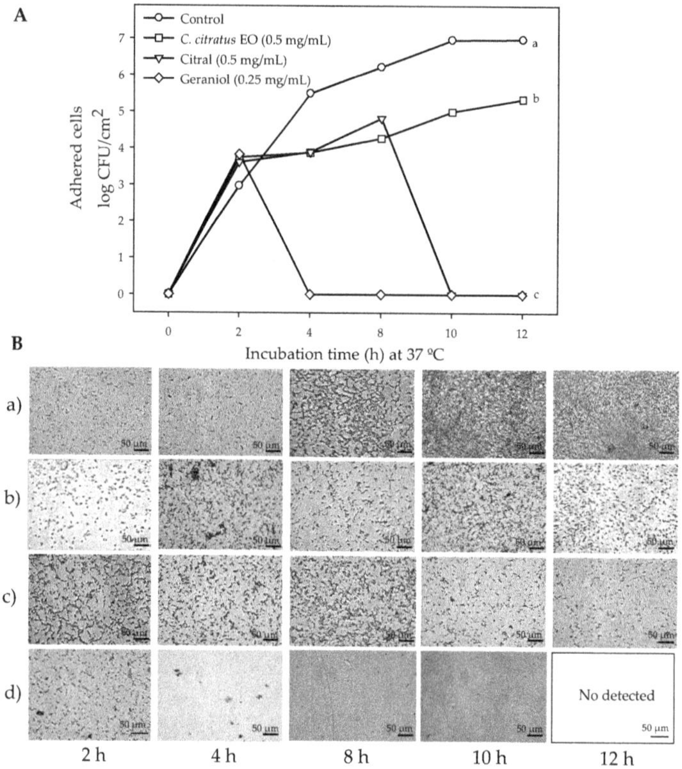

Figure 2. (**A**) Viability changes of biofilm embedded *E. coli* O157:H7 cells exposed to non-lethal concentrations of *C. citratus* EO, citral, and geraniol; different letters indicate significant differences among average of treatments ($p < 0.05$). The values are means ± SD, $n = 3$. (**B**) Light microscopy analysis of *E. coli* O157:H7 biofilms: (a) control, (b) *C. citratus* EO, (c) citral, (d) geraniol. Microphotographs were captured at 600x magnification in an Axio-Vert Microscope.

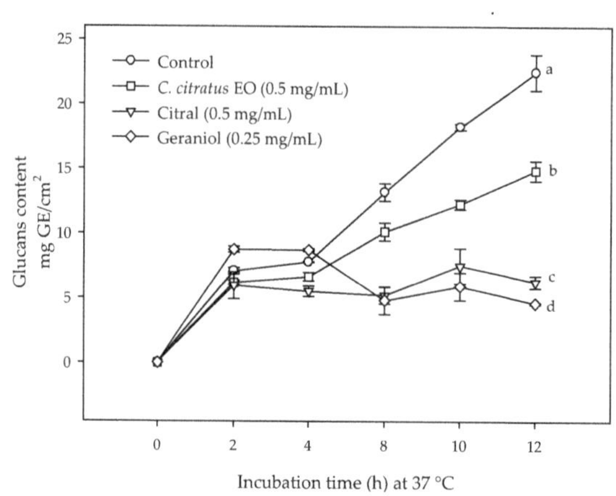

Figure 3. Glucans content on stainless steel coupons produced by *E. coli* O157:H7 biofilms exposed to non-lethal concentrations of *C. citratus* EO, citral, and geraniol; different letters indicate significant differences among treatments ($p < 0.05$). The values are means ± SD, $n = 3$.

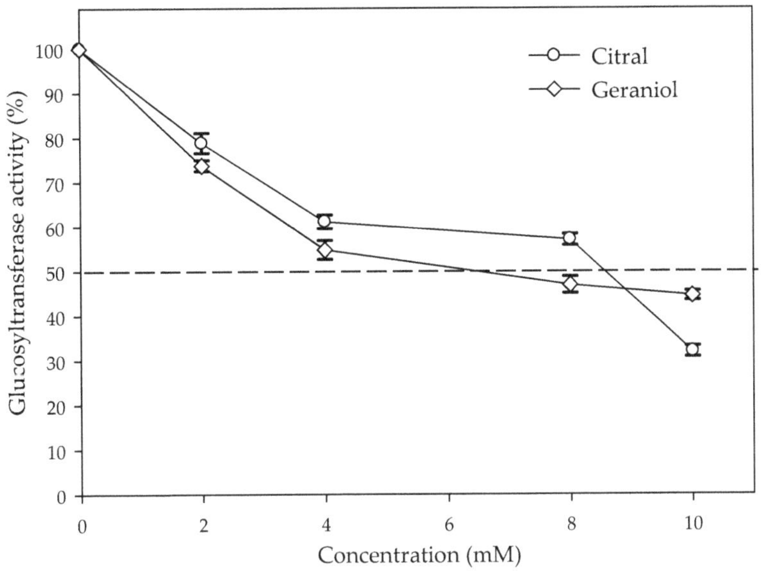

Figure 4. Glucosyltransferase inhibition by the presence of citral and geraniol at different. concentrations ($p < 0.05$). The values are means ± SD, $n = 3$.

Table 1. Kinetic parameters of glucosyltransferase exposed to citral and geraniol.

Agent	Concentration (μM)	K_m * (μM)	V_{max} * (μmol UDP**-glucose min/mL)	Ki (μM) *
Citral	0	3.42	714.28	
	8	2.66	476.19	7
	10	2.66	303.03	
Geraniol	0	3.42	714.28	
	8	2.20	416.66	6.5
	10	?	256.41	

* Values are means of three replicated experiments. ** UDP: Uridine diphosphate.

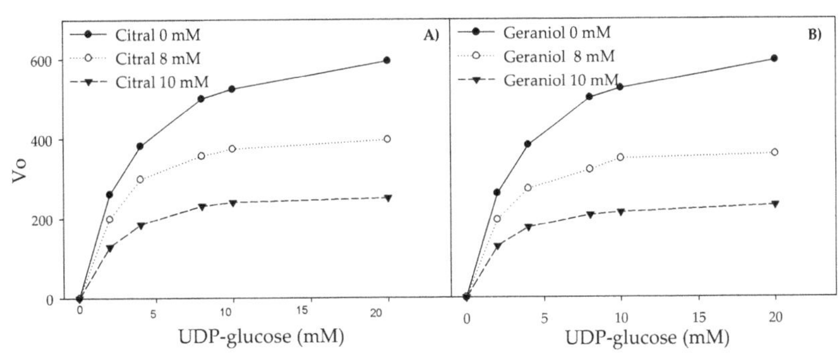

Figure 5. *Cont.*

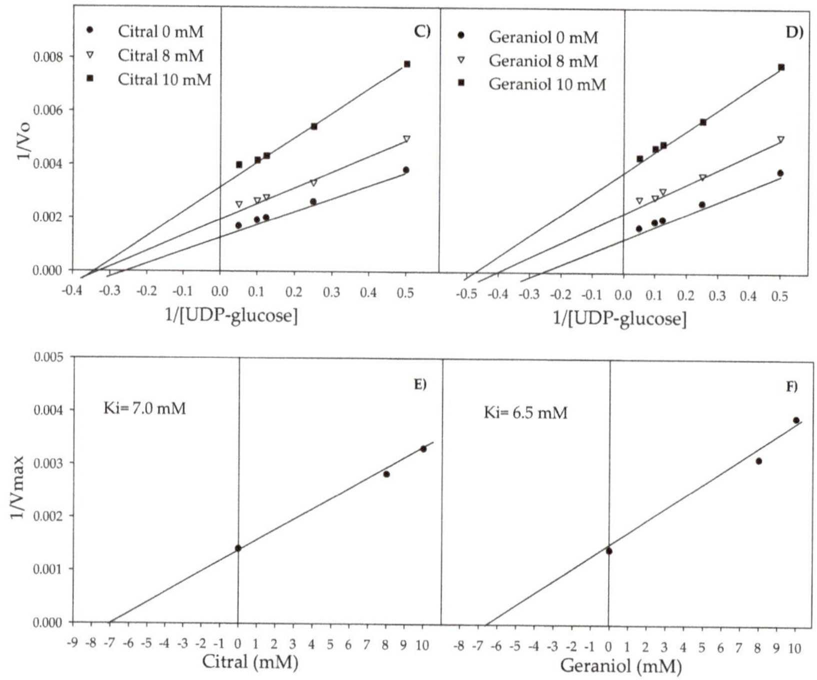

Figure 5. Reaction velocity of glucosyltransferase as a function of substrate concentration in the presence of citral (**A**) and geraniol (**B**). Lineweaver–Burk double reciprocal plot of the glucosyltransferase activity in the presence of citral (**C**) and geraniol (**D**). The double reciprocal plot of the glucosyltransferase activity as a function of citral (**E**) and geraniol (**F**) as a graphical method to calculate K_i. Every point is a mean of three replicated experiments.

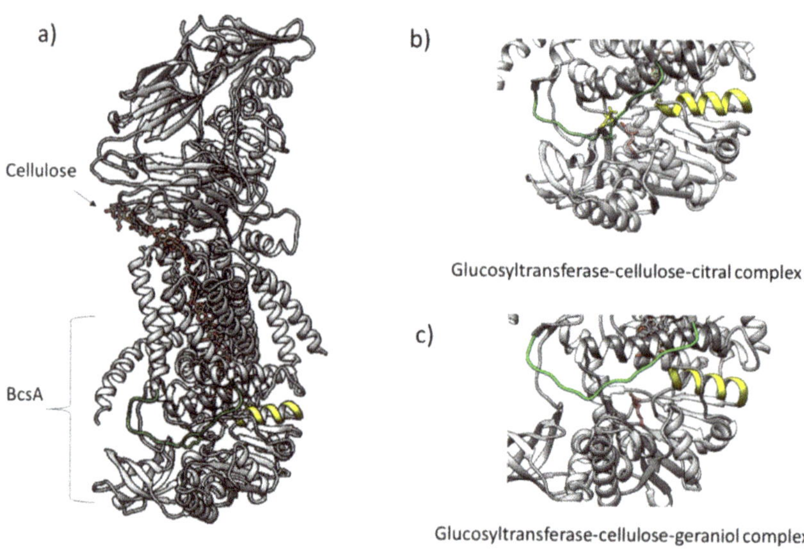

Figure 6. Glucosyltransferase (**a**) interactions with citral (**b**) and geraniol (**c**) blocking the gating loop (green) and the helix finger (yellow) movements during cellulose processing.

3. Discussion

The contamination of food contact surfaces and the resistance of *E. coli* O157:H7 to disinfection processes are associated with its ability to form biofilms. The important role of glucosyltransferase producing glucans to strengthen the *E. coli* O157:H7 biofilms makes its inhibition an attractive target to reduce the biofilm formation. In this regard, *C. citratus* EO, citral, and geraniol have shown antimicrobial activities against many Gram-positive and Gram-negative bacteria, including *E. coli*. However, their effect on glucosyltransferase activity in relation with the biofilm formation has not been previously evaluated.

C. citratus EO, citral, and geraniol inhibited the planktonic growth of *E. coli* O157:H7, and this effect could be attributed to their abilities to degrade membrane proteins and cell permeability. The higher antibacterial activity of citral and *C. citratus* EO compared with geraniol could be related to their hydrophobic characteristics, since they have partition coefficients (Log P) of 3 and 3.5, respectively [16,17], and these values could reflect a higher rate of interaction with the bacterial membrane. On the other hand, geraniol showed the lowest antibacterial activity against *E. coli* O157:H7, which may be explained considering its relatively lower lipophilic character (Log P = 2.9) [16] given by its hydroxyl group, which makes it more difficult to pass through non-polar environments such as the cell membrane [18] compared with citral and *C. citratus* EO. A similar situation was described for thymol (possess a hydroxyl group), which showed a lower efficacy against *E. coli* (MIC = 5 mg/mL) compared to p-cymene (absence of hydroxyl groups), which showed higher antibacterial activity (MIC = 2.5 mg/mL) [18].

Previously, Ortega-Ramirez et al. [10] reported the inhibitory effect of *C. citratus* EO against planktonic *E. coli* at 2.21 mg/mL. On the other hand, other EOs also showed efficacy to inhibit *E. coli* O157:H7; for example, Kim et al. [19] reported that concentrations of 0.001 to 0.01mg/mL of bay, clove, and pimento berry EO significantly inhibited the biofilm formation of *E. coli* O157:H7. Bazargani and Rohloff [20] reported an inhibition of *E. coli* O157:H7 adhesion of 72.3, 56.2, and 98.4% by coriander (1.6 mg/mL), anise (12.5 mg/mL), and peppermint EO (6.3 mg/mL), respectively. These results showed that *C. citratus* EO and its terpenes, citral and geraniol, showed efficacy as antibacterial agents inhibiting planktonic growth of *E. coli* O157:H7 even at low doses. It is important to mention that no previous reports of MBICs of these treatments were found in the revised literature; however, few mechanistic studies have been proposed. It is possible that the lower adhesion of the treated bacteria could be related to the interference in the adhesion process. Therefore, it has to be highlighted that the interest of this study was to evaluate the effect of *C. citratus* EO, citral, and geraniol on glucosyltransferase activity, glucan production, and biofilm development to propose a more complete mechanisms against cell communities that are the natural way of bacterial organization instead individual planktonic cells. For this reason, lower doses than MICs and MBICs were used to only affect the production of glucans without affecting cell viability.

Glucan production during *E. coli* O157:H7 biofilm formation was significantly reduced by citral and geraniol. Among the factors regulating the production of glucans in biofilms are the intercellular communication and the biosynthetic pathways [21]. Intercellular communication in *E. coli* occurs throughout the detection of acyl-homoserine lactones [22]; this process triggers the expression of virulence genes and the enzymatic production of glucans [23]. Thus, within the potential mechanisms of action of terpenes inhibiting glucans production are: (i) down-regulation of glucans synthase genes or a (ii) direct effect on the activity of such system [24]. Both approaches have been tested in other bacterial systems; however, most of the evidence has been directed to a possible effect on the enzymatic production of this polymer, as was done in the present study [24,25].

The ability of bacteria to adhere and form biofilms on different surfaces has substantial implications in the food industry due to safety, quality, and economic issues [26]. As mentioned above, the presence of glucans protects cells from the action of disinfectants and physical cleaning processes. In this sense, it is possible to use *C. citratus* EO, citral, and geraniol as alternative disinfectants to inhibit biofilm formation as well as to help enhance the effect of other cleaning methods. These data can be compared with previous studies that showed the efficacy of plant extracts and their active constituents to inhibit the production of water-insoluble glucans and biofilms of plaque-forming bacteria. Extracts of *Plectranthus barbatus*, *Plectranthus ecklonii*, and *Rheum undulatum* were effective in inhibiting the production of glucans in crude extracts of *Streptococcus sobrinus* and *S. mutans* [24].

For the same bacteria, Koo et al. [25] reported IC_{50} of 0.35 and 0.28 mg/mL for apigenin and farnesol, respectively. Also, epigallocatechin gallate, epigallocatechin, tannic acid, and catechol at 0.1 mg/mL inhibited the production of water-insoluble glucans of 73.1, 68.5, 68, and 67.6%, respectively [27]. In these studies, the reduction of glucans production was related with biofilm inhibition; however, most of them were done on dental plaque and tooth decay bacteria, not in a foodborne pathogen such as *E. coli* O157:H7. From the obtained results, it was observed that *C. citratus* EO, citral, and geraniol were effective in inhibiting the glucans production at non-lethal concentrations, maintaining their effect during the biofilm formation process.

C. citratus terpenes affected glucosyltransferase activity and, based on the obtained kinetic constants, this suggested an uncompetitive inhibition mechanism of glucosyltransferase by citral and geraniol, indicating that both terpenes bound reversibly to the enzyme–substrate complex, forming a ternary complex catalytically inactive. Citral and geraniol are molecules capable of accepting and donating hidrogens atoms and possess non-polar properties to establish hydrophobic interactions [16]. The interaction of terpenes within the hydrophobic pocket below the gating loop and the helix finger could affect the consequent UDP-glucose binding and glucan synthesis [6]. Cellulose synthase is activated by the presence of c-di-GMP, specifically by conformational changes caused by binding c-di-GMP, leading to an open state of the gating loop away from the active site cleft and near the water–lipid interface, where the loop is stabilized by the hydrophobic interactions with the BcsA's amphipathic interface helices forming a transmembrane channel [6]. In this sense, the interruption of the helix finger movement by the presence of citral or geraniol affected the glucan polymerization by influencing the retraction and the insertion of the gating loop [6].

Although there is no evidence of the effect of plant extracts on the glucosyltransferase activity of *E. coli*, there are studies with the dental bacteria *Streptococcus* [24,25]. Plant extracts of *P. barbatus*, *P. ecklonii*, and *R. undulatum* inhibited the activity of glucosyltransferase in crude extracts of *S. sobrinus* (IC_{50} = 1.0, 1.2 and 0.142 mg/mL, respectively) and *S. mutans* (IC_{50} = 3.1, 1.6 and 0.079 mg/mL, respectively) [24]. Within the same study, rosmarinic acid, one of the main components of these plants, showed IC_{50} of 2.1 and 3.9 mg/mL for *S. sobrinus* and *S. mutans* enzyme extracts, respectively. However, these studies did not propose any inhibition mechanism. On the other hand, oleic and linoleic acids showed to be uncompetitive inhibitors of glucosyltransferase; these fatty acids interacted with the substrate–enzyme complex, decreasing the velocity reaction in a similar way to that observed with *C. citratus* EO terpenes [28].

C. citratus EO and its components also inhibited the activity of other enzymes; for example, *C. citratus* EO inhibited MARK4, a kinase enzyme involved in apoptosis, inflammation, and many other regulatory pathways [14]. In another study, seven monoterpenes of *C. citratus* EO were evaluated on pentoxyresorufin activity, obtaining IC_{50} of 0.087 mM for (-)-α-pinene, 0.089 mM for (+)-α-pinene, 0.76 mM for α-terpinene, and 1.19 mM for citral [29]. For this reason, it is important to consider the effect of the rest of the EO components against glucosyltransferase activity, glucan production, and biofilm inhibition of *E. coli* O157:H7. As shown in previous studies, there is evidence that *C. citratus* EO and its compounds were capable of inhibiting different enzymes, but there was no evidence of the effect of this EO against *E. coli* O157:H7 biofilm-glucans-glucosyltransferase, which is the contribution of this study.

4. Material and Methods

4.1. Susceptibility of Planktonic and Biofilm E. coli O157:H7 Cells to C. citratus EO, Citral, and Geraniol

The antibacterial efficacies of *C. citratus* EO (W523100), citral (W230316), and geraniol (W250716) (Sigma-Aldrich, St. Louis, MO, USA) were evaluated against the growth of planktonic and biofilm *E. coli* O157:H7 (ATCC 43890). MIC experiments were performed by the broth microdilution method reported by the Clinical and Laboratory Standards Institute or CLSI [30] with some modifications. Briefly, 5 μL of an overnight inoculum of *E. coli* O157:H7 (1×10^6 CFU/mL) diluted in sterile saline solution were added to a sterile 96-well microtitre plate (Costar 96, Sigma-Aldrich, St. Louis, MO, USA), followed by 295 μL of EO, citral, and geraniol diluted in Luria Bertani o MH (LB) broth at concentrations from 1 to 20 mg/mL, obtaining 2-fold dilutions, respectively. The microplate was incubated at 37 °C for 24 h, and the MICs were determined as the lowest concentrations of each agent that completely inhibited the visible growth of planktonic cells.

For inhibiting biofilm bacteria, MBICs were determined as the lowest dose of each compound inhibiting the bacterial adhesion on stainless steel coupons ($1 \times 1 \times 0.1$ cm, grade 304) during 24 h of incubation at 37 °C [31]. Different concentrations of natural compounds (0–20 mg/mL) were added into test tubes with 10 mL of MH broth containing stainless steel coupons. Then, the tubes were inoculated with *E. coli* O157:H7 (1×10^6 CFU/mL, diluted in sterile saline solution) and incubated at 37 °C for 24 h under static conditions; then, the coupons were removed from the culture medium and washed with sterile distilled water to remove weakly adhered cells. Afterward, the coupons were placed in 5 mL of sterile peptone water and subjected to an ultrasonic bath (40 kHz) for 5 min to release the strongly adhered cells and were counted by plating on MH agar after 24 h of incubation at 37 °C (log CFU/cm^2). Both inhibitory concentrations were obtained by triplicate from three independent experiments, and the obtained results were expressed as mg/mL [31].

4.2. Effect of C. citratus EO, Citral, and Geraniol on the Glucans Content in E. coli O157:H7 Biofilms

Lower doses than MICs and MBICs were used to only affect the production of glucans without affecting cell viability. The conditions used for biofilm formation were as described above; applying *C. citratus* EO (0.5 mg/mL), citral (0.5 mg/mL), and geraniol (0.25 mg/mL), viable cells were counted at different times (0, 2, 4, 8, 10, 12 h) at 37 °C. Biofilm cells adhered to stainless steel coupons as well as planktonic cells in the culture medium were determined as described above, expressing results as log CFU/cm^2 and log CFU/mL, respectively. Also, biofilms were stained with 0.1% crystal violet solution for 10 min and fixed with Lugol to observe morphological changes during the exposure to the treatments using an inverted microscope (Zeiss Axio Vert A1 Inverted, Carl Zeiss, NY, USA), viewing with phase contrast at 600× [32].

The glucans production by treated bacteria was expressed as glucose equivalents (GE) per area of stainless steel (cm^2) [32]. Coupons were removed from the culture medium after incubation and then washed with water to remove weakly adhered cells. Subsequently, they were placed into tubes containing 5 mL of water and 30 μL of formaldehyde (33%) (Sigma Aldrich, St. Louis, MO, USA) and left at 4 °C for 1 h. Subsequently, 2 mL of NaOH (1 M) (Sigma Aldrich, St. Louis, MO, USA) were added to the tubes, sonicated for 5 min, and stored for 3 h at 4 °C. The final volume (7 mL) was filtered (millipore 0.22 μm) and dialyzed with Milli-Q water using a dialysis membrane (3500 Da) (Sigma Aldrich, St. Louis, MO, USA) at 4 °C for 24 h, and the > 3500 Da fraction was lyophilized. The lyophilized sample was diluted in 300 μL of Milli-Q water for the subsequent quantification of glucans adhered to the stainless steel coupons. The glucans were determined with the phenol/sulfuric acid method [33] using glucose as standard and expressing results as mg of glucose equivalents per area, GE/cm^2.

4.3. *Inhibition of Glucosyltransferase Activity by Citral and Geraniol*

Glucosyltransferase (SRP0416, Sigma Aldrich, St. Louis, MO, USA) activity was measured in the presence of citral and geraniol at 0, 8, and 10 μM; lemongrass EO was excluded from this assay considering the variety of chemical structures in its content, making it difficult to establish a molar relation. This was measured in 300 μL of buffer solution (40 mM Tris-HCl, pH 8, 15 mM $MgCl_2$, 1 mM $CaCl_2$, and 5 mM UDP-glucose) containing each concentration of terpenes; this mixture was pre-incubated at 30 °C for 10 min, and the reaction was initiated by adding the glucosyltransferase (EC 2.4.1.). The enzyme activity was measured using the fluorometric assay [34] that monitored the release of UDP-fluorescein (λ_{ex} 490 nm; λ_{em} 514 nm) as a product of the UDP-glucose hydrolysis (the absence of terpenes in the reaction was taken as 100% activity).

The initial reaction velocities (V_o) were obtained using 2 mM of glucosyltransferase, substrate at 2, 4, 8, 10, and 20 μM, and the individual terpenes at 8 and 10 μM, respectively. The experimental data were fitted to a non-linear model, applying the equation of Michaelis–Menten for K_m and V_{max} calculation, and then these values were fitted to the Lineweaver–Burk equation. The type of inhibition was determined analyzing the Lineweaver–Burk graph, and the K_i values of the individual terpenes were taken from the x-intercepts of $1/V_{max}$ versus the terpene concentration [35]; this assay was performed three times to assure reproducibility.

4.4. *Molecular Docking of Glucosyltransferase with Citral and Geraniol*

Molecular docking was used to identify possible interactions between the individual terpenes (citral and geraniol, respectively) with the glucosyltransferase crystallographic model (PDB 5EIY) [6]; the used citral and geraniol models were PubChem 638011 and PubChem 637566. This analysis was done using the AutoDoc Vina application in the UCSF Chimera version 1.13 software (Resource for Biocomputing, Visualization, and Informatics, University of California, San Francisco, CA, USA) to obtain affinity energies (kcal/mol) with the lowest root-mean-square deviation (RMSD, Å) between glucosyltransferase and each terpene. Ten binding modes with a 3 level of exhaustiveness search and a 3 kcal/mol level of maximum energy difference were set as basic parameters during the analysis.

4.5. *Statistical Analysis*

A completely randomized experimental design was done for all assays. The effect of *C. citratus* EO, citral, and geraniol, as well as the exposure time (0, 2, 4, 8, 10, and 12 h) were evaluated on the count of viable planktonic and biofilm cells and the glucans production. In addition, a Pearson correlation was done between the secreted glucans and the biofilm formation. All experiments were done by triplicate, expressing the results as means ± standard deviation. An analysis of variance (ANOVA) was done for all the assays to estimate significant differences among treatments, and the means were compared by the Tukey–Kramer test. All experiments were performed at $p \leq 0.05$ using the statistical software NCSS 2007 (NCSS, LLC, Utah, USA).

5. Conclusions

C. citratus EO, citral, and geraniol were capable of inhibiting *E. coli* O157:H7 biofilm formation, decreasing cell adhesion and glucans production on stainless steel surfaces. This inhibitory effect could be related to an uncompetitive inhibition of glucosyltransferase activity caused by the presence of citral and geraniol. These results suggest a possible inhibition mechanism of terpenes on biofilm formation of *E. coli* O157:H7.

Author Contributions: Conceptualization, I.V.-A., G.A.G.-A., M.A.M.-T. and J.F.A.-Z.; Data curation, L.A.O.-R.; Funding acquisition, J.F.A.-Z.; Investigation, L.A.O.-R.; Methodology, L.A.O.-R.and M.M.G.-P.; Resources, J.F.A.-Z.; Supervision, J.F.A.-Z.; Writing—original draft, L.A.O.-R. and M.M.G.-P.; Writing—review & editing, I.V.-A., G.A.G.-A., M.A.M.-T. and J.F.A.-Z. All authors have read and agreed to the published version of the manuscript.

Funding: The authors are thankful for the financial support of the Mexican Council for Science and Technology CONACYT (Project number 60224). Ortega-Ramirez and Gutierrez-Pacheco acknowledge CONACYT for the received Doctoral Fellowships.

Conflicts of Interest: The authors declare no conflict of interest.

References

1. Gutiérrez-Pacheco, M.M.; Bernal-Mercado, A.T.; Vázquez-Armenta, F.J.; Martínez-Tellez, M.A.; González-Aguilar, G.A.; Lizardi-Mendoza, J.; Madera-Santana, T.J.; Nazzaro, F.; Ayala-Zavala, J.F. Quorum sensing interruption as a tool to control virulence of plant pathogenic bacteria. *Physiol. Mol. Plant P* **2019**, *106*, 281–291. [CrossRef]
2. CDC. Reports of Selected *E. coli* Outbreak Investigations. Available online: https://www.cdc.gov/ecoli/outbreaks.html (accessed on 18 June 2018).
3. Ryu, J.H.; Beuchat, L.R. Biofilm formation by *Escherichia coli* O157:H7 on stainless steel: Effect of exopolysaccharide and Curli production on its resistance to chlorine. *Appl. Environ. Microbiol.* **2005**, *71*, 247–254. [CrossRef]
4. Beloin, C.; Roux, A.; Ghigo, J.-M. Escherichia coli biofilms. In *Bacterial Biofilms*; Springer: Berlin/Heidelberg, Germany, 2008; pp. 249–289.
5. Lim, E.S.; Koo, O.K.; Kim, M.-J.; Kim, J.-S. Bio-enzymes for inhibition and elimination of *Escherichia coli* O157: H7 biofilm and their synergistic effect with sodium hypochlorite. *Sci. Rep.* **2019**, *9*, 9920. [CrossRef] [PubMed]
6. Morgan, J.L.; McNamara, J.T.; Fischer, M.; Rich, J.; Chen, H.M.; Withers, S.G.; Zimmer, J. Observing cellulose biosynthesis and membrane translocation in crystallo. *Nature* **2016**, *531*, 329–334. [CrossRef] [PubMed]
7. Zulfa, Z.; Chia, C.; Rukayadi, Y. In vitro antimicrobial activity of *Cymbopogon citratus* (lemongrass) extracts against selected foodborne pathogens. *Int. Food Res. J.* **2016**, *23*, 1262–1267.
8. Singh, B.R.; Singh, V.; Singh, R.K.; Ebibeni, N. Antimicrobial activity of lemongrass (*Cymbopogon citratus*) oil against microbes of environmental, clinical and food origin. *Int. Res. J. Pharm. Pharmacol.* **2011**, *1*, 228–236.
9. Ahmad, A.; Viljoen, A.M.; Chenia, H.Y. The impact of plant volatiles on bacterial *Quorum sensing*. *Lett. Appl. Microbiol.* **2015**, *60*, 8–19. [CrossRef] [PubMed]
10. Ortega-Ramirez, L.A.; Silva-Espinoza, B.A.; Vargas-Arispuro, I.; Gonzalez-Aguilar, G.A.; Cruz-Valenzuela, M.R.; Nazzaro, F.; Ayala-Zavala, J.F. Combination of *Cymbopogon citratus* and *Allium cepa* essential oils increased antibacterial activity in leafy vegetables. *J. Sci. Food Agric.* **2017**, *97*, 2166–2173. [CrossRef] [PubMed]
11. Tofiño-Rivera, A.; Ortega-Cuadros, M.; Galvis-Pareja, D.; Jiménez-Rios, H.; Merini, L.J.; Martínez-Pabón, M.C. Effect of Lippia alba and *Cymbopogon citratus* essential oils on biofilms of *Streptococcus mutans* and cytotoxicity in CHO cells. *J. Ethnopharmacol.* **2016**, *194*, 749–754. [CrossRef]
12. Adukwu, E.C.; Allen, S.C.; Phillips, C.A. The anti-biofilm activity of lemongrass (*Cymbopogon flexuosus*) and grapefruit (*Citrus paradisi*) essential oils against five strains of *Staphylococcus aureus*. *J. Appl. Microbiol.* **2012**, *113*, 1217–1227. [CrossRef]
13. Oliveira, M.A.C.; Borges, A.C.; Brighenti, F.L.; Salvador, M.J.; Gontijo, A.V.L.; Koga-Ito, C.Y. *Cymbopogon citratus* essential oil: Effect on polymicrobial caries-related biofilm with low cytotoxicity. *Braz. Oral Res.* **2017**, *31*, e89. [CrossRef] [PubMed]
14. Naz, F.; Khan, F.I.; Mohammad, T.; Khan, P.; Manzoor, S.; Hasan, G.M.; Lobb, K.A.; Luqman, S.; Islam, A.; Ahmad, F.; et al. Investigation of molecular mechanism of recognition between citral and MARK4: A newer therapeutic approach to attenuate cancer cell progression. *Int. J. Biol. Macromol.* **2018**, *107*, 2580–2589. [CrossRef] [PubMed]
15. de Oliveira, E.R.; Alves, D.S.; Carvalho, G.A.; de Oliveira, B.M.R.G.; Aazza, S.; Bertolucci, S.K.V. Toxicity of Cymbopogon flexuosus essential oil and citral for Spodoptera frugiperda. *Ciência Agrotecnol.* **2018**, *42*, 408–419. [CrossRef]
16. Pubchem PubChem Compound Database. Available online: https://pubchem.ncbi.nlm.nih.gov/ (accessed on 21 November 2018).

17. Treesuwan, W.; Neves, M.A.; Uemura, K.; Nakajima, M.; Kobayashi, I. Preparation characteristics of monodisperse oil-in-water emulsions by microchannel emulsification using different essential oils. *LWT* **2017**, *84*, 617–625. [CrossRef]
18. Cristani, M.; D'Arrigo, M.; Mandalari, G.; Castelli, F.; Sarpietro, M.G.; Micieli, D.; Venuti, V.; Bisignano, G.; Saija, A.; Trombetta, D. Interaction of four monoterpenes contained in essential oils with model membranes: Implications for their antibacterial activity. *J. Agric. Food Chem.* **2007**, *55*, 6300–6308. [CrossRef] [PubMed]
19. Kim, Y.-G.; Lee, J.-H.; Gwon, G.; Kim, S.-I.; Park, J.G.; Lee, J. Essential Oils and Eugenols Inhibit Biofilm Formation and the Virulence of *Escherichia coli* O157:H7. *Sci. Rep.* **2016**, *6*, 36377. [CrossRef]
20. Bazargani, M.M.; Rohloff, J. Antibiofilm activity of essential oils and plant extracts against *Staphylococcus aureus* and *Escherichia coli* biofilms. *Food Control* **2016**, *61*, 156–164. [CrossRef]
21. Gupta, P.; Pruthi, P.A.; Pruthi, V. Role of Exopolysaccharides in Biofilm Formation. In *Introduction to Biofilm Engineering*; ACS Publications: Washington, DC, USA, 2019; pp. 17–57.
22. Ha, J.H.; Hauk, P.; Cho, K.; Eo, Y.; Ma, X.; Stephens, K.; Cha, S.; Jeong, M.; Suh, J.Y.; Sintim, H.O.; et al. Evidence of link between quorum sensing and sugar metabolism in *Escherichia coli* revealed via cocrystal structures of LsrK and HPr. *Sci. Adv.* **2018**, *4*, eaar7063. [CrossRef]
23. Da Re, S.; Ghigo, J.M. A CsgD-independent pathway for cellulose production and biofilm formation in *Escherichia coli*. *J. Bacteriol.* **2006**, *188*, 3073–3087. [CrossRef]
24. Figueiredo, N.L.; de Aguiar, S.R.M.; Falé, P.L.; Ascensão, L.; Serralheiro, M.L.M.; Lino, A.R.L. The inhibitory effect of *Plectranthus barbatus* and *Plectranthus ecklonii* leaves on the viability, glucosyltransferase activity and biofilm formation of *Streptococcus sobrinus* and *Streptococcus mutans*. *Food Chem.* **2010**, *119*, 664–668. [CrossRef]
25. Koo, H.; Pearson, S.K.; Scott-Anne, K.; Abranches, J.; Cury, J.A.; Rosalen, P.L.; Park, Y.K.; Marquis, R.E.; Bowen, W.H. Effects of apigenin and tt-farnesol on glucosyltransferase activity, biofilm viability and caries development in rats. *Oral Microbiol. Immunol.* **2002**, *17*, 337–343. [CrossRef]
26. Galie, S.; Garcia-Gutierrez, C.; Miguelez, E.M.; Villar, C.J.; Lombo, F. Biofilms in the food industry: Health aspects and control methods. *Front. Microbiol.* **2018**, *9*, 898. [CrossRef] [PubMed]
27. Kim, Y.; Jang, S.; Kim, H.; Kim, S. Deodorizing, antimicrobial and glucosyltransferase inhibitory activities of polyphenolics from biosource. *Korean J. Chem. Eng.* **2017**, *34*, 1400–1404. [CrossRef]
28. Osawa, K.; Miyazaki, K.; Shimura, S.; Okuda, J.; Matsumoto, M.; Ooshima, T. Identification of cariostatic substances in the cacao bean husk: Their anti-glucosyltransferase and antibacterial activities. *J. Dent. Res.* **2001**, *80*, 2000–2004. [CrossRef] [PubMed]
29. De-Oliveira, A.C.; Ribeiro-Pinto, L.F.; Paumgartten, F.J. In vitro inhibition of CYP2B1 monooxygenase by β-myrcene and other monoterpenoid compounds. *Toxicol. Lett.* **1997**, *92*, 39–46. [CrossRef]
30. Clsi, C. *Performance Standards for Antimicrobial Susceptibility Testing; Twenty-Fourth Informational Supplement*; American National Standards Institute: Washington, DC, USA, 2014.
31. Tapia-Rodriguez, M.R.; Hernandez-Mendoza, A.; Gonzalez-Aguilar, G.A.; Martinez-Tellez, M.A.; Martins, C.M.; Ayala-Zavala, J.F. Carvacrol as potential quorum sensing inhibitor of *Pseudomonas aeruginosa* and biofilm production on stainless steel surfaces. *Food Control* **2017**, *75*, 255–261. [CrossRef]
32. Gutierrez-Pacheco, M.M.; Gonzalez-Aguilar, G.A.; Martinez-Tellez, M.A.; Lizardi-Mendoza, J.; Madera-Santana, T.J.; Bernal-Mercado, A.T.; Vazquez-Armenta, F.J.; Ayala-Zavala, J.F. Carvacrol inhibits biofilm formation and production of extracellular polymeric substances of *Pectobacterium carotovorum* subsp. *carotovorum*. *Food Control* **2018**, *89*, 210–218. [CrossRef]
33. Dubois, M.; Gilles, K.; Hamilton, J.; Rebers, P.; Smith, F. Colorimetric method based on phenol sulfuric acid. *Anal. Chem.* **1956**, *28*, 356.
34. Palmerini, C.A.; Datti, A.; Vanderelst, I.E.; Minuti, L.; Orlacchio, A. An approach for fluorometric determination of glycosyltransferase activities. *Glycoconj. J.* **1996**, *13*, 631–636. [CrossRef]
35. Copeland, R.A. *Evaluation of Enzyme Inhibitors in Drug Discovery: A Guide for Medicinal Chemists and Pharmacologists*; John Wiley & Sons: Hoboken, NJ, USA, 2013.

antibiotics

MDPI

Brief Report

Is the Antimicrobial Activity of Hydrolates Lower than That of Essential Oils?

Maura Di Vito [1,2,*], Antonina Smolka [2], Maria Rita Proto [1], Lorenzo Barbanti [1], Fabrizio Gelmini [3], Edoardo Napoli [4], Maria Grazia Bellardi [1], Paola Mattarelli [1], Giangiacomo Beretta [3], Maurizio Sanguinetti [2,5,†] and Francesca Bugli [2,5,†]

1 Dipartimento di Scienze e Tecnologie Agro-Alimentari, Università of Bologna, Viale G. Fanin 42, 40127 Bologna, Italy; mariarita.proto2@unibo.it (M.R.P.); lorenzo.barbanti@unibo.it (L.B.); mariagrazia.bellardi@unibo.it (M.G.B.); paola.mattarelli@unibo.it (P.M.)
2 Dipartimento di Scienze Biotecnologiche di Base, Cliniche Intensivologiche e Perioperatorie, Università Cattolica del Sacro Cuore, Largo A. Gemelli 8, 00168 Rome, Italy; smolka2015@gmail.com (A.S.); maurizio.sanguinetti@unicatt.it (M.S.); francesca.bugli@unicatt.it (F.B.)
3 Dipartimento di Scienze e Politiche Ambientali, Università degli Studi di Milano, via Celoria, 2, 20133 Milano, Italy; fabrizio.gelmini@unimi.it (F.G.); giangiacomo.beretta@unimi.it (G.B.)
4 Istituto di Chimica Biomolecolare, Consiglio Nazionale delle Ricerche, via P. Gaifami 18, 95126 Catania, Italy; edoardo.napoli@icb.cnr.it
5 Dipartimento di Scienze di Laboratorio e Infettivologiche, Fondazione Policlinico Universitario "A. Gemelli" IRCCS, 00168 Rome, Italy
* Correspondence: wdivit@gmail.com; Tel.: +39-051-209-6267 or +39-06-30154964
† Equally contributed.

Abstract: Among the top five human infections requiring medical treatment is dermatitis. Treatment of bacterial and fungal skin infections is usually based on antibiotic therapy, which is often ineffective due to the involvement of antibiotic-resistant microbial strains. The aim of this study was to compare the antimicrobial activity of essential oils (EOs) and hydrolates (Hys) extracted from six aromatic plants grown in Italy (*Lavandula angustifolia, Lavandula intermedia, Origanum hirtum, Satureja montana, Monarda didyma,* and *Monarda fistulosa*) towards fungal (*Candida albicans, Candida parapsilosis, Candida glabrata* and *Candida tropicalis; Trichophyton soudanense, Trichophyton tonsurans, Trichophyton rubrum, Trichophyton violaceum* and *Microsporum canis*) and bacterial strains (*Staphylococcus aureus* MRSA, *Staphylococcus aureus* MSSA, *Streptococcus pyogenes, E. faecalis, Enterococcus faecalis* VRE, and *Enterococcus faecium*) potentially pathogenic for human skin. The composition and antimicrobial activity of EOs and Hys were evaluated using the Gas-chromatography mass spectrometry and micro dilution-broth test, respectively. The volatiles' conversion factors (CFs) were calculated to compare the activity of Hys with that of the corresponding EOs. Data show that, although the minimum inhibitory concentration values of EOs are lower than the corresponding Hys, the volatiles contained in Hys are more effective at inhibiting microbial growth because they are active at lower concentrations.

Keywords: *Satureja montana; Lavandula angustifolia; Lavandula intermedia; Origanum hirtum; Monarda didyma; Monarda fistulosa*

Citation: Di Vito, M.; Smolka, A.; Proto, M.R.; Barbanti, L.; Gelmini, F.; Napoli, E.; Bellardi, M.G.; Mattarelli, P.; Beretta, G.; Sanguinetti, M.; et al. Is the Antimicrobial Activity of Hydrolates Lower than That of Essential Oils?. *Antibiotics* **2021**, *10*, 88. https://doi.org/10.3390/antibiotics10010088

Received: 14 November 2020
Accepted: 15 January 2021
Published: 18 January 2021

1. Introduction

Among the top five human infections requiring medical treatment is dermatitis [1]. Treatment of bacterial and fungal skin infections is usually based on antibiotic therapy, which is often ineffective due to the involvement of antibiotic-resistant microbial strains such as methicillin-resistant *Staphylococcus aureus* (MRSA) [2] and *Candida* sp. [3]. In recent decades, given the poor innovation in the discovery of new antimicrobials and the frequency of recalcitrant skin infections, the need for innovative anti-infective therapeutics is becoming more and more urgent. In this field, great interest in the last 20 years has been focused on the potential of natural products.

In recent years, there has been growing interest in natural products obtained from aromatic plant distillation: essential oils (EOs) and hydrolates (Hys). As such, there are many scientific articles about the effectiveness of EOs in various contexts: antimicrobials, immunomodulatory, antioxidants, anti-inflammatory, pain-relievers, etc., but there is little evidence on the activities of Hys.

Official Pharmacopoeias well define the two natural products. The EO is considered to be a complex odorous product obtained by steam distillation, hydro-distillation, or by the dry distillation of a plant, some of its parts, or, in the case of OEs obtained from *Citrus* spp., through appropriate mechanical cold processes [4]. Similarly, starting from 2012, the French Pharmacopoeia defines the Hy as a product obtained through the distillation of different parts of aromatic plants, which separates from the essential oil at the end of the distillation [5].

While they originate from the same process, the two distillation products are quite different in terms of chemical composition and effectiveness.

EOs are hydrophobic mixtures mainly characterized by terpene molecules that, on the contrary, are extremely diluted in Hys. In fact, the Hys are hydrophilic solutions characterized, up to a maximum of 1 g/L, by the terpene components present in the corresponding EO [6]. Furthermore, in the Hy, the relative ratio of each terpenic molecule will be conditioned by its hydrophilic characteristics. Owing to this, the major components of an EO may not be the same that is present in the corresponding Hy.

Due to the high oxicity of many terpene compounds [7], essential oils require special warnings when used *per os* or in topical applications [8]. On the contrary, Hys resulting from dilution of terpenic solutions are less toxic and can be used more easily for the same applications.

However, only few studies have been carried out on EOs and Hys obtained from the same distillation process in order to compare their chemical composition [9–11], or study some of their activities such as psychopharmacological and anti-cancer activities [12,13], or larvicidal and nematodicidal ones [14,15]. Our group participated in these early investigations, assessing the chemical composition and the antimicrobial activity of the EO and Hy obtained from *Monarda citriodora* in a recent research. The study showed that, to achieve the same inhibitory effect of EO, a higher volume of Hy was necessary; however, in this volume, the concentration of active components was lower than that present in the corresponding EO, i.e., the EO from the same plant source [16]. Therefore, data indicate a higher likelihood for the active compounds isolated from *M. citriodora* Hy to be more active in the aqueous phase, because they can more easily reach their target, or because they are not contrasted with antagonistic compounds present only in the OE.

Given this background and in view of improving the knowledge on Hy potential uses, the first aim of this study was to evaluate the antimicrobial activity of six EOs and the companion Hys isolated from the same aromatic plant cultivated in Italy, towards fungal and bacterial strains potentially pathogenic for human skin. The following microorganisms isolated from patients with skin infections included the following. Six bacteria: methicillin-resistant *Staphylococcus aureus* (MRSA), methicillin- susceptible *Staphylococcus aureus* (MSSA), *Streptococcus pyogenes*, vancomycin-resistant enterococci (VRE) *Enterococcus faecalis* and *Enterococcus faecium*. Four drug-resistant yeasts: *Candida albicans*, *Candida parapsilosis*, *Candida glabrata* and *Candida tropicalis*. Five dermatophytes: *Trichophyton soudanense*, *Trichophyton tonsurans*, *Trichophyton rubrum*, *Trichophyton violaceum* and *Microsporum canis*. The second aim was to compare the relative concentration of active volatiles present in EOs and Hys obtained from the same plant by using the volatiles' conversion factor (CF).

2. Results

2.1. GC-MS and Gravimetric Analyses

The chromatographic analysis of EOs shows phytocomplexes that are quite different (Table 1). *Lavandula angustifolia* has linalyl acetate and β-linalool at respective concentrations of 33.35% and 28.36%, while *L. intermedia* EO has the same components at concentrations of 36.47% and 27.99%, respectively. The EO of *Origanum hirtum* is mainly characterized by thymol, γ-terpinene and p-cymene at 36.3%, 23.81% and 18.83%, respectively, while the EO of *Satureja montana* has carvacrol as a major compound (concentration of 63.1%), followed by γ-terpinene (concentration of 13.44%). Both *Monarda didyma* and *M. fistulosa* EOs show carvacrol (20.59% and 35.18%, respectively) and γ-terpinene (13.07% and 16.85%, respectively) as major compounds, while thymol and *p*-cymene are the third most concentrated components in the respective *M. didyma* and *M. fistulosa*. The rest of the components present in EOs show concentrations lower than 10%.

The analysis of Hy (Table 2) shows β-linalool, α-terpinen-4-ol and α-terpineol (42.5%, 20.33 and 19.1%, respectively) as major chemical compounds of *L. angustifolia* Hy. *L. intermedia* Hy is characterized by β-linalool, camphor and 1,8-cineol (34.17%, 22.12% and 19.08%, respectively) as major compounds, while *S. montana* has carvacrol and thymol as the major compounds (85.79% and 13.88%, respectively). *O. hirtum* Hy has only one component, thymol (100% concentration).

M. didyma has carvacrol and thymol (48.44% and 34.03%, respectively) as major compounds, while *M. fistulosa* has only carvacrol (84.68%) at a concentration above 10%. All the other components show a concentration lower than 10%. It is important to remember that the concentrations of chemicals identified in the Hys are referred at most to 1 g/L, which is the maximum terpenes concentration present in Hy. Results of the gravimetric analyses are shown in Table 2. The qualitative and quantitative analyses of the extract obtained for the gravimetric analysis are not shown because they are redundant and perfectly superimposable to those obtained from the gas-chromatographic analysis.

Table 1. Chemical composition of EOs.

			Average (% *n* = 3)					
Components	E-RI	L-RI	*L. angustifolia*	*L. intermedia*	*O. hirtum*	*S. montana*	*M. didyma*	*M. fistulosa*
2,3-Dimethyl-3-buten-2-ol	741	746	-	-	-	0.05	-	-
Thujene	923	928	0.11	0.08	1.30	1.11	1.81	3.48
α-Pinene	931	936	0.29	0.63	0.76	0.76	0.57	0.79
Camphene	945	950	0.10	0.33	0.07	0.22	0.23	0.15
Sabinene	967	973	0.06	0.13	-	0.07	1.12	0.28
1-Octen-3-ol	974	980	0.08	0.08	0.10	0.47	4.50	4.08
3-Octanone	979	985	0.27	-	-	-	0.20	0.13
β-Pinene	972	978	0.14	0.51	0.09	0.11	0.27	0.26
Myrcene	983	989	3.56	1.39	1.12	0.95	2.28	3.62
α-Phellandrene	998	1004	0.10	0.04	0.23	0.21	0.40	0.66
Hexyl acetate	1004	1010	-	0.03	-	-	-	-
3-Carene	1005	1011	0.15	0.09	0.06	0.06	0.20	0.32
α-Terpinene	1011	1017	0.07	0.05	3.04	1.98	3.69	5.69
p-Cymene	1018	1024	0.13	0.05	18.83	9.82	8.08	13.85
Limonene	1024	1030	1.53	-	0.39	0.66	0.88	1.06
1,8-Cineole	1026	1032	1.54	9.20	0.04	0.20	1.36	-
(Z)-β-Ocimene	1031	1038	5.44	0.60	1.29	0.04	-	-
(E)-β-Ocimene	1041	1048	3.13	0.63	0.22	0.02	-	-
γ-Terpinene	1053	1060	0.19	0.12	23.81	13.44	13.07	16.85
cis- Linalool oxide (f)	1069	1075	0.13	0.06	-	-	-	-
Terpinolene	1080	1087	0.26	0.29	0.12	0.05	0.22	0.21
β−Linalool	1092	1099	28.36	27.99	0.40	0.48	8.71	1.24
No Match	1197	1203	0.04	-	-	-	-	-
1-Octen-3-ol, acetate	1103	1110	0.61	0.09	-	-	-	-
Neo-allo-ocimene	1122	1130	3.28	-	-	-	-	-
Camphor	1136	1143	0.25	7.27	-	-	-	-
n-Hexyl isobutyrate	1144	1151	-	0.05	-	-	-	-

Table 1. *Cont.*

			Average (% *n* = 3)					
Components	**E-RI**	**L-RI**	*L. angustifolia*	*L. intermedia*	*O. hirtum*	*S. montana*	*M. didyma*	*M. fistulosa*
Borneol	1159	1167	0.77	3.40	0.05	0.50	0.56	0.24
Lavandulol	1161	1168	0.14	-	-	-	-	-
p-Cymen-8-ol	1176	1184	-	-	-	0.01	-	-
Cryptone	1181	1189	0.11	-	-	-	-	-
α-Terpineol	1182	1190	0.31	0.50	0.06	0.06	0.94	0.23
n-Hexyl n-butyrate	1184	1192	0.23	-	-	-	-	-
cis-Sabinene hydrate	1212	1219	0.10	0.10	-	0.07	-	-
Isobornyl formate	1231	1239	0.03	-	-	-	-	-
Thymol methyl ether	1226	1234	-	-	5.37	-	4.47	0.40
Pulegone	1226	1234	-	-	4.05	-	-	-
Hexyl 3-methylbutyrate	1236	1244	-	0.07	-	-	-	-
Carvacrol methyl ether	1235	1243	-	-	-	-	7.36	6.74
Tymoquinone	1244	1252	-	-	-	0.03	-	-
Geraniol	1247	1255	-	-	-	-	-	0.47
Linalyl acetate	1247	1255	33.35	36.47	-	-	-	-
Bornyl acetate	1275	1284	0.07	-	-	-	-	-
Lavandulol acetate	1281	1289	1.28	2.31	-	-	-	-
Thymol	1282	1290	-	-	36.30	1.21	15.40	1.87
Carvacrol	1292	1300	-	-	0.13	63.16	20.59	35.18
L-Terpinen-4-ol	1295	1302	5.50	2.93	0.27	0.29	-	-
δ-Elemene	1328	1337	-	0.06	-	0.06	-	-
Neryl acetate	1354	1362	0.46	0.16	-	-	-	-
Carvacrol acetate	1364	1373	-	-	-	0.13	-	-
β-Copaene	1367	1376	0.04	-	-	0.04	-	-
α-Copaene	1367	1376	-	-	-	0.05	-	-
Geranyl acetate	1371	1380	0.78	0.30	-	-	-	-
β-Bourbonene	1375	1384	-	-	0.09	0.04	-	-
β-Elemene	1381	1390	-	-	-	0.01	-	-
Humulene	1397	1407	0.07	0.03	-	0.03	0.03	0.06
β-Caryophillene	1411	1420	5.75	1.71	0.67	1.53	1.00	1.20
cis-α-Bergamotene	1425	1430	0.22	0.09	-	-	-	-
trans-α-Bergamotene	1425	1434	0.05	0.05	-	-	-	-
γ-Elemene	1426	1436	-	-	-	0.09	-	-
(Z)-β-Farnesene	1436	1446	0.21	0.45	-	-	-	-
(E)-β-Farnesene	1446	1456	0.07	-	-	-	-	-
Geranyl propionate	1467	1477	-	0.24	-	-	-	-
γ-Muurolene	1466	1476	-	0.04	0.07	-	-	-
Germacrene D	1471	1481	0.21	0.28	-	0.28	-	-
Zingiberene	1485	1495	-	0.03	-	-	-	-
β-Bisabolene	1498	1508	-	-	0.41	0.88	-	-
γ-Cadinene	1503	1513	-	0.30	0.15	0.02	-	-
δ-Cadinene	1513	1523	0.04	-	0.23	0.07	-	-
β-Sesquiphellandrene	1513	1524	-	0.07	-	-	-	-
Caryophyllene oxide	1570	1581	0.08	-	-	0.07	-	-
Cadinol T	1629	1640	-	0.14	-	-	-	-
α-Bisabolol	1671	1683	-	0.14	-	-	-	-

Note. RI = Retention Indices. SD < 5%, RI-E = RI experimentally determined, RI-L = RI determined through Libraries.

Table 2. Chemical composition of volatile compounds in hydrolate.

			Average (%)					
Components	**E-RI**	**L-RI**	*L. angustifolia*	*L. intermedia*	*O. hirtum*	*S. montana*	*M. didyma*	*M. fistulosa*
3-Methyl-4-penten-1-ol	781	786	-	0.11	-	-	-	-
3-Hexen-1-ol	852	857	0.10	-	-	-	0.03	0.16
5,5-Dimethyl-2(5H)-furanone	946	952	0.52	-	-	-	-	-
1-Octen-3-ol	976	980	-	0.19	-	-	6.64	5.59
3-Octanone	979	985	-	-	-	-	0.05	0.04
1,8-Cineole	1026	1032	0.90	19.08	-	-	0.33	-
cis-Linalool oxide(f)	1069	1075	0.78	0.76	-	-	-	-
trans-Linalool oxide(f)	1077	1083	2.40	-	-	-	-	-
β-Linalool	1092	1099	42.15	34.17	-	-	6.94	0.63
Camphor	1136	1143	0.32	22.12	-	-	-	-
Eucarvone	1142	1150	0.15	-	-	-	-	-
Sabina ketone	1148	1156	0.14	-	-	-	-	-
Isopulegol	1152	1159	1.42	-	-	-	-	-
Borneol	1159	1166	2.50	3.17	-	-	0.77	0.22
α-Terpineol	1182	1190	19.01	5.20	-	-	1.56	0.30
Verbenone	1198	1206	-	-	-	0.05	-	-
Not identified	1209	1215	0.42	0.15	-	-	-	-
Cumin aldehyde	1230	1238	0.07	-	-	-	-	-
6,7-Dihydro-7-hydroxylinalool	1229	1237	3.58	1.17	-	-	-	-
2-Hydroxycineol	1239	1247	-	0.26	-	-	-	-
Geraniol	1247	1255	0.77	0.07	-	-	-	0.61
Thymol	1282	1290	-	-	100	13.88	34.03	6.66
Cumin alcohol	1282	1290	0.18	-	-	-	-	-
Not identified	1287	n.d.	0.52	-	-	-	-	-
Carvacrol	1292	1300	-	-	-	85.79	48.44	84.68
L-Terpinen-4-ol	1295	1302	20.23	7.63	-	-	1.22	1.11
Not identified	1406	n.d.	-	1.16	-	-	-	-
Not identified	1493	n.d.	2.99	-	-	-	-	-
Cadinol T	1629	1640	-	0.63	-	-	-	-
α-Cadinol	1641	1652	-	0.16	-	-	-	-
α-Bisabolol	1671	1682	-	0.77	-	-	-	-
Palmitic acid, ethyl ester	1981	1993	0.10	0.79	-	0.06	-	-
Stearic acid, ethyl ester	2183	2196	0.05	0.65	-	-	-	-
Squalene	2776	2790	0.03	1.44	-	0.21	-	-
Gravimetric analysis [a]			0.09	0.05	0.04	0.06	0.03	0.04

Note. RI = Retention indices. [a] Values are expressed as % (*w*/*w*). SD < 5%, RI-E = RI experimentally determined, RI-L = RI determined through Libraries.

2.2. Broth Microdilution Susceptibility Test

Table 3 shows the Minimum Inhibitory Concentration (MIC) and Minimum Lethal Concentration (MLC) of the tested EOs. The table also displays the values of Inhibition Rate or Lethal Rate of 90% (IR90 and LR90, respectively) of strains. The EOs of *S. montana* and *O. hirtum* are the most active, showing IR90 values of 0.25% and 1 % *v*/*v*, respectively, and LR90 values of 0.25% *v*/*v* and 1% *v*/*v*, respectively. All the other EOs have IR90 and LR90 values greater than or equal to 2% *v*/*v*, except *M. didyma* EO showing IR90 and LR90 values equal to 1% *v*/*v* and > 2% *v*/*v*, respectively. Specifically, while the EO of *S. montana* acts in equal measure on all three microbial types (bacteria, yeasts, and dermatophytes), the EO of *O. hirtum* acts primarily on bacteria and yeasts, while that of *M. fistulosa* on dermatophytes.

Table 3. Inhibitory and lethal activities of EOs.

		EOs (% *v*/*v*)											
	Clinical Strains	**LA**		**LI**		**OH**		**SM**		**MD**		**MF**	
D	Bacteria	MIC	MLC	MIC	MLC	MIC	MLC	MIC	MLC	MIC	MLC	MIC	MLC
0.1SA(R)	*S. aureus* MRSA	>2	>2	2	>2	≤0.06	<0.06	≤0.06	≤0.06	1	2	0.5	1
0.2SP	*S. pyogenes*	>2	>2	1	2	0.125	0.125	0.125	0.125	0.25	0.5	2	2
0.3EF(R)	*E. faecalis* VRE	>2	>2	2	2	0.125	0.125	0.125	0.125	0.25	0.5	2	2
0.4EF	*E. faecium*	>2	>2	2	2	0.125	0.125	0.125	0.125	0.5	0.5	2	1
0.5SA	*S. aureus* MSSA	>2	>2	2	>2	0.125	0.25	0.125	0.125	0.5	1	2	2
0.6EF	*E. faecalis*	>2	>2	2	2	≤0.06	0.25	≤0.06	0.125	0.25	>2	2	2
	Yeasts	MIC	MLC	MIC	MLC	MIC	MLC	MIC	MLC	MIC	MLC	MIC	MLC
3.1CA	*C. albicans*	>2	>2	2	>2	0.25	0.25	0.25	0.25	0.25	0.25	1	2
0.1CP (R)	*C. parapsilosis*	>2	>2	2	>2	0.25	0.5	0.25	0.25	0.5	0.5	1	2
0.2CG (R)	*C. glabrata*	>2	>2	2	>2	0.25	0.25	0.25	0.25	0.25	0.25	1	2
0.3CT (R)	*C. tropicalis*	>2	>2	2	>2	0.25	0.5	0.25	0.25	0.25	0.5	1	2
	Dermatophytes	MIC	MLC	MIC	MLC	MIC	MLC	MIC	MLC	MIC	MLC	MIC	MLC
0.1TS	*T. soudanense*	2	2	1	2	1	1	0.125	0.125	0.5	0.5	0.5	0.5
0.2TS	*T. tonsurans*	1	>2	1	2	0.5	0.5	0.125	0.125	0.25	0.5	0.25	0.25
0.3TS	*T. rubrum*	2	2	2	2	1	1	0.25	0.25	2	2	0.5	0.5
0.4TS	*T. violaceum*	0.125	0.06	0.125	0.06	0.125	0.06	0.125	0.125	0.25	0.125	0.125	0.06
0.5TS	*M. canis*	>0.5	>2	0.25	0.25	1	1	0.25	0.25	2	2	0.5	0.5
	IR90/LR90	>2	>2	2	>2	1	1	0.25	0.25	1	2	2	2
	IR50/LR50	>2	>2	2	2	0.25	0.25	0.125	0.125	0.25	0.5	1	2

Note. D = Designation, IR90= Inhibition Rate of 90% of strains, LR90 = Lethal Rate of 90% of strains, IR50 = Inhibition Rate of 50% of strains, LR50 = Lethal Rate of 50% of strains, LA = *Lavandula angustifolia*, LI=*Lavandula intermedia*, OH = *Origanum hirtum*, SM = *Satureja montana*, MD = *Monarda didyma*, MF = *Monarda fistulosa*.

As shown in Table 4, values obtained from the analysis of the antimicrobial effectiveness of the Hys indicate the Hys of *O. hirtum* and *M. didyma* (IR90 value 50% *v*/*v*) as more active than the others against bacteria, yeast and dermatophytes. However, it was not possible to study Hys concentrations greater than 50% *v*/*v*, as this would have introduced a significant methodological bias by reducing the amount of nutrient broth necessary for microbial growth.

Table 4. Inhibitory and lethal activities of Hys.

		Hys (% *v*/*v*)											
	Clinical Strains	**LA**		**LI**		**OH**		**SM**		**MD**		**MF**	
D	Bacteria	MIC	MLC	MIC	MLC	MIC	MLC	MIC	MLC	MIC	MLC	MIC	MLC
0.1SA(R)	*S. aureus* MRSA	>50	>50	>50	>50	6.25	50	>50	>50	>50	>50	>50	>50
0.2SP	*S. pyogenes*	>50	>50	>50	>50	50	>50	>50	>50	50	>50	>50	>50
0.3EF(R)	*E. faecalis* VRE	>50	>50	>50	>50	50	>50	>50	>50	50	>50	>50	>50
0.4EF	*E. faecium*	>50	>50	>50	>50	50	50	>50	>50	>50	>50	>50	>50
0.5SA	*S. aureus* MSSA	>50	>50	>50	>50	50	>50	>50	>50	50	>50	>50	>50
0.6EF	*E. faecalis*	>50	>50	>50	>50	50	>50	>50	>50	50	>50	>50	>50
	Yeasts	MIC	MLC	MIC	MLC	MIC	MLC	MIC	MLC	MIC	MLC	MIC	MLC
3.1CA	*C. albicans*	>50	>50	>50	>50	50	50	50	50	50	50	25	50
0.1CP (R)	*C. parapsilosis*	>50	>50	>50	>50	50	>50	50	>50	50	50	25	50
0.2CG (R)	*C. glabrata*	>50	>50	>50	>50	50	50	50	50	50	>50	25	50
0.2CT (R)	*C. tropicalis*	>50	>50	>50	>50	50	50	50	50	50	50	25	50
	Dermatophytes	MIC	MLC	MIC	MLC	MIC	MLC	MIC	MLC	MIC	MLC	MIC	MLC
0.1TS	*T. soudanense*	50	>50	50	>50	50	50	50	50	25	50	25	50
0.2TS	*T. tonsurans*	50	>50	50	>50	25	50	50	50	25	50	25	>50
0.3TS	*T. rubrum*	>50	>50	>50	>50	50	50	50	50	50	50	25	25

Table 4. *Cont.*

		Hys (% v/v)											
	Clinical Strains	LA		LI		OH		SM		MD		MF	
0.4TS	*T. violaceum*	50	50	12.5	25	6.25	12.5	25	25	12.5	12.5	≤6.25	6.25
0.5TS	*M. canis*	>50	>50	>50	>50	50	50	50	50	50	50	25	25
	IR90/ LR90	>50	>50	>50	>50	50	>50	>50	>50	>50	>50	>50	>50
	IR50/LR50	>50	>50	>50	>50	50	50	50	50	50	50	25	50

Note: D = Designation, IR90 = Inhibition Rate of 90% of strains, LR90 = Lethal Rate of 90% of strains, IR50 = Inhibition Rate of 50% of strains, LR50 = Lethal Rate of 50% of strains, LA = *Lavandula angustifolia*, LI = *Lavandula intermedia*, OH = *Origanum hirtum*, SM = *Satureja montana*, MD = *Monarda didyma*, MF = *Monarda fistulosa*.

In particular, the *O. hirtum* Hy at a concentration of 50% *v/v* is the only one that can inhibit all bacteria growth but is unable to exert cytocidal effect at the same concentration, while fungi (yeast and dermatophytes) show greater sensitivity to Hys (Table 3). Specifically, the Hys of *S. montana*, *O. hirtum* and *M. didyma* have inhibitory and cytocidal effect against most dermatophytes at a concentration equal to 50% *v/v*, and only *M. fistulosa* is able to inhibit all strains at a concentration of 25% *v/v*, but it is not capable of having cytocidal effects for values <50% *v/v*.

2.3. Comparison Between EOs and Hys

Table 5 shows the values of the peaks' total areas of the chemicals of both EOs (EOTA) and Hys (HYTA), the volatiles' Conversion Factor (CF) obtained as EOTA/ HYTA, and the value of the $IR50_{Hy}/CF$ ratio. This last parameter indicates the value that the $IR50_{Hy}$ would have if the Hy were concentrated as the EO. As shown in Table 5, the value of the $IR50_{Hy}/CF$ ratio is lower than that of $IR50_{EO}$ for all the EOs.

Table 5. Volatile concentrations in EOs and HYs, their relationships, and IR50 comparison at equivalent volatile concentrations.

Heading	Total Area			
	O. hirtum	*S. montana*	*M. didyma*	*M. fistulosa*
EOTA	2.23×10^{13}	7.13×10^{13}	8.21×10^{11}	7.53×10^{11}
HYTA	2.03×10^{10}	1.61×10^{11}	5.03×10^{8}	4.25×10^{8}
CF	1.13×10^{3}	4.42×10^{2}	1.63×10^{3}	1.77×10^{3}
$IR50_{Hy}/CF$ (% v/v)	0.044	0.113	0.031	0.014
$IR50_{EO}$ (% v/v)	0.25	0.125	0.25	1
$IR50_{EO}/(IR50_{Hy}/CF)$	5.68	1.11	8.33	71.43

Note: EOTA = Essential Oil Total volatiles Area, HYTA = Hydrolate Total volatiles Area, CF= volatiles' Conversion Factor.

This means that, to have the same antimicrobial activity in the EO, a relative concentration of volatiles between 1.11 (*S. montana*) and 71.43 (*M. fistulosa*) times as high as that contained in the Hy is required.

The same difference is evidenced in the activity of EO and Hy against each microbial strain. Table 6 shows the concentration of EOs and Hys necessary to obtain the Inhibitory concentration of the 50% (IC50) of the initial inoculum, and the $IC50_{Hy}/CF$ ratio that is the $IC50_{Hy}$ value normalized according to the volatiles' concentration. IC50 values were obtained, starting from the inhibition curve calculated using OD450 values obtained from the micro-broth dilution test. In Table 6, values of dermatophytes are not reported. In fact, due to the inhomogeneity of their growth, they were only evaluated by visual reading, as specified in "Material and Methods section". Additionally, in this case, the visual exam points out that $IC50_{Hy}/CF$ ratios are significantly lower than the respective $IC50_{EO}$ values.

Table 6. Comparison of the IC50 values of each Hy vs. the corresponding EO.

		% *v*/*v*											
	Clinical Strains	**OH**			**SM**			**MD**			**MF**		
D	**Bacteria**	**$IC50_{Hy}$**	**$IC50_{Hy}/CF$**	**$IC50_{EO}$**	**$IC50_{Hy}$**	**$IC50_{Hy}/CF$**	**$IC50_{EO}$**	**$IC50_{Hy}$**	**$IC50_{Hy}/CF$**	**$IC50_{EO}$**	**$IC50_{Hy}$**	**$IC50_{Hy}/CF$**	**$IC50_{EO}$**
0.1SA(R)	*S. aureus* MRSA	1.94 ± 3.92	0.00 ± 0.00	n.c.	24.41 ± 2.60	0.06 ± 0.00	n.c.	52.77 ± 6.36	0.03 ± 0.00	1.17 ± 0.71	119.03 ± 17.50	0.06 ± 0.01	0.33 ± 0.04
0.2SP	*S. pyogenes*	31.23 ± 20.32	0.03 ± 0.02	0.14 ± 0.02	33.10 ± 1.80	0.07 ± 0.00	0.30 ± 0.17	28.87 ± 2.78	0.02 ± 0.00	0.36 ± 0.05	102.82 ± 54.31	0.06 ± 0.03	0.84 ± 0.01
0.3EF(R)	*E. faecalis* VRE	25.05 ± 9.28	0.02 ± 0.00	0.01 ± 0.02	38.27 ± 20.00	0.09 ± 0.04	0.10 ± 0.01	22.82 ± 0.22	0.01±0.00	0.18 ± 0.02	59.76 ± 10.71	0.03 ± 0.01	0.88 ± 0.01
0.4EF	*E. faecium*	21.67 ± 0.77	0.02 ± 0.00	0.04 ± 0.03	28.66 ± 3.12	0.07 ± 0.00	0.07 ± 0.01	35.28 ± 10.40	0.02±0.01	0.41 ± 0.04	40.88 ± 20.10	0.02 ± 0.01	0.41 ± 0.03
0.5SA	*S. aureus* MSSA	24.95 ± 10.50	0.02 ± 0.00	0.12 ± 0.02	29.78 ± 6.84	0.07 ± 0.01	0.10 ± 0.02	18.79 ± 0.26	0.01±0.00	0.45 ± 0.00	32.16 ± 14.41	0.02 ± 0.01	0.58 ± 0.03
0.6EF	*E. faecalis*	29.74 ± 3.98	0.03 ± 0.00	n.c.	68.10 ± 22.00	0.15 ± 0.05	n.c.	17.35 ± 0.01	0.01±0.00	0.21 ± 0.01	22.38 ± 7.69	0.01 ± 0.00	0.79 ± 0.03
	Yeast	$IC50_{Hy}$	$IC50_{Hy}/CF$	$IC50_{EO}$	$IC50_{Hy}$	$IC50_{Hy}/CF$	$IC50_{EO}$	$IC50_{Hy}$	$IC50_{Hy}/CF$	$IC50_{EO}$	$IC50_{Hy}$	$IC50_{Hy}/CF$	$IC50_{EO}$
3.1CA	*C. albicans*	11.60 ± 0.32	0.01 ± 0.00	0.15 ± 0.04	25.29 ± 4.57	0.06 ± 0.01	0.19 ± 0.01	27.57 ± 17.16	0.02 ± 0.01	0.01 ± 0.05	11.29 ± 5.04	0.01 ± 0.00	0.49 ± 0.04
0.1CP (R)	*C. parapsilosis*	20.75 ± 3.63	0.02 ± 0.00	0.16 ± 0.04	26.08 ± 1.86	0.06 ± 0.00	0.15 ± 0.01	20.78 ± 1.10	0.01 ± 0.00	0.13 ± 0.07	13.70 ± 0.24	0.01 ± 0.00	0.53 ± 0.04
0.2CG (R)	*C. glabrata*	27.53 ± 1.36	0.02 ± 0.00	0.22 ± 0.00	27.59 ± 0.92	0.06 ± 0.00	0.19 ± 0.01	28.92 ± 1.31	0.02 ± 0.00	0.29 ± 0.00	16.41 ± 0.03	0.01 ± 0.00	0.62 ± 0.11
0.3CT (R)	*C. tropicalis*	28.79 ± 2.24	0.03 ± 0.00	0.11 ± 0.03	24.33 ± 0.72	0.05 ± 0.01	0.18 ± 0.12	27.86 ± 3.87	0.02 ± 0.00	0.16 ± 0.01	15.14 ± 1.10	0.01 ± 0.00	0.38 ± 0.06

Note. n.c. = This value cannot be calculated because the corresponding MIC value is lower than the minimum dilution tested. IC50 = Inhibitory Concentration of the 50% of initial inoculum, CF = volatiles' Conversion Factor, OH = *Origanum hirtum*, SM = *Satureja montana*, MD = *Monarda didyma*, MF = *Monarda fistulosa*.

More generally, the average values of $IC50_{Hy}$/CF and $IC50_{EO}$, calculated on four bacterial strains (excluding 01SA(R) and 0.6EF strains) and four yeasts, indicate that the two distillation products (EO and Hy) from *S. montana* show the smallest differences in terms of effectiveness related to volatiles concentrations: in the average of the eight cases, an amount of the EO 8.2 times as concentrated as that of the Hy is needed to attain the same inhibition of microbial growth. However, products obtained from the *O. hirtum* and *Monarda* genus illustrate the greatest difference in terms of the biological activity related to the volatiles' concentration. In fact, a quantity of *O. hirtum*, *M. didyma* and *M. fistulosa* EOs, respectively, 5.7, 16 and 42.3 times as concentrated as the corresponding Hys is necessary. In this respect, the IC50 comparison between EOs and Hys outlines the same ranking as the IR50 comparison between EOs and Hys (Table 5), strengthening the differences in efficacy between the two distillation products.

3. Discussion

For more than half a century, humans have relied primarily on antibiotics and vaccines to treat and prevent microbial infections. In recent decades, despite the great progress in the medical and pharmaceutical fields, the traditional treatment of infectious diseases is often ineffective due to the increased resistance of microbial strains to antibiotics. To date, one fifth of global deaths is due to infectious diseases [17], as the uncontrolled use of antibiotics in the clinical, veterinary, and agricultural fields has led to the spread of multidrug-resistant microbial strains. While the pharmaceutical industry has addressed this problem by modifying existing antibiotics and developing new ones, microbial strains respond to the pharmaceutical industry by inactivating these new strategies with the development of antibiotic resistance. This scenario clearly highlights the need for new antimicrobial agents with different modes of action than those of traditional antibiotics.

Natural products are among the most promising candidates because they have low toxicity, low environmental impact, and a broad spectrum of action when compared to synthetic antimicrobial substances.

Many studies have shown the antimicrobial activity of various EOs [18,19] also regarding muti-drug resistant bacteria and fungi, due to a broad spectrum of cytocidal activity [20,21]. For example, the EO of *S. montana*, in addition to anti-oxidant activity, proved effective against bacteria and dermatophytes; especially *T. violaceum*, *T. rubrum*, *T. tonsurans*, *T. mentagrophytes* and *P. oryzae* [22,23], while the EO obtained from *O. hirtum* showed antimicrobial activity against both Gram+ and Gram- strains [24,25]. The EOs belonging to the *Lavandula* genus, in addition to having an antimicrobial activity against a broad spectrum of microorganisms [26–28], show sedative properties on the central nervous system, as well as anti-inflammatory and re-epithelializing properties [29–31]. Furthermore, EOs and Hys derived from non-native plants belonging to the *Monarda* genus grown in Italy, have shown interesting antimicrobial activities towards Gram+, Gram- yeasts and environmental fungi [32–34].

The effectiveness of active ingredients was also studied. β-Linalool is a non-toxic alcohol most common in nature. It is present in the phytocomplexes of lavender EOs but also of many other EOs. In the EO of *Cinnamomum camphora* (Ho wood) it can reach concentrations higher than 90%. Literature data show its comprehensive range of bioactive properties including antimicrobial activity [35]. The main component of both EO and Hy of *O. hirtum* is the thymol, a phenol monoterpene isomer of carvacrol, particularly present in EOs obtained from species belonging to the *Thymus* genus. This natural compound has an antimicrobial spectrum wider than that of β-linalool, including Gram-positive, Gram-negative bacteria (especially pathogens of the airways), and fungi. Finally, it shows the ability to interfere with the fungal transformation process from the cellular form to the hyphal form [36]. The antimicrobial activity of carvacrol, main component of both *S. montana* and *Monarda* spp. natural products, is higher than that of the other volatile compounds due to the free hydroxyl group, hydrophobicity, and the phenol moiety. In particular, it shows a great activity against Gram- food-borne pathogens [37].

Among the main active compounds analyzed, it is possible to identify an activity gradient (linalool < thymol < carvacrol). This gradient is consistent with the data of antimicrobial efficacy actually observed, as the least active natural compounds are those obtained from the *Lavandula* genus, while the others show stronger antimicrobial activities.

Moreover, several EOs have been shown to interfere with the ability of microorganisms to form biofilm, which is often linked to chronic, difficult-to-treat infections such as skin and wound infections [38,39]. *S. montana* EO was shown to be able to inhibit biofilm formation and interfere with preformed biofilms of Gram+ bacteria, including *S. aureus* [23].

Despite the high antimicrobial activity of EOs, use as such is not recommended due to their high concentration of hydrophobic active ingredients with a toxic potential. Therefore, to avoid toxic effects, EOs need to be used in low concentrations by diluting them in an appropriate vehicle before use.

On the contrary, Hys are hydrophilic solutions containing up to a maximum 1g/L of the EOs active compounds. Although more perishable than EOs, they are generally safe and do not need to be diluted in a vehicle before use. This feature of Hys makes them interesting both for oral intake and skin applications. The latter use becomes especially important in the presence of skin infections.

However, the antimicrobial activity of Hys would certainly appear to be milder than that of the corresponding EOs. In fact, the simple comparison of MIC values obtained from the antimicrobial analysis of the EOs and Hys used in this study evidence that the first are more effective at a lower concentration. Tables 1 and 2 show that the EOs active on at least the 50% of the strains have inhibitory and cytocidal actions at concentrations ranging between 0.125% v/v and 2% v/v. Whereas, the Hys must be used at concentrations between 25% v/v and 50% v/v to reach the same antimicrobial activity, i.e., they need to be from 25 to 200 times more concentrated than EOs.

However, if we consider the relative concentration of active chemicals, can we say that Hys really have milder antimicrobial actions than the corresponding EOs? Tables 5 and 6 show that this cannot be said. In fact, the calculated $IR50_{Hy}/CF$ is lower than the $IR50_{EO}$, as well as the $IC50_{Hy}/CF$ calculated for each microbial strain is lower than the $IC50_{EO}$. This means that, to obtain the inhibition of 50% of growth of both the initial inoculum of each strain and total microbial strains, a concentration of EOs' volatiles greater than that of the corresponding Hys is required. It results, therefore, in the Hys' volatiles being relatively more effective than those of EOs. This activity could be due to the hydrophilic environment of Hy, which provides a greater bioavailability of volatiles for the interaction with bacteria and fungi [40], or to the antagonistic action present among chemical components of the EO phytocomplex.

These data are interesting because they show the antimicrobial activity of Hys from another point of view, especially as it concerns potential clinical applications for the treatment of skin infections. In fact, in these pathologies, local applications that are simultaneously effective for the patient and safe for intact or damaged skin are indispensable.

Potential applications encompass all small skin infections that need daily local treatments with antimicrobial creams and ointments, but also of more serious pathologies such as *Tinea capitis* generated by dermatophytes that essentially afflicts children, or antibiotic resistant/sensitive infections of sores or wounds whose treatment becomes important for skin re-epithelialization, or chronic vaginal infections induced by yeasts in which the topical use of concentrated EOs is absolutely contraindicated due to their toxicity.

In all cases, the use of Hys with antimicrobial activity compatible with a cutaneous or mucosal treatment would be of great interest. In fact, Hys are already on the market, and they can be used on the skin of non-allergic subjects without inducing adverse effects. Currently, Hys in Italy are used in formulations of cosmetic products for body care, or they are sold pure for cosmetic and food use. As is well known, the Italian market is a famous perfume and fragrance hub that is constantly looking for new products and is able to influence the Hys production of primary producers. Globally, the Hys market in Europe has been growing for several years, attaining, in 2018, a 40% share of the world market [41].

From 2019 to 2024, this share is set to increase by an additional 5.2% [42]. Owing to these reasons and in light of our preliminary data, it becomes more and more interesting to deepen the studies on Hys.

4. Materials and Methods

4.1. Clinical Strains

Fifteen clinical strains (six Gram-positive bacterial strains and nine fungal strains), which are potential skin pathogens provided by the UOC of Microbiology of Policlinico Universitario A. Gemelli of Rome, Italy, were used. Two of the six bacterial strains were resistant (R) to antibiotics. Bacterial strains were: *Staphylococcus aureus* MRSA (0.1R), *Streptococcus pyogenes* (0.2), *Enterococcus faecalis* VRE (0.3R), *Enterococcus faecium* (0.4), *Staphylococcus aureus* MSSA (0.5), *Enterococcus faecalis* (0.6). Whereas, four of the nine fungal strains were yeasts (*Candida albicans* (3.1), *Candida parapsilosis* (0.1R), *Candida glabrata* (0.2R), and *Candida tropicalis* (0.3R)), three of which were resistant to common antifungals, and five dermatophytes (*Trichophyton rubrum*, *Trichophyton tonsurans*, *Trichophyton soudanense*, *Trichophyton violaceum*, and *Microsporum canis*). Mueller Hinton medium (Becton Dickinson and Company, Cockeysville, MD, USA) was used to grow bacterial strains at 37 °C for 24 h, while fungal strains were grown on RPMI broth and Sabouraud agar medium (Oxoid, Wade Road, Basingstoke, Hants, UK). In particular, yeasts were grown at 37 °C for 24 h, and dermatophytes at 30 °C for 7 days.

4.2. Essential Oils and Related Hydrolates

EOs and Hy from six aromatic plants grown and processed in Italy were studied (*S. montana*, *L. angustifolia*, *L. intermedia*, *O. hirtum*, *M. didyma*, and *M. fistulosa*). All EOs and Hys were kindly granted by FX Laboratorio Benessere srl (Arzignano, Vicenza, Italy), except for those isolated from *M. didyma* and *M. fistulosa* species, which were provided by DISTAL, University of Bologna.

4.3. Gas Chromatography Mass Spectrometry Analysis

Analyses were performed on a Bruker ScionSQ gas chromatograph, coupled with a single quadrupole mass-spectrometer (GC-MS) (Bruker, Milan, Italy). Compounds were separated BD-5 a semi-standard non-polar column (30 m × 0.25 mm, i.d. 0.25 µm) (Phenomenex, Bologna, Italy). EOs were diluted 1:1000 (*v*/*v*) in ethyl acetate, and 1 µL of this dilution was injected into GC-MS. Samples of hydrolate were diluted 1:5 (*v*/*v*) in ethanol (99.8%), and 1 µL of this dilution was injected into GC-MS. The percentage (*w*/*w*) of the amount of the compounds of EO present in Hy was carried out gravimetrically. Peaks were identified by comparing the retention times with those of authentic standard MS fragmentation patterns and final confirmation by matching with the components of the commercial library NIST mass spectral database (vers. 6.41). The percentage composition of the oils was computed by the normalization method from the GC peak areas. R.I. were generated by using a series of n-alkanes from C7 to C40 (Sigma-Aldrich, Milan, Italy) and compared with data reported in the literature [43–46]. All analyses were repeated in triplicate.

4.4. Gravimetric Analysis

Five mL of each Hy were subjected to liquid/liquid isolation with 5 mL of CH_2Cl_2 (n = 3). The organic phases were pooled, and the solvent evaporated by means of a rotary evaporator at reduced pressure. The residue obtained was weighed and the percentage (*w*/*v*) content of volatiles in the hydrolate evaluated.

4.5. Broth Microdilution Susceptibility Test

The broth microdilution (BMD) susceptibility test according to the European Committee on Antimicrobial Susceptibility Testing (EUCAST) international guidelines were performed. The BMD test was performed on a 96-well plate by adding 100 µL of a cell suspension equal to 5×10^5 CFU/mL to a final volume of 200 μL. Scalar dilutions, between 50% *v*/*v* (500 μL/mL) and 3.125% *v*/*v* (31.25 μL/mL) of Hy and between 2% (20 μL /mL) and 0.06% (0.6 μL/mL) of EO were tested. EOs and Hys were dissolved in a suitable nutrient agar (as specified in paragraph 4.1) and 0.5% *v*/*v* of Tween 80 was used to deliver the EOs into the hydrophilic medium. Plates were incubated overnight at 37 °C. After this period, MIC values were determined by spectrophotometric reading at 450 nm (EL808, Biotek, Winooski, VT, USA), except for MICs values of the dermatophytes, which were assessed by visual reading. To evaluate the MLC, 5 μL of the content of each well was seeded on Muller Hilton or Sabouraud agar plates, which were incubated for 24 h at 37 °C. The MIC is defined as the lowest concentration that completely inhibits the organism's growth when compared to the growth of control. Whereas, the MLC is defined as the lowest concentration corresponding to the death of 99.9% or more of the initial inoculum. Each test was performed in triple, and both negative and positive controls were included. Values corresponding to the IR or LR of 50% and 90% of all strains were calculated. As discussed in the "Data management" paragraph, the value corresponding to a concentration of EOs or Hys necessary to obtain the inhibition of 50% of the initial inoculum was extrapolated for each strain analyzed.

4.6. Comparison Between EO and Hy

Hy and EO comparison was made, as described in Di Vito M et al. [16]. Comparison was based on comparing the total volatiles content of EO with that of the corresponding Hy. Briefly, the Essential Oil Total volatiles Area (EOTA) and the Hydrolate Total volatiles Area (HYTA) were calculated by evaluating areas covered by the total volatiles in the chromatograms multiplied by EO and Hy respective dilutions prior to GC–MS (1000 and 5, respectively). The semi-quantitative volatiles' Conversion Factor (CF) between the EO and the Hy was assumed to be the EOTA/HYTA ratio. Comparison between an EO and its corresponding Hy was made by dividing the IC50 or IR50 of each Hy by its CF. If the value of this ratio corresponds to the value of IC50 or IR50 of the EO, it means that the two natural products are equivalent in terms of relative antimicrobial activity, as the same amount of volatiles is needed in both EO and Hy to inhibit the growth of 50% of the initial inoculum. Whereas, values of this ratio lower or higher than the IC50 or IR50 of the OE show a relative antimicrobial activity of volatiles contained in the Hy higher or lower than that of the EO, respectively.

4.7. Data Management

The IC50 value of each natural substance (*O. hirtum, S. montana, M. didyma* and *M. fistulosa*) and distillation product (EO and Hy) vs. each microbial strain was obtained by interpolating the OD450 values corresponding to the tested dilutions with a regression line, and calculating the dilution value (% *v*/*v*) corresponding to half of the OD450 value of the positive control. All the values obtained from both the microbiological and chemical analyzes were processed obtaining mean and standard deviation values.

5. Conclusions

An intrinsic and intriguing question that emerges from this study is to establish which topical application (hydrophobic EOs or hydrophilic Hys) is most suitable for healing different skin infections. Our short communication highlights an aspect still unexplored by the scientific literature regarding the real antimicrobial effectiveness of the active ingredients contained in Hys compared to the EOs from the same plant source. The use of odorous aqueous solutions with low concentrations of active ingredients in the treatment of minor and chronic skin infections is certainly interesting for the fight against antibiotic resistance. Furthermore, since the terpenic active ingredients are not very soluble in water, most Hys have a low number still present; *O. hirtum*, has only one. This makes these natural products also interesting for pharmaceutical companies who are looking for new natural products with antimicrobial action, but need "standardizable" products to be tested in clinical trials conducted according to scientific rigor.

Author Contributions: Conceptualization, M.D.V., F.B.; Methodology, M.D.V., F.G., P.M.; Formal Analysis, M.D.V., A.S., F.G.; Investigation, M.D.V., A.S.; Resources, P.M., M.S.; Data Curation, M.D.V., L.B., M.R.P.; Writing—Original Draft Preparation, M.D.V.; Writing—Review & Editing, P.M., F.B., E.N., M.G.B., G.B. All authors have read and agreed to the published version of the manuscript.

Funding: This research received no external funding.

Institutional Review Board Statement: Not applicable.

Informed Consent Statement: Not applicable.

Data Availability Statement: *M. didyma* and *M. fistulosa* plants were used in this study. Vouchers are deposited at the Herb garden of Casola valsenio (Ravenna, Italy) (Sauro Biffi).

Acknowledgments: Lori Morrison is gratefully acknowledged for manuscript proofreading.

Conflicts of Interest: The authors declare no conflict of interest.

References

1. Millikan, L.E. Complementary medicine in dermatology. *Clin. Dermatol.* **2002**, *20*, 602–605. [CrossRef]
2. Halcón, L.; Milkus, K. Staphylococcus aureus and wounds: A review of tea tree oil as a promising antimicrobial. *Am. J. Infect. Control* **2004**, *32*, 402–408. [CrossRef]
3. Kozics, K.; Bučková, M.; Puškárová, A.; Kalászová, V.; Cabicarová, T.; Pangallo, D. The Effect of Ten Essential Oils on Several Cutaneous Drug-Resistant Microorganisms and Their Cyto/Genotoxic and Antioxidant Properties. *Molecules* **2019**, *24*, 4570. [CrossRef]
4. *European Pharmacopeia*, 6th ed.; European Directorate for the Quality of Medicines and Healthcare, Supplement 5.8; Council of Europe: Strasbourg, France, 2007.
5. AA. VV. *French Pharmacopeia*; Lyon, France. 2012. Available online: https://ansm.sante.fr/Mediatheque/Publications/Pharmacopee-francaise-Plan-Preambule-index (accessed on 17 January 2021).
6. D'Amato, S.; Serio, A.; Chaves Lopez, C.; Paparella, A. Hydrosols: Biological activity and potential as antimicrobials for food applications. *Food Control* **2018**, *86*, 126–137. [CrossRef]
7. Zárybnický, T.; Boušová, I.; Ambrož, M.; Skálová, L. Hepatotoxicity of monoterpenes and sesquiterpenes. *Arch. Toxicol.* **2018**, *92*, 1–13. [CrossRef]
8. Herman, A.; Herman, A.P. Essential oils and their constituents as skin penetration enhancer for transdermal drug delivery: A review. *J. Pharm. Pharmacol.* **2015**, *67*, 473–485. [CrossRef]
9. Sainz, P.; Andrés, M.F.; Martínez-Díaz, R.A.; Bailén, M.; Navarro-Rocha, J.; Díaz, C.E.; González-Coloma, A. Chemical composition and biological activities of Artemisia pedemontana subsp. assoana essential oils and hydrolate. *Biomolecules* **2019**, *9*, 558. [CrossRef] [PubMed]
10. Lei, G.; Li, J.; Zheng, T.; Yao, J.; Chen, J.; Duan, L. Comparative Chemical Profiles of Essential Oils and Hydrolate Extracts from Fresh Flowers of Eight Paeonia suffruticosa Andr. Cultivars from Central China. *Molecules* **2018**, *23*, 3268. [CrossRef] [PubMed]
11. Wajs-Bonikowska, A.; Sienkiewicz, M.; Stobiecka, A.; Maciąg, A.; Szoka, Ł.; Karna, E. Chemical Composition and Biological Activity of Abies alba and A. koreana Seed and Cone Essential Oils and Characterization of Their Seed Hydrolates. *Chem. Biodivers.* **2015**, *12*, 407–418. [CrossRef] [PubMed]
12. Carlini, E.A.; De Oliveira, A.B.; De Oliveira, G.G. Psychopharmacological effects of the essential oil fraction and of the hydrolate obtained from the seeds of Licaria puchury-major. *J. Ethnopharmacol.* **1983**, *8*, 225–236. [CrossRef]
13. Guesmi, F.; Tyagi, A.K.; Prasad, S.; Landoulsi, A. Terpenes from essential oils and hydrolate of *Teucrium alopecurus* triggered apoptotic events dependent on caspases activation and PARP cleavage in human colon cancer cells through decreased protein expressions. *Oncotarget* **2018**, *9*, 32305. [CrossRef] [PubMed]
14. Carvalho, A.F.U.; Melo, V.M.M.; Craveiro, A.A.; Machado, M.I.L.; Bantim, M.B.; Rabelo, E.F. Larvicidal activity of the essential oil from Lippia sidoides cham. against Aedes aegypti linn. *Mem. Inst. Oswaldo Cruz* **2003**, *98*, 569–571. [CrossRef] [PubMed]
15. Andrés, M.F.; González-Coloma, A.; Muñoz, R.; De la Peña, F.; Julio, L.F.; Burillo, J. Nematicidal potential of hydrolates from the semi industrial vapor-pressure extraction of Spanish aromatic plants. *Environ. Sci. Pollut. Res.* **2018**, *25*, 29834–29840. [CrossRef] [PubMed]
16. Di Vito, M.; Bellardi, M.G.; Mondello, F.; Modesto, M.; Michelozzi, M.; Bugli, F.; Sanguinetti, M.; Sclocchi, M.C.; Sebastiani, M.L.; Biffi, S.; et al. Monarda citriodora hydrolate vs essential oil comparison in several anti-microbial applications. *Ind. Crops Prod.* **2019**, *128*, 206–212. [CrossRef]
17. WHO (World Health Organization). Antimicrobial Resistance: Global Report on Surveillance. Available online: https://apps.who.int/iris/bitstream/handle/10665/112642/9789241564748_eng.pdf;jsessionid=2EF9B2B5D4C9757B7C4A3433C8413F7B?sequence=1 (accessed on 18 January 2021).

18. Wińska, K.; Mączka, W.; Łyczko, J.; Grabarczyk, M.; Czubaszek, A.; Szumny, A. Essential Oils as Antimicrobial Agents—Myth or Real Alternative? *Molecules* **2019**, *24*, 2130. [CrossRef]
19. Kalemba, D.; Kunicka, A. Antibacterial and Antifungal Properties of Essential Oils. *Curr. Med. Chem.* **2003**, *10*, 813–829. [CrossRef] [PubMed]
20. Bucková, M.; Puškárová, A.; Kalászová, V.; Kisová, Z.; Pangallo, D. Essential oils against multidrug resistant gram-negative bacteria. *Biologia* **2019**, *73*, 803–808. [CrossRef]
21. Mayaud, L.; Carricajo, A.; Zhiri, A.; Aubert, G. Comparison of bacteriostatic and bactericidal activity of 13 essential oils against strains with varying sensitivity to antibiotics. *Lett. Appl. Microbiol.* **2008**, *47*, 167–173. [CrossRef]
22. Entela Haloc, I.; Toska, V.; Baldisserotto, A.; Goci, E.; Vertuani, S.; Manfredini, S. Evaluation of antifungal activity of Satureja Montana essential oil before and after inclusion in beta-cyclodextrine. *Int. J. Pharm. Pharm. Sci.* **2014**, *6*, 187–191.
23. Vitanza, L.; Maccelli, A.; Marazzato, M.; Scazzocchio, F.; Comanducci, A.; Fornarini, S.; Crestoni, M.E.; Filippi, A.; Fraschetti, C.; Rinaldi, F.; et al. Satureja montana L. essential oil and its antimicrobial activity alone or in combination with gentamicin. *Microb. Pathog.* **2019**, *126*, 323–331. [CrossRef]
24. Karakaya, S.; El, S.N.; Karagözlü, N.; Sahin, S. Antioxidant and antimicrobial activities of essential oils obtained from oregano (*Origanum vulgare* ssp. hirtum) by using different extraction methods. *J. Med. Food* **2011**, *14*, 645–652. [CrossRef] [PubMed]
25. Grondona, E.; Gatti, G.; López, A.G.; Sánchez, L.R.; Rivero, V.; Pessah, O.; Zunino, M.P.; Ponce, A.A. Bio-efficacy of the essential oil of oregano (Origanum vulgare Lamiaceae. Ssp. Hirtum). *Plant Foods Hum. Nutr.* **2014**, *69*, 351–357. [CrossRef] [PubMed]
26. Bajalan, I.; Rouzbahani, R.; Pirbalouti, A.G.; Maggi, F. Chemical Composition and Antibacterial Activity of Iranian Lavandula × hybrida. *Chem. Biodivers.* **2017**, *14*, e1700064. [CrossRef] [PubMed]
27. Tardugno, R.; Serio, A.; Pellati, F.; D'Amato, S.; Chaves López, C.; Bellardi, M.G.; Di Vito, M.; Savini, V.; Paparella, A.; Benvenuti, S. Lavandula × intermedia and Lavandula angustifolia essential oils: Phytochemical composition and antimicrobial activity against foodborne pathogens. *Nat. Prod. Res.* **2019**, *33*, 3330–3335. [CrossRef]
28. D'Auria, F.D.; Tecca, M.; Strippoli, V.; Salvatore, G.; Battinelli, L.; Mazzanti, G. Antifungal activity of Lavandula angustifolia essential oil against Candida albicans yeast and mycelial form. *Med. Mycol.* **2005**, *43*, 391–396. [CrossRef]
29. Cardia, G.F.E.; Silva-Filho, S.E.; Silva, E.L.; Uchida, N.S.; Cavalcante, H.A.O.; Cassarotti, L.L.; Salvadego, V.E.C.; Spironello, R.A.; Bersani-Amado, C.A.; Cuman, R.K.N. Effect of Lavender (*Lavandula angustifolia*) Essential Oil on Acute Inflammatory Response. *Evid. Based Complement. Alternat. Med.* **2018**, *2018*, 1413940. [CrossRef]
30. Pérez-Recalde, M.; Ruiz Arias, I.E.; Hermida, É.B. Could essential oils enhance biopolymers performance for wound healing? A systematic review. *Phytomedicine* **2018**, *38*, 57–65. [CrossRef]
31. Kasper, S. An orally administered lavandula oil preparation (Silexan) for anxiety disorder and related conditions: An evidence based review. *Int. J. Psychiatry Clin. Pract.* **2013**, *17* (Suppl. 1), 15–22. [CrossRef]
32. Li, H.; Yang, T.; Li, F.-Y.; Yao, Y.; Sun, Z.-M. Antibacterial activity and mechanism of action of Monarda punctata essential oil and its main components against common bacterial pathogens in respiratory tract. *Int. J. Clin. Exp. Pathol.* **2014**, *7*, 7389–7398.
33. Ricci, D.; Epifano, F.; Fraternale, D. The Essential Oil of *Monarda didyma* L. (Lamiaceae) Exerts Phytotoxic Activity in vitro against Various Weed Seed. *Molecules* **2017**, *22*, 222. [CrossRef]
34. Mattarelli, P.; Epifano, F.; Minardi, P.; Di Vito, M.; Modesto, M.; Barbanti, L.; Bellardi, M.G. Chemical Composition and Antimicrobial Activity of Essential Oils from Aerial Parts of Monarda didyma and Monarda fistulosa Cultivated in Italy. *J. Essent. Oil Bearing Plants* **2017**. [CrossRef]
35. Pereira, I.; Severino, P.; Santos, A.C.; Silva, A.M.; Souto, E.B. Linalool bioactive properties and potential applicability in drug delivery systems. *Colloids Surf. B. Biointerfaces* **2018**, *171*, 566–578. [CrossRef] [PubMed]
36. Salehi, B.; Mishra, A.P.; Shukla, I.; Sharifi-Rad, M.; Contreras, M.D.M.; Segura-Carretero, A.; Fathi, H.; Nasrabadi, N.N.; Kobarfard, F.; Sharifi-Rad, J. Thymol, thyme, and other plant sources: Health and potential uses. *Phytother. Res.* **2018**, *32*, 1688–1706. [CrossRef] [PubMed]
37. Sharifi-Rad, M.; Varoni, E.M.; Iriti, M.; Martorell, M.; Setzer, W.N.; Del Mar Contreras, M.; Salehi, B.; Soltani-Nejad, A.; Rajabi, S.; Tajbakhsh, M.; et al. Carvacrol and human health: A comprehensive review. *Phytother. Res.* **2018**, *32*, 1675–1687. [CrossRef] [PubMed]
38. Malone, M.; Bjarnsholt, T.; McBain, A.J.; James, G.A.; Stoodley, P.; Leaper, D.; Tachi, M.; Schultz, G.; Swanson, T.; Wolcott, R.D. The prevalence of biofilms in chronic wounds: A systematic review and meta-analysis of published data. *J. Wound Care* **2017**, *26*, 20–25. [CrossRef] [PubMed]
39. Vázquez-Sánchez, D.; Galvão, J.A.; Mazine, M.R.; Gloria, E.M.; Oetterer, M. Control of Staphylococcus aureus biofilms by the application of single and combined treatments based in plant essential oils. *Int. J. Food Microbiol.* **2018**, *286*, 128–138. [CrossRef]
40. Van de Vel, E.; Sampers, I.; Raes, K. A review on influencing factors on the minimum inhibitory concentration of essential oils. *Crit. Rev. Food Sci. Nutr.* **2019**, *59*, 357–378. [CrossRef]
41. Anonymous Hydrosols Market. *Research Report-Global Forecast. Till 2024*; Market Research Future®. 2020. Available online: https://www.marketresearchfuture.com/reports/hydrosols-market-4789 (accessed on 18 January 2021).
42. *Channel (Store-Based and Non-Store-Based), and Region (North America, Europe, Asia-Pacific, and Rest of the World)—Forecast till Hydrosols Market Global Research Report Information by Source (Rose, Roman Chamomile, Neroli, Lavender, and Others), Category (Organic and Conventional), Distribution 2024, Maharashtra, India*. Maharashtra, India. 2020. Available online: https://www.marketresearchfuture.com/reports/hydrosols-market-4789 (accessed on 17 January 2021).

43. Kilic, A.; Hafizoglu, H.; Kollmannsberger, H.; Nitz, S. Volatile Constituents and Key Odorants in Leaves, Buds, Flowers, and Fruits of *Laurus nobilis* L. *J. Agric. Food Chem.* **2004**, *52*, 1601–1606. [CrossRef]
44. Takaku, S.; Haber, W.A.; Setzer, W.N. Leaf essential oil composition of 10 species of Ocotea (Lauraceae) from Monteverde, Costa Rica. *Biochem. Syst. Ecol.* **2007**, *35*, 525–532. [CrossRef]
45. Babushok, V.I.; Linstrom, P.J.; Zenkevich, I.G. Retention Indices for Frequently Reported Compounds of Plant Essential Oils. *J. Phys. Chem. Ref. Data* **2011**, *40*. [CrossRef]
46. Dötterl, S.; Wolfe, L.M.; Jürgens, A. Qualitative and quantitative analyses of flower scent in Silene latifolia. *Phytochemistry* **2005**, *66*, 203–213. [CrossRef] [PubMed]

MDPI
St. Alban-Anlage 66
4052 Basel
Switzerland
Tel. +41 61 683 77 34
Fax +41 61 302 89 18
www.mdpi.com

Antibiotics Editorial Office
E-mail: antibiotics@mdpi.com
www.mdpi.com/journal/antibiotics

www.ingramcontent.com/pod-product-compliance
Lightning Source LLC
Chambersburg PA
CBHW061453210926
51850CB00008B/1703